BIOMIMETIC AND SUPRAMOLECULAR SYSTEMS RESEARCH

BIOMIMETIC AND SUPRAMOLECULAR SYSTEMS RESEARCH

ARTURO H. LIMA
EDITOR

Nova Biomedical Books
New York

For permission to use material from this book please contact us:
Telephone 631-231-7269; Fax 631-231-8175
Web Site: http://www.novapublishers.com

Library of Congress Cataloging-in-Publication Data

Biomimetic and supramolecular systems research / Arturo H. Lima (editor).
p. ; cm.
Includes bibliographical references and index.
ISBN 978-1-60456-405-1 (hardcover)
1. Biomimetics. 2. Biomimetic polymers. 3. Nanostructured materials. 4. Supramolecular chemistry. I. Lima, Arturo H.
[DNLM: 1. Biomimetic Materials--chemical synthesis. 2. Biocompatible Materials. 3. Macromolecular Substances. 4. Nanostructures. 5. Nanotechnology--methods. QT 37 B61485 2008] QP517.B56B48 2008
610.82'4--dc22 *2008006474*

Published by Nova Science Publishers, Inc. New York

Contents

Preface

This new book is dedicated to new research on conceiving, designing, constructing and testing materials, structures, devices and systems which are based on supramolecular aggregates and on larger-scale assemblies which replicate biological structures and processes.

Chapter I - Nanostructured organic materials can be constructed from unit molecules through appropriate assembling processes in which supramolecular chemistry plays an important role. Certain parts of the structures obtained can be referred to as nanostructured supermolecules. Various potential functions can be expected for such nanostructured supermolecules. In particular, biomimetic functionality could be sought for supramolecular systems since these functions cannot be obtained using current nanotechnologies involving inorganic materials. In this chapter, nanostructured supermolecules, hybridized with various materials, are described with biomimetic functions such as regulated permeation, controlled release, and materials conversion. The research appearing in this chapter is roughly categorized into areas of composites of lipid thin films and polymers, layer-by-layer assembly, and mesoporous materials. Lipid thin films are known to undergo reversible changes in their fluidity, usually known as phase transitions often between crystalline and liquid crystalline phases. Immobilization of these lipid thin films onto appropriate supports can lead to materials that co-possess mechanical stability and capabilities of controlled release. Layer-by-layer assembly provides both soft thin films and polymer coated hollow capsules whose structures can be frustrated by external stimuli, resulting in controllable permeability changes of entrapped drugs. Mesoporous materials have nanochannels with controlled structural dimensions and extremely high surface areas. Mesoporous materials modified with operating valves can be used for drug storage and release under control by external stimuli

Chapter II - So-called ElectroActive Polymers (EAP) currently offer attractive properties as "smart" materials for supramolecular electromechanical transduction, suitable for actuation devices capable of mimicking functional characteristics of biological muscles and structures. Within the broad EAP family, dielectric elastomers represent one of the most performing classes of materials, able to show superior electromechanical properties. This chapter presents the ongoing research activities and the latest results achieved in the authors' lab for the development of soft actuators made of dielectric elastomers, along with their possible uses for biomimetic and biomedical systems. Devices here described exploit the basic principle of operation of any dielectric elastomer actuator, i.e. an electrical squeezing of an elastomeric insulator, although specifically implemented by means of different structural configurations with tailored active features. In particular, new actuators shaped as contractile linear devices are emphasised, with reference to possible applications currently considered.

Chapter III - Silicon microelectronics and MOS technology are among the most important fields of development in modern science and technology. On the other hand, the biotechnology community is feeling the need to implement fast, portable, low cost and user friendly biosensors to monitor different species in the everyday life (e.g. air and water pollution, biomedical applications, etc.). A promising approach to evolve biosensor technology appears to take advantage of the enormous potential offered by the silicon microelectronic integration technology.

In this chapter the authors discuss MOS based Si biosensors in which the sensitive elements (both proteins and DNA strands) are covalently immobilized on the transistor gate surface and are able to detect either organic or inorganic molecules.

To further improve both the biosensor responsivity and selectivity, the use of nanostructures is a viable tool. The authors describe how a supramolecular approach based on diblock copolymers self-assembling can be used for nanoscale definition of mono and two dimensional anchoring sites of the bio-elements, with control on size and position at nanometer level.

Chapter IV - Photonic crystals are composite materials possessing a periodical modulation of their dielectric constant on the scale length of the visible wavelength. This dielectric periodicity affects light propagation creating allowed and forbidden photon energy regions. Among different types of photonic crystals, the authors addressed their attention to artificial opals i.e. three dimensional self-assembled arrays of spheres packed in a face centred cube crystal lattice. In order to tune the optical properties of such photonic crystals, the authors inserted gold nanoparticles in the interstices between the spheres. They call this process doping. In this work the authors report on the optical properties of artificial opals doped at various levels with gold nanoparticles (NpAu) stabilized with citrate molecules. By increasing the NpAu doping level, the opal photonic stop band was bathochromically shifted as well as its full width half maximum and intensity were reduced. No effects of nanoparticle coalescence was observed as previously detected for NpAu prepared by laser ablation. In order to improve the mechanical stability of NpAu doped opals, the authors started a preliminary thermal annealing study with in situ monitoring of optical properties. They found that thermal degradation of doped samples was reduced with respect to bare opals. In addition, the annealing temperature of NpAu doped opals was much higher than that observed for bare ones.

Chapter V - Micro- and nano-patterned materials are of great importance for the design of new nanoscale electronic, optical and mechanical devices, ranging from sensors to displays. A prospective system that can support a designed functionality is elastomeric polyurethane thin films with nano- or micro-modulated surface structures ("wrinkles"). These wrinkles can be induced on different lengthscales by mechanically stretching the films, without the need for any sophisticated lithographic techniques. In the present article the authors focus on the experimental control of the wrinkling process. A simple model for wrinkle formation is also discussed, and some preliminary results reported. Hierarchical assembly of these tunable structures paves the way for the development of a new class of materials with a wide range of applications, from electronics to biomedicine.

Chapter VI - Apatite is the main inorganic component of natural bone. It is both bioactive and osteoconductive. Apatite has been used extensively as a coating on the surface of bioinert materials, especially metals, to enhance the bioactivity of these materials. There are many coating techniques available for applying apatite onto different substrates, including plasma spraying, chemical vapor deposition (CVD), sputtering coating, sol-gel coating, electrochemical deposition, electrophoresis, and biomimetic coating. Among these, plasma spraying is the most widely used technique for applying apatite coating on metal substrates. However, the high temperature employed in the coating process can result in decomposition and reduced crystallinity of apatite coating, and in situ incorporation of drugs and proteins becomes impossible with this coating technique. Although the substrate is remained relatively cool (approximately 200°C) during plasma spraying, the temperature applied is still too high for most polymeric substrates.

In the 1980s, Professor Kokubo from the Kyoto University and his colleagues proposed a biomimetic method to form a bone-like apatite layer at a physiological temperature and pH using simulated body fluid (SBF). SBF was initially used to assess the bioactivity of different biomaterials. It was found that bioactive materials could induce apatite formation, but bioinert materials could not. Later, the biomimetic method was more widely used as a coating technique to apply bone-like apatite onto different biomaterials. However, a coating could only be formed on the surface of bioactive materials, such as silica gel, titania gel, tantalum, and alkaline-treated titanium and its alloys. It is generally believed that certain surface functional groups are responsible for the apatite formation on these materials. A recent study has evaluated the effect of the four most promising surface functional groups, including -OH, $-SO_3H$, $-PO_4H_2$ and –COOH, on apatite formation, and found that carboxyl end groups yield more apatite deposition than any other functional groups. In order to form an apatite coating on relatively inert materials, a nucleation aid, normally $CaO\text{-}SiO_2$-based glass, has been used. First, the substrate is placed in contact with glass particles in SBF thus forming apatite nuclei on its surface. The substrate is then transferred to another SBF with relatively higher calcium and phosphate concentrations (e.g. 1.5×SBF) to allow the further growth from the apatite nuclei in situ. Using this approach, bone-like apatite has been successfully deposited onto the surface of polymers, alumina and 316 stainless steel to name but a few materials. An alternative approach to form apatite coating on inert substrates is the utilization of concentrated SBF alone. Not only does apatite form on the surface of inert materials, but also its formation rate is significantly improved by using concentrated SBF. In recent years, biomimetic apatite coating has been applied to the surface of bioinert tissue-engineered scaffolds to enhance the osteogenic potential of the material. Homogenous apatite coatings have been obtained on different types of scaffolds, such as titanium foam and poly(D,L-lactic-co-glycolic acid) (PLGA) scaffold. These coatings have been found to support osteoblast attachment and proliferation, which are important properties of bone regenerative materials.

Chapter VII - The ability of living organisms to control the structure and morphology of biominerals has attracted a lot of attention over the past 20 years. In parallel to biological studies devoted to *in vivo* processes, so-called "biomimetic" approaches have been widely developed. These approaches aim at reproducing *in vitro* the biomineral morphology and/or structure in biologically-relevant conditions. First steps in this direction have extensively relied on the identification of biomolecules extracted from living organisms and their evaluation in mineralization experiments in diluted solutions. Further insights on biomineralization processes were gained by using biomimetic synthetic molecules. Recently, more accurate model systems that take into account confinement effects and question the role of mineral precursors have been developed. However, in many instances, such biomimetic approaches have so far failed to accurately reproduce the complexity of natural materials. This raises the question whether *in vitro* models will allow the complete elucidation of biomineralization processes.

This commentary discusses the recent achievements, actual limitations and possible perspectives of such biomimetic approaches. After a general introduction aiming at presenting the main concepts involved in current biomineralization mimetic studies, the authors will illustrate the state-of-art of this very active field of research through the examples of some popular models in this field. The authors will show that, despite the differences that exist in the associated mineral phases and in the conditions of their formation, common trends in biomimetic strategies and resulting materials can be identified. Finally, the authors will try to identify to what extent *in vitro* experiments may allow the elucidation of the *in vivo* mechanisms of biomineralization. In parallel, they will also emphasize that an accurate replication of the biomineral features may not be required to obtain synthetic materials useful for future applications.

Chapter VIII - Phenyleneethynylenes (PEs) are a promising class of conjugated materials for optoelectronic devices due to their high electronic delocalization and strong emission properties. Depending on the type of optoelectronic application, specific molecular and supramolecular organization is required. For electroluminescent diodes, it is desirable to deposit the emitting layer with the phenyleneethynylene molecules oriented with the conjugated backbone, responsible of the electroluminescent properties, in the same direction of the electrons-holes injection from the electrodes. On the other hand, when the final application is a solar cell, it is preferred that the photoconductive part of the molecule is exposed to the incident light as much as possible. These requirements must also match other physicochemical characteristics that are necessary for any optoelectronic device, such as solubility, thermal and chemical stability, among others. With this idea in mind, in this chapter, the authors report on the molecular engineering of phenyleneethynylenes by changing the chemical structure either in the main conjugated backbone or with proper lateral substituents. Different molecular architectures: totally rod-like, semirigid, dendrimers or functionalized nanoparticles can be synthesized and specific supramolecular organization can be induced by varying the film deposition technique.

Chapter IX - In this study, the microstructures of a ceromer (Artglass®) reinforced by either glass fibers (GlasSpan®) or polyethylene fibers (Connect®) were characterized and compared and the influence of the fiber reinforcement on the flexural strength of the resulting products evaluated. With this objective, seven bars of each material were produced. One bar of each material was separated for microstructural analysis. The microstructural samples were subjected to metallographic polishing and finishing, and then analyzed using optical microscopy at different magnifications. The images obtained were treated using an image processing computer program (Image Pro Plus) in order to quantify the microstructure by calculating the mean diameter and mean volume fraction of fibers. The flexure tests were made by three-point bending, using six samples of each material. After statistical analysis, the results showed that the mean diameter of the glass fibers (4μm) was smaller than the polyethylene ones (23.6 μm). The mean volume fraction of glass fibers (0.42) was larger than that of the polyethylene fibers (0.28) and the mean center-to-center distance between fibers was smaller in the glass fibers material (33 μm) than in the polyethylene fibers material (61 μm). The flexural strength of both glass and polyethylene fiber-reinforced materials was statistically equal, despite the fiber volume fraction being statistically larger in the fiber glass material.

Chapter X - Fabrication of an adequate scaffolding material having apt architecture, mechanical properties and biocompatibility for optimal cell-material recognition and biosynthesis is one of the great challenges of tissue engineers. In this study, the authors introduced a novel porous biomaterial of marine origin, the skeleton of *Tetraclita rufotincta* (Crustacea, Cirripedia) as a potent scaffolding material for bioengineering applications. They chose this biomaterial due to its porous architecture, chemical composition and since it is an abundant species along the tidal zone of the Red Sea and other oceans. The authors showed that the main inorganic constitute of the *Tetraclita rufotincta* (*T. rufotincta*) skeleton is $CaCO_3$ in the form of calcite. Light microscopy, scanning electron microscopy (SEM) and XTT cell proliferation assays of mesenchymal stem cells (MSCs) that were seeded on *T. rufotincta* biomatrices have shown fast adhesion and high proliferation rates. Histochemical, immunoflourescence and quantitative real time polymerase chain reaction (qPCR) surveys revealed that MSCs differentiated into chodrogenic lineage and subsequently secreted chondorgenic extra cellular matrix constitutes without addition of any inducer. The authors suggest that crystalline skeletal deriviatives of *T. rufotincta* are bioactive, enhance chondrogenesis of MSCs and therefore could be exploited for tissue engineering applications in particularly for *ex-vivo* cartilage bioengineering.

Chapter XI - Biomimetic based bioelectronics have emerged as a breakthrough with great potential capable of generating new concepts and technologies for the development of the next generation electronics. The authors have developed several biomimicking devices consisting of protein such as biomemory and image sensors, and so on. Bioelectronic device for information storage composed of ferredoxin was based on the electron transport function of plant photosynthesis. The information storage function of the fabricated device was investigated by applying the reduction potential as a write function and measuring the stored reducing charge as a read function. And a biomolecular photoreceptor consisting of bacteriorhodopsin(bR)/flavin complex Lamgmuir-blodgett(LB) films was developed for color recognition. By mimicking the early process of color perception in the human visual system, the light-sensitive biological pigments with photoelectronic conversion function were chosen and used as the main components of the device. bR and flavin were deposited onto patterned ITO glass, and a 9-pixel photoreceptor with a sandwich-type structure of ITO/bR/flavin/electrolyte gel/Pt was constructed. When simple color images were projected onto the photoreceptor, the shapes and the colors of projected images were successfully recognized by the knowledge-based algorithm, and the processed images were visualized onto the monitor connected to the photoreceptor through a data acquisition system and personal computer. And a bioelectronic device consisting of bR as photoreceptor and 6-nitrospiropyran as noise filter was developed for the extraction of desired image. In the constructed bioelectronic device, the increment of the intensity ratio of transmitted lights in comparison with that of input lights was obtained when various input lights of the different light intensity were irradiated onto the films. The proposed devices here can be combined and extended to the development of biorobots with the fundamental researches in the fields of brain/neuroscience, material science, and mechanics, which joint activities will create a new field of biomechatronics. With the whole new world of electronics, the concept of a biocomputer or a biorobot will no longer appeared as a science fiction, but will be realized in this century.

Chapter XII - The study of the linked processes related to the chemical precipitation from supersaturated solutions is required to understand the mechanisms involved in numerous natural processes, such as the biomineralization of biofunctional structures and the precipitation of chemical sedimentary rocks. Also, it offers us an exceptional opportunity to investigate how the matter self-assemblies into hierarchy ordered structures. On the other hand, the phenomenon of the chemical precipitation can be applied in the laboratory as an efficient and versatile synthetic route to prepare novel materials of both fundamental and practical interest.

In this chapter, the authors present an overview of the fundamentals of the crystallization from solutions, drawing especial attention to their application in the production of uniform nanocrystalline materials with engineered features.

Chapter XIII - The objective of present investigation is to systematically study the crystallization effect of autoclaving hydrothermal treatment at 100-200°C, vacuum and atmospheric heat treatments at 400-800°C on the microstructural evolution of plasma-sprayed HA coatings (HACs), especially focus on clarifying the effect of hydrothermal crystallization with lower heating temperatures. This study will also clarify the HA crystallization on the fluctuation of bonding strength and evaluate the failure mechanism of coatings through the Weibull model.

Post-heat treatments can promote the crystallization of HACs, and it is worth noting that just 150°C is required to see a significant rise in crystallinity for the hydrothermally-treated HACs compared with the other heat treatments. The ambient saturated steam atmosphere of hydrothermal treatment is favorable to eliminate the TCP, TP, CaO and amorphous calcium phosphate phases within the HACs. Low-temperature hydrothermal treatment shows a superior crystallization effect than high-temperature heat treatments in a vacuum or in an atmospheric environment.

According to the reaction kinetics of Arrhenius equation, the heating temperature, especially higher than 500°C, is a control factor for promoting the HA crystallization. Considering the hydrothermal treatment, experimental evidence confirmed that the saturated steam pressure of hydrothermal treatment can promote and accelerate the HA crystallization. The ambient saturated steam pressure is thought of as another control factor that not only reduces the activation energy of HA crystallization calculated from the Arrhenius equation, but also plays an important role in lowering crystallization temperatures. The finely-crystallized HA crystalline observed within hydrothermally-treated HACs is recognized as a nucleation and grain growth phenomenon by the replenishment of OH^- groups into the hydroxyl-deficient HA structure. A self-healing effect can be recognized to diminish the spraying defects with these crystalline HA.

Referring to the results of tensile adhesion tests, the HA crystallization and the thermal-induced microstructural degradation are concluded as two effects to dominate the bonding strength evolution and data fluctuation. Hydrothermally-treated HACs with higher bonding strength (38.9 MPa) than vacuum and atmospheric heat-treated ones (36.0 MPa and 37.1 MPa, respectively) can be recognized for its enhanced fracture toughness and much denser microstructure from the self-healing effect to reduce the strength degradation. Through the statistical analysis of Weibull model, the strengthening HACs are generally reliable materials with a wear-out failure model. It means that the crystallization effect promotes plasma-sprayed HACs to have a better reliability for further biological applications.

The bonding strength degradation occurred at heating temperatures above 600°C is mainly due to the effect of its crystallization-induced contraction through the thermal dilatometry analysis. This effect generates a thermal stress within the HACs because the mismatch of the thermal expansion coefficient between the crystallized HACs and Ti-6Al-4V substrate, and the HACs will be cracked in the form of significant contraction-induced cracks. The cracked microstructural feature will induce the bonding strength degradation and further increase the bonding strength data fluctuation. This phenomenon even induces a low Weibull modulus and a decreased reliability of the HACs. In contrast, the hydrothermally-treated HACs with low-temperature crystallization and a self-healing effect will be effective to prevent the above-mentioned microstructural defect and bonding strength degradation from the high-temperature crystallization process.

In: Biomimetic and Supramolecular Systems Research ISBN: 978-1-60456-405-1
Editor: Arturo H. Lima

Chapter I

Controlled Release and Materials Conversion Using Nanostructured Supermolecules

Katsuhiko Ariga[1], Jonathan P. Hill, Qingmin Ji, and Ajayan Vinu
World Premier International (WPI) Research Center for Materials Nanoarchitectonics (MANA), National Institute for Materials Science (NIMS), Japan

1. Abstract

Nanostructured organic materials can be constructed from unit molecules through appropriate assembling processes in which supramolecular chemistry plays an important role. Certain parts of the structures obtained can be referred to as nanostructured supermolecules. Various potential functions can be expected for such nanostructured supermolecules. In particular, biomimetic functionality could be sought for supramolecular systems since these functions cannot be obtained using current nanotechnologies involving inorganic materials. In this chapter, nanostructured supermolecules, hybridized with various materials, are described. with biomimetic functions such as regulated permeation, controlled release, and materials conversion. The research appearing in this chapter is roughly categorized into areas of composites of lipid thin films and polymers, layer-by-layer assembly, and mesoporous materials. Lipid thin films are known to undergo reversible changes in their fluidity, usually known as phase transitions often between crystalline and liquid crystalline phases. Immobilization of these lipid thin films onto appropriate supports can lead to materials that co-possess mechanical stability and capabilities of controlled release. Layer-by-layer assembly provides both soft thin films and polymer coated hollow capsules whose structures can be frustrated by external stimuli, resulting in controllable permeability changes of entrapped drugs. Mesoporous materials have nanochannels with controlled structural dimensions and extremely high surface areas. Mesoporous materials modified with

[1] Corresponding address: World Premier International (WPI) Research Center for Materials Nanoarchitectonics (MANA) and Supermolecules Group, National Institute for Materials Science (NIMS), 1-1 Namiki, Tsukuba 305-0044, Japan. Phone: +81-29-860-4597, Fax: + 81-29-860-4832, E-mail: ARIGA.Katsuhiko@nims.go.jp.

operating valves can be used for drug storage and release under control by external stimuli

2. Introduction

Without doubt a variety of the activities in our daily lives are supported by the rapid development of nanotechnology. Machines and devices with high information capacities have become ubiquitous and they are constructed mostly through advanced techniques in micro-fabrication of inorganic substances especially semiconductors and metals. On the other hand, nanoscience and nanotechnology based on organic and biological substances is a substantial emerging field. The soft, flexible, and dynamic functions expected for organic nanostructures should be distinct from those possessed by nanostructures composed of inorganic materials.

Fabrication strategies of these two types of nanomaterials differ in that those for inorganic nanostructures often proceed by creation of minature structures from objects of larger dimensions, is the so-called top-down approach, while nanostructures of organic materials can be initiated from unit molecules through appropriate assembling processes. The latter strategy is known as the bottom-up approach, where methods for assembly of organic units are the main research targets. Therefore, supramolecular chemistry, which deals with molecular interaction, molecular recognition, and molecular assembly, plays an important role in nanofabrication of organic and biological materials. Certain parts of the structures obtained can be regarded as nanostructured supermolecules.

The practical uses of inorganic micro- and nanostructures are already realized and are based on well-matured technology of inorganic materials. In contrast, the potential of nanostructured supermolecules is still relatively unexplored. In particular, biomimetic functions is an unique target for supramolecular nanotechnologies because these functions cannot be achieved by using conventional inorganic material nanotechnologies. For example, selective permeation and controlled release of target drugs can be investigated using nanostructured supermolecules. These functions can be regarded as mimics of biomembranes or biospecific receptors and have potential in practical applications as drug delivery systems. Another especially attractive potential function is that of specific material conversion, which can be considered mimicry of the activity of enzymes and is occasionally useful for sensor application.

However, we know that simple supramolecular assemblies are often very fragile and are not always strong enough to be used in practical applications. Therefore, hybridization of the nanostructured supermolecules with mechanically strong materials such as polymers, silicates, or carboniferous materials is sometimes considered as a method to circumvent this disadvantage. Hybrid nanostructured supermolecules comprising various materials are described in this chapter with emphasis on those concerned with functions of regulated permeation, controlled release, and material conversions. The subject of this chapter is roughly categorized into (a) composites of lipid thin films and polymers, (b) layer-by-layer assembly, and (c) mesoporous materials. Examples have been selected widely from the most recent thirty year period to illustrate the unique characteristics of nanostructured supermolecules.

3. Composites of Lipid Thin Films and Polymers

Spherical assemblies based on lipid bilayer structures are called liposomes (from phospholipids) or vesicles (from other amphiphiles) and can be regarded as cell membrane mimics (Figure 1). These structures are capable of trapping drug molecules in their inner aqueous cavities. Permeation of materials through lipid bilayer structures can be regulated by using the fluidity of the lipid bilayer membranes [1,2]. In many cases, permeation through the lipid bilayer is suppressed when in its crystalline phase and accelerated upon entering a disordered, liquid-crystalline state. Such changes in material permeation can be usually controlled by controlling temperature. The surface of liposomes and vesicles can be modified, and other components are easily incorporated so that liposomes and vesicles have been widely studied for delivery of various drugs to specific cellular targets [3,4]. They can be modified easily to incorporate targeting moieties that allow for guided delivery of the drug. For example, Fields and coworkers incorporated specifically designed triple-helical peptide sequences into liposomes for use against metastatic melanoma cells that overexpress the CD44 cell surface receptor (Figure 2) [5]. Their experimental results suggested that these liposomes can be constructed and their structures fine-tuned rationally for drug delivery applications based on lipid composition.

From engineering points of view, liposomes and vesicles are themselves too fragile to be utilized for a wide range of applications. Hybridization of lipid bilayer structures with mechanically stable materials such as polymers could provide a practical solution to this problem. Some ideal examples of this method can be seen in pioneering research on lipid-corked nylon capsules completed by Okahata and coworkers in the 1980s [6]. They 'corked' semipermeable nylon capsules with bilayer membranes of several dialkyldimethylammonium bromides (Figure 3) [7]. Permeation of NaCl from the inner aqueous phase through the lipid-immobilized nylon layer was reduced by factors of 8-50 relative to that of the uncorked capsule.

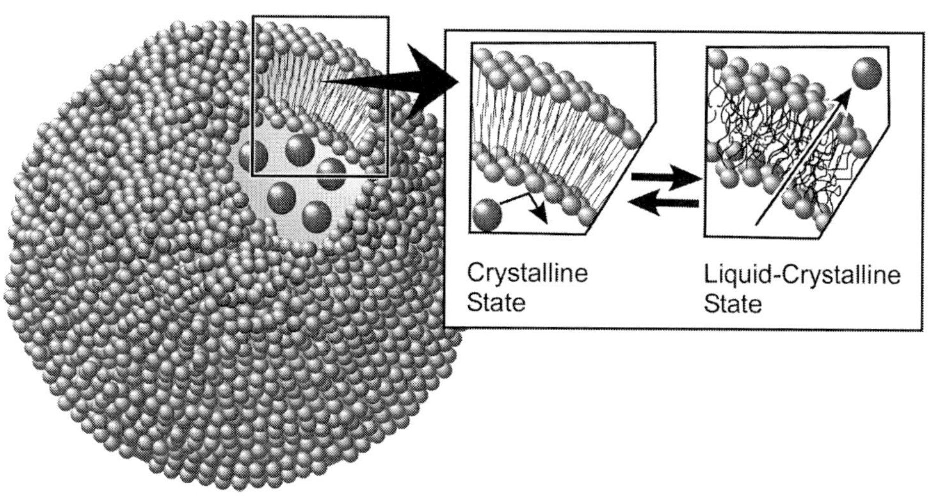

Figure 1. Liposome or vesicle structures. Inserted figures represent phase transition behavior of lipid bilayer membrane.

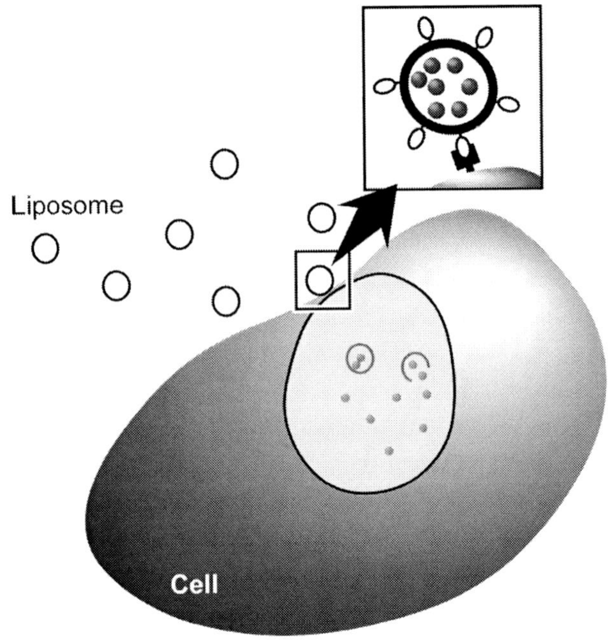

Figure 2. Drug delivery by functionalized liposome to cell.

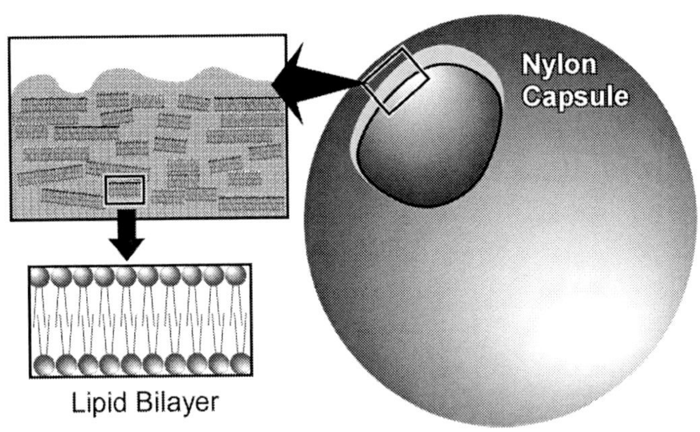

Figure 3. A lipid-corked nylon capsule.

The permeability of NaCl was drastically increased above the phase transition temperature (T_c) because of an increase in membrane fluidity. The NaCl flux above T_c was enhanced with a decrease in the alkyl chain length of the bilayers. Since the structure of bilayer-forming lipids can be varied greatly, different types of chemical and physical stimuli can be selected for controlling permeability through the nylon capsules. Thus, corked synthetic bilayers containing a dissociative ammonium head group for the permeation of a fluorescent probe trapped in the inner aqueous phase was investigated under various pH conditions [8]. The permeability could be increased or decreased reversibly depending on pH. Such pH-control of material permeation was achieved only at temperatures above T_c of the coating bilayers. Material permeation through the nylon ultrathin capsules coated with synthetic bilayers containing acid-dissociative head groups were also investigated [9]. In those cases, permeability was reduced under neutral conditions with an increase in the acidic medium. Control of permeability by using external chemical stimuli was also investigated. For example, permeation of NaCl from the inner aqueous phase of nylon ultrathin capsules coated with sodium didodecyl phosphate bilayer membrane was increased reversibly by treatment with Ca^{2+}, but decreased to the original rate by external washing with ethylenediamine-*N,N,N',N'*-tetraacetic acid (EDTA) [10]. A similar permeation control was also realized using lipid bilayers having phosphatidylethanolamine head groups [11].

Physical stimuli other than temperature are also applicable to permeation control of the lipid-corked nylon capsules. If an azobenzene-containing lipid is used as the bilayer component then photoresponsive permeation control can be achieved [12]. The same research group also demonstrated reversible regulation of NaCl permeation through the lipid-coated capsule membrane by using ultrasonic irradiation [13]. The activation energies of permeation suggested that the lipid bilayer structures became fluid-like liquid crystalline state upon ultrasonic irradiation resulting in an increase in NaCl permeation. Permeation control upon they electrical breakdown of the lipid bilayer structures was also reported [14]. Permeation of NaCl, glucose, and water-soluble dyes through the lipid-corked nylon membranes was enhanced by 5 - 176 times under a 20 - 80 V electric field relative to that in the absence of an electric field. When exposure to the electric field ceased even after a long charging duration (3 - 5 min), the permeability returned to the original slow rate. The latter fact suggests that this permeability control could be reproduced repeatedly without damaging either corking bilayers or nylon capsule membranes. Signal receptive permeability control may be useful as a model of the synaptic systems in which a nerve electrical impulse initiates the rapid release of a chemical intermediary.

The lipid-corked nylon capsule may not be perfectly stable under harsh conditions. Stability of the hybrids can be effectively increased by polymerization of lipids within a nylon matrix. Okahata and coworkers examined the effect of lipid polymerization on the barrier effect, phase transition behaviors, and stability of the hybrid structure against lipid leakage [15]. The results obtained are summarized schematically in Figure 4. Monomeric bilayers showed a large barrier effect and clear permeation changes at the phase transition point (T_c), but a serious drawback is that they are liable to release of the lipid components from the capsule under harsh conditions at high temperature (Figure 4a). When lipids are polymerized at the hydrophobic alkyl chain moiety, the polymerization enhanced the stability of the corking lipid, but largely decreased both the barrier effect and the bilayer property because of shrinkage and disordering in the bilayer region (Figures 4b and 4c). Phase transition behavior upon melting of alkyl chains was maintained by polymerization of the lipids at the hydrophilic group with a long spacer chain, and was accompanied by an enhanced stability of the bilayer structures (Figure 4d). However, the barrier effect was not significantly improved. In the case of corking bilayers polymerized at the counterion, the bilayer-corked capsule showed improved stability and a large barrier effect without loss of the bilayer properties (Figure 4e). The latter method seems to be the best strategy to ensure permeation control with enhanced stability of the hybrids.

In an alternative to polymerization of the lipid components within the nylon matrix, grafting of stimuli-responsive polymers onto the nylon capsule membranes also leads to materials possessing effective permeation control.

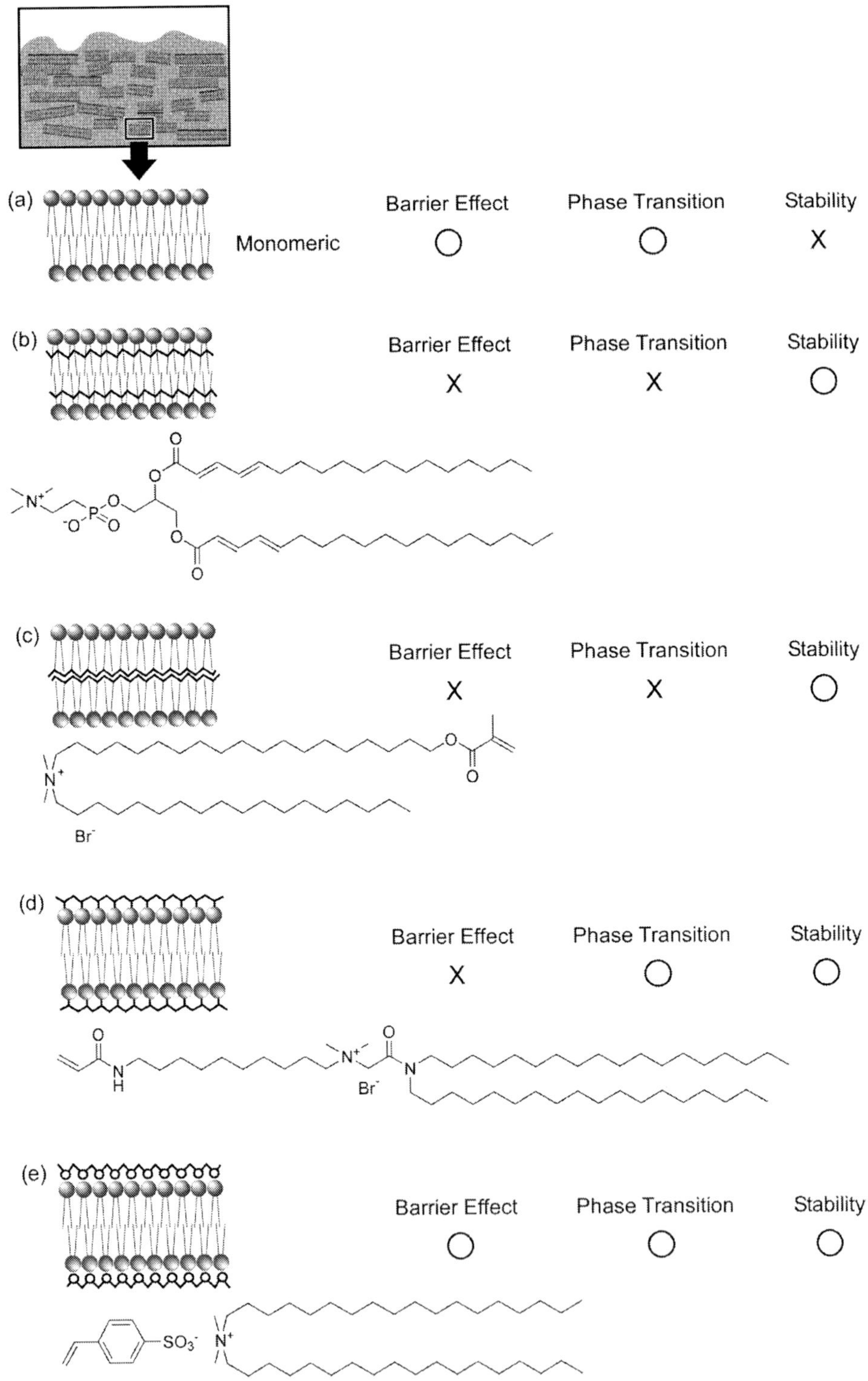

Figure 4. Effect of lipid polymerization on the barrier effect, phase transition behaviors, and stability: (a) monomeric bilayer; (b) and (c) bilayer polymerized at alkyl chains; (d) bilayer polymerized at the hydrophilic group; (e) bilayer polymerized at the counterion.

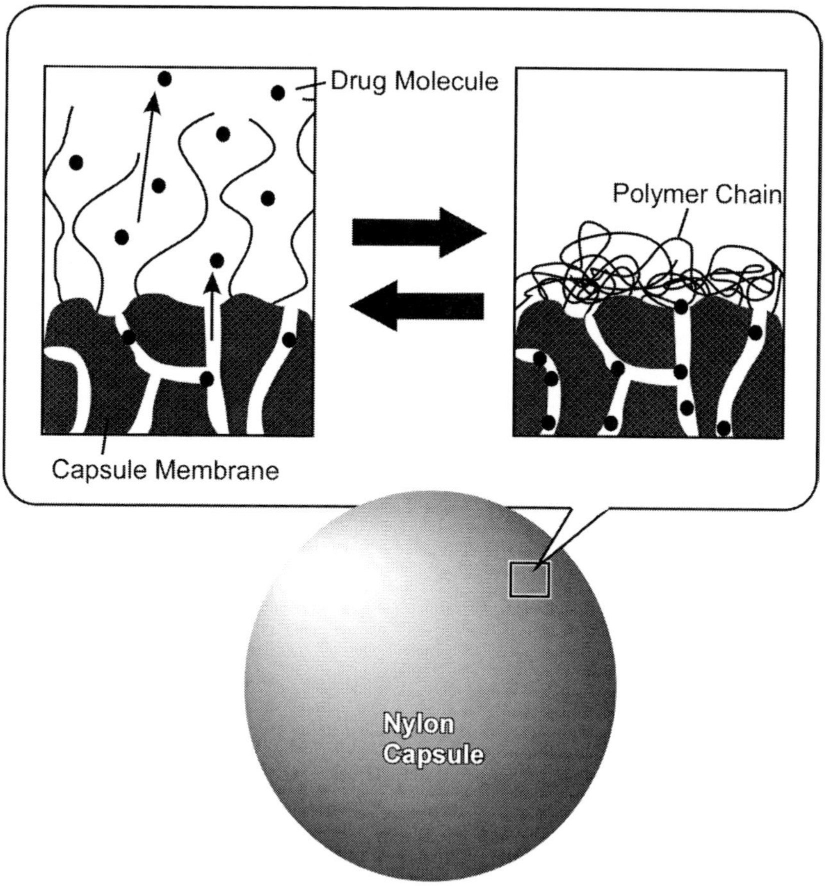

Figure 5. Permeation control by polymer-grafted nylon capsule.

As illustrated in Figure 5, conformational changes of the grafted polymers can be used to regulate drug release from the capsule interior. For example, use of dissociative polymers such as poly(methacrylic acid) enabled control of drug release upon pH change [16]. The poly(methacrylic acid) chains have negative charge at higher pH and are expanded due to electrostatic repulsion. In contrast, neutralization of the chains at low pH made them shrink. From this work an effective permeation of the drug molecules was realized at higher pH. Some imide-type polymers change their conformation and shrink considerably above a particular temperature. Grafting of these polymers led to permeation control upon temperature variation [17]. In order to similarly apply biospecific interactions to this system, synthetic glycopolymers were grafted onto the nylon capsule [18]. Specific interaction with lectins such as concanavalin A regulated the assembly behavior of the glycopolymer chains, resulting in permeation control of drugs from the capsule interior.

Polymer-grafted capsules could be also applied to materials conversion. In the example shown in Figure 6, a nylon capsule membrane with surface-grafted polymers bearing hydrophobic -onium salts (ammonium and phosphonium) was used [19,20], where the -onium salt acted as a phase-transfer catalyst across the porous capsule membrane, accelerating reactions between benzyl bromide in the inner organic phase and sodium azide in the outer aqueous phase.

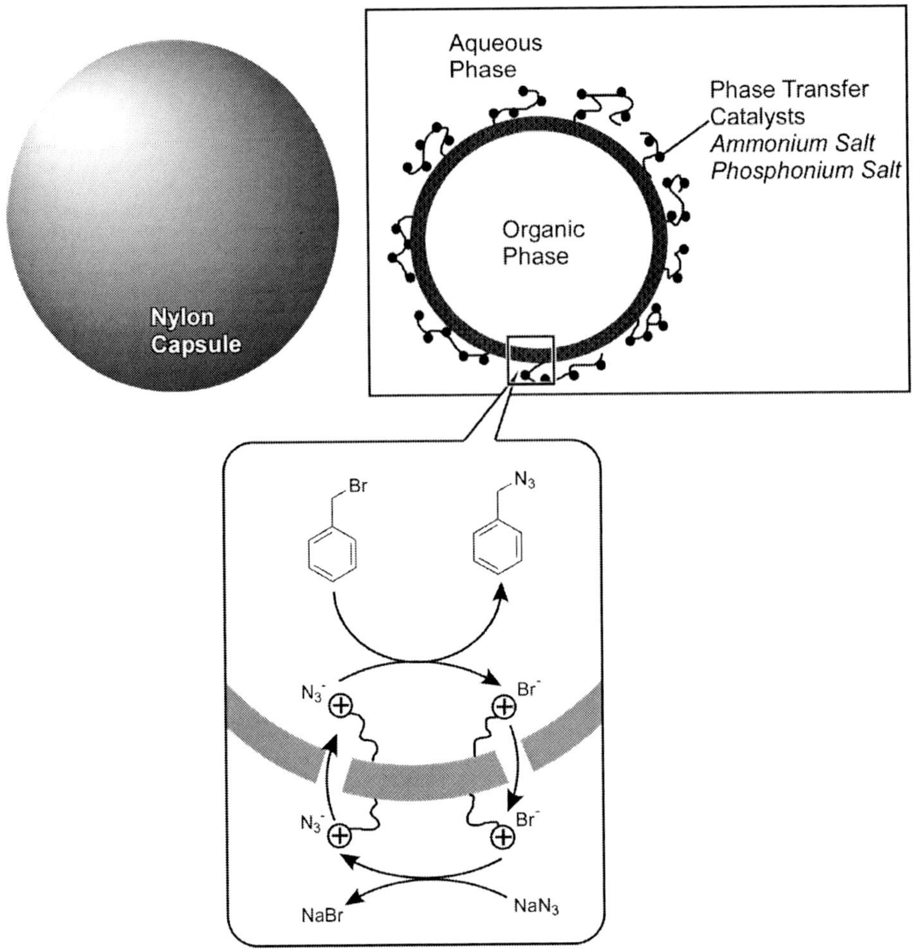

Figure 6. Phase transfer catalysis by polymer-grafted nylon capsule.

The first step is the exchange of the counteranions of the -onium salts from Br^- to N_3^- in the outer aqueous phase. The phase transfer of N_3^- from the outer aqueous to the inner organic phase across the capsule membrane next occurs. The third, rate-limiting step is the nucleophilic substitution of benzyl bromide by N_3^- anions in the inner organic phase or the oil/water interface. Because catalyst moieties always localize at the interfacial region, the reaction between reagents in oil and water phases smoothly proceeds. Redox-type phase transfer catalysis was similarly demonstrated using capsules with grafted chains of viologen-containing polymers [21,22].

As stronger support, porous glass plate was used for immobilization of lipid films. Okahata and coworkers reported permeation control through self-assembled monolayers (SAM) of dialkylsilane amphiphiles that had been covalently immobilized within glass pores (Figure 7) [23]. The permeation rate of NaCl across the porous glass plate was regulated by phase transitions from the solid state to fluid liquid crystalline state of the immobilized monolayers in the pores. Other more meaningful research was performed using Langmuir-Blodgett (LB) technique. The same research group demonstrated permeation control through a single monolayer structure [24,25].

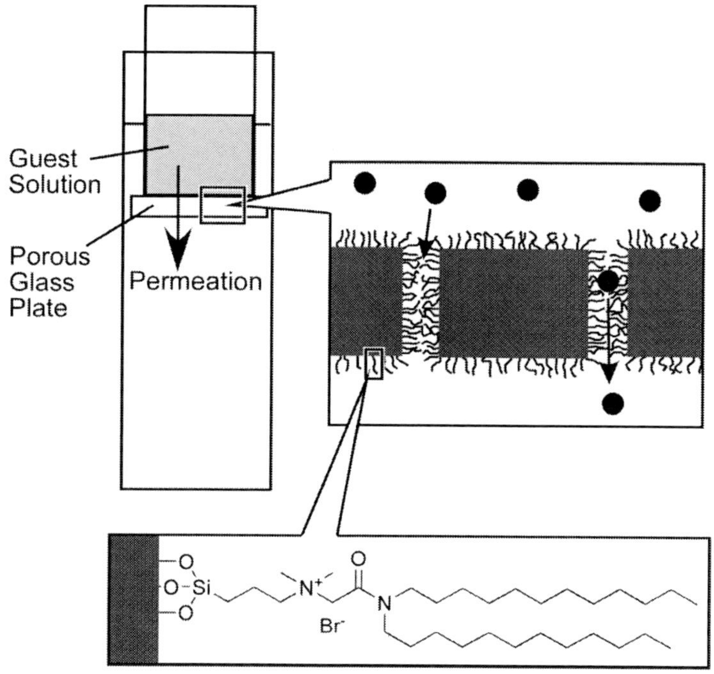

Figure 7. Permeation control through self-assembled monolayers (SAM) of dialkylsilane amphiphiles.

As shown in Figure 8A, in order to immobilize only a single monolayer on a porous glass surface, they combined the SAM concept with the LB technique, i.e., an organosilane monolayer was first prepared and polymerized at the air–water interface and subsequently transferred for immobilization onto a solid substrate. In the surface pressure-molecular area (π-A) isotherm of dialkylalkoxysilane, the condensed phase was only observed at pH 2, while on pure water the monolayer was relatively expanded with a phase transition to a condensed phase. FT-IR spectra of a 30-layer LB film prepared under the acidic condition showed peaks assignable to Si-O-Si at 1026 and 1095 cm^{-1}, while Si-O-C stretching bands at 1080 and 1103 cm^{-1} were observed for the LB films transferred from pure water. These facts imply that the monolayer of dialkylalkoxysilane was polymerized to form well a condensed phase on the acidic subphase. The polymerized monolayer was subsequently transferred and covalently immobilized on the porous glass plate. Permeation of a water-soluble fluorescent probe through the glass plate containing the immobilized monolayer was evaluated. Permeation coefficients at different temperatures are shown in the form of an Arrhenius plot (Figure 8B). The permeation through the polymerized monolayer exhibited a discontinuity at around 45 °C, which is very close to the phase transition temperature as determined by differential scanning calorimetry (DSC). This confirmed that the abrupt permeation change was caused by the phase transition of the polymerized monolayer from a crystalline state to a liquid crystalline state. This is the first example of permeation control using an only 2 nm-thick monolayer.

The LB technique can also be useful for immobilization of biological components such as proteins, although proteins sometimes suffer denaturation caused by the high surface tension on a water surface. Okahata et al. prepared a lipid-coated glucose oxidase (GOD) that could be obtained as a precipitate by mixing aqueous solutions of the enzyme [26,27].

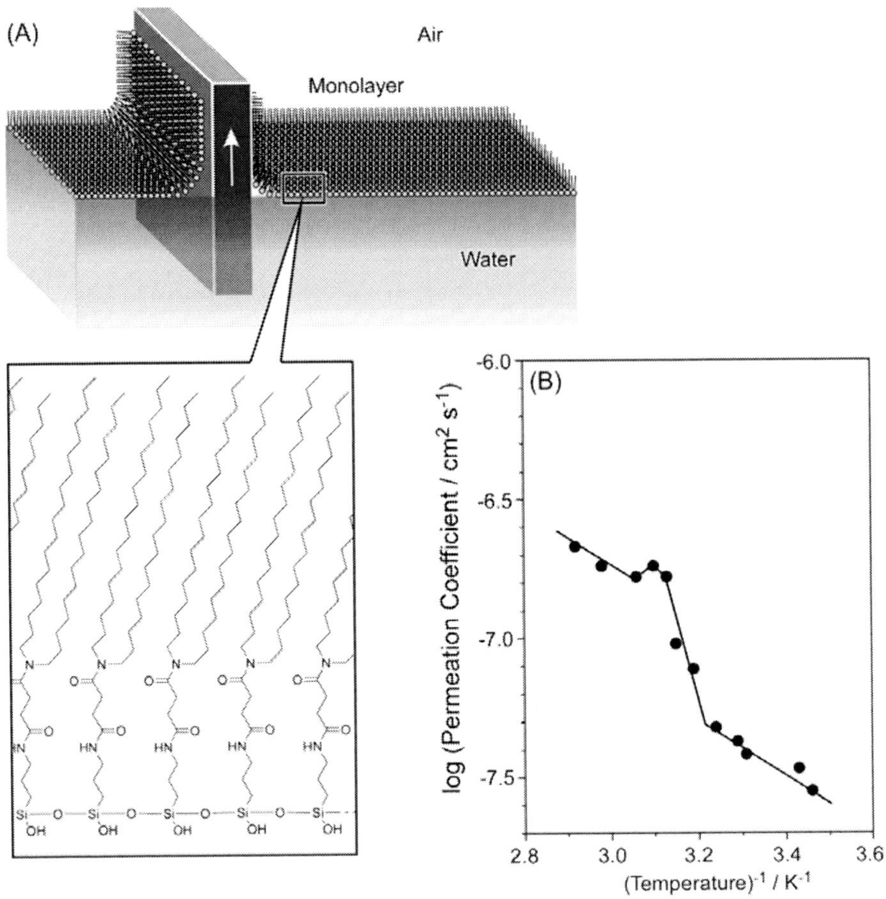

Figure 8. (A) Polymerized monolayer prepared using Langmuir-Blodgett (LB) technique. (B) Arrhenius plot of permeation coefficient through the immobilized monolayer.

The complex formed can be spread on water without causing serious denaturation, and could be transferred onto a Pt electrode (Figure 9). The electrode obtained was used as a glucose sensor in amperometric measurements.

4. Layer-By-Layer Assembly

Although SAM and LB methods provide organized film structures, these techniques are not always applicable to many materials. This incompatibility can be circumvented by using the alternate layer-by-layer (LbL) adsorption technique, which has been paid much attention as a versatile method for thin film preparation [28,29]. Figure 10A illustrates the method of electrostatic layer-by-layer assembly where a cationic polyelectrolyte is first adsorbed on a negatively charged surface of a solid support where over-adsorption usually occurs thus reversing the surface charge. The next adsorption of an anionic particle again reverses the surface charge. This alternation in the surface charges permits continuous fabrication of the layered structure.

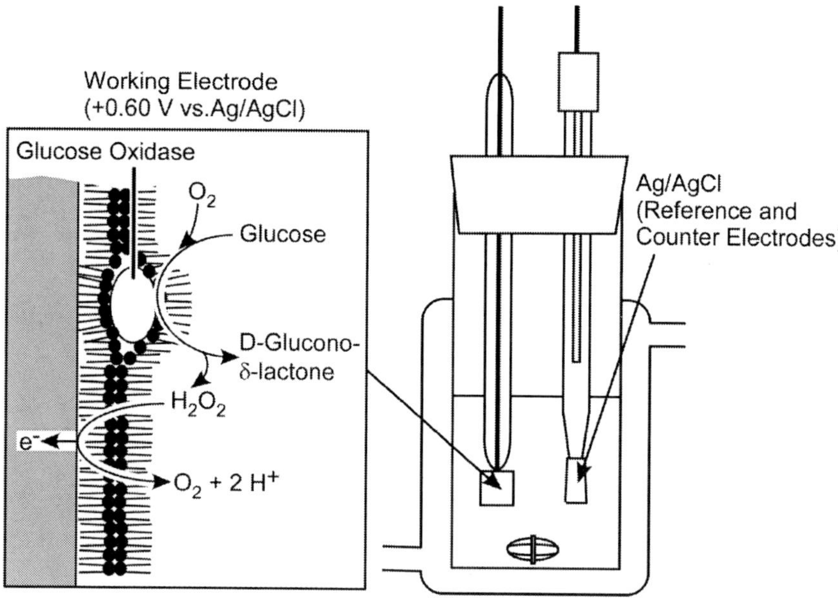

Figure 9. Glucose sensor composed from an LB film of a lipid-coated glucose oxidase.

Because of the simplicity of the assembling process, there is a vast choice of materials available including polymers [30,31], biomaterials [32-34], dye assemblies [35], and inorganic substances [36,37]. In addition, other molecular interactions such as hydrogen bonding [38], metal interaction [39], charge transfer complexation [40], stereocomplex formation [41,42], supramolecular inclusion [43], and bio-specific recognition [44] can be exploited. Therefore, the LbL technique has become a powerful methodology for the preparation of nanostructured supermolecules. This concept can be extended to preparation of three-dimensional nano-objects, using a colloidal core as a supporting material (Figure 10B) [45]. LbL assembly on the surface of the colloidal core is followed by destruction of the central core resulting in hollow capsules.

Nanostructured supermolecules prepared by LbL method possess soft and flexible properties, and therefore, they are appropriate for the development of dynamic functions such as controlled release and materials conversion. For example, Advincula and coworkers developed pH-sensitive bipolar ion-permoselective films prepared by LbL assembly and photocrosslinking of benzophenone-modified poly(acrylic acid) (PAA-BP) and poly(allylamine hydrochloride) (PAH-BP) (Figure 11) [46]. These films can change permeation selectivity between cationic molecular probe ($Ru(NH_3)_6^{3+}$) and anionic molecular probe ($Fe(CN)_6^{3-}$) depending on ambient pH conditions. At high pH (for example, pH 10) the LbL film is permeable to the cationic probe but impermeable to the anionic probe because deprotonation of free carboxylic acid ($-COO^-$) promotes and suppresses permeability of cationic and anionic probes, respectively. This film reverses its permeation selectivity at low pH region (for example, at pH 3.2), i.e., the LbL film is permeable to an anionic probe but less permeable to a cationic probe due to the protonation of free amine ($-NH_3^+$).

Rubner, Cohen, and coworkers introduced soft LbL films within the pores of a polymer support to prepare a pH-sensitive permeation valve (Figure 12) [47].

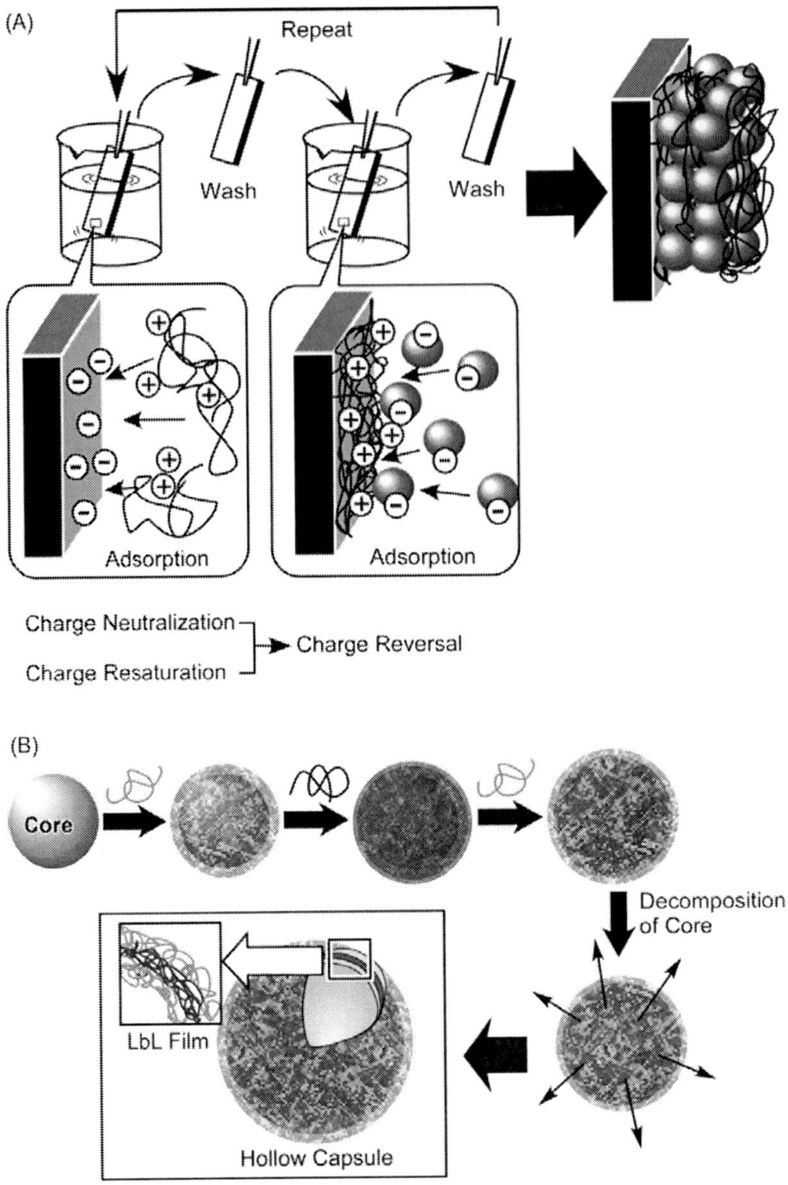

Figure 10. Layer-by-layer assembly; (A) planar film preparation; (B) hollow capsule formation.

Track-etched polycarbonate membranes (TEPC) were filled with the LbL assembled polyelectrolyte multi-layers comprising poly(allylamine hydrochloride) (PAH) and poly(styrenesulphonic acid sodium salt). Gating properties of the multilayer-modified TEPC membranes upon their swelling/deswelling were studied by measuring the flux of pH-adjusted deionized water. Large discontinuous changes in the transmembrane flux were observed, indicating that the multilayers within the cylindrical pores of TEPC membranes exhibit discontinuous swelling/deswelling behavior. This hybrid membrane showed reversible gating properties as the pH condition of feed solution was alternated between pH 2.5 and 10.5.

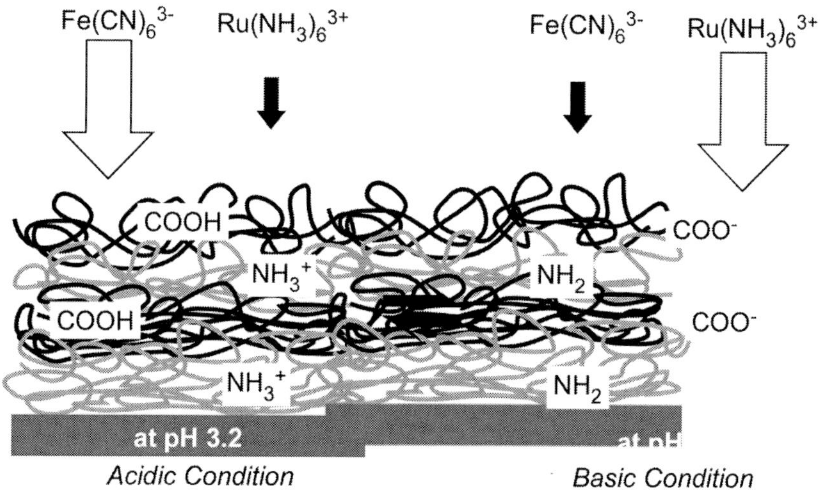

Figure 11. pH-Sensitive bipolar ion-permoselective films prepared by LbL assembly.

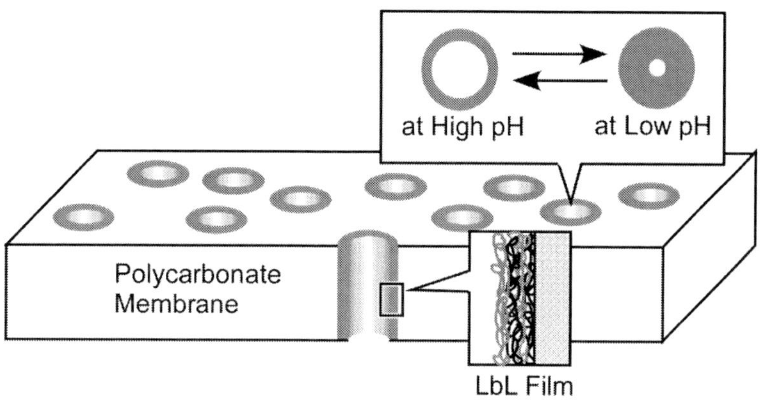

Figure 12. pH-Sensitive permeation valve of LbL films within pores of polycarbonate membranes.

Because this LbL valve performs hysteretic gating, either a "closed" or "open" state at one pH condition can be realized depending on the pretreatment history. Selection of retention or transmission by high molecular weight polymers by varying the membrane pretreatment condition becomes possible. Such gate properties of water would be useful in microfluidic channel devices.

Control of gas permeation through specially designed LbL assemblies was also investigated by Shiratori and coworkers [48]. They deposited amine-based PAH and carboxylic-acid-based poly(acrylic acid) PAA on the surface of fiberglass cloths that were subsequently used for toxic gas removal. Chemically arranged ultrathin films might be useful for removal of gaseous substances from the environment, which is of current importance with respect to serious environmental problems such as passive smoking and "sick building syndrome." The gas-removal performance of the LbL filter was evaluated using static and gas-flow systems. The filter showed remarkably high performance for both ammonia and acetaldehyde removal through both chemical and physical adsorptions.

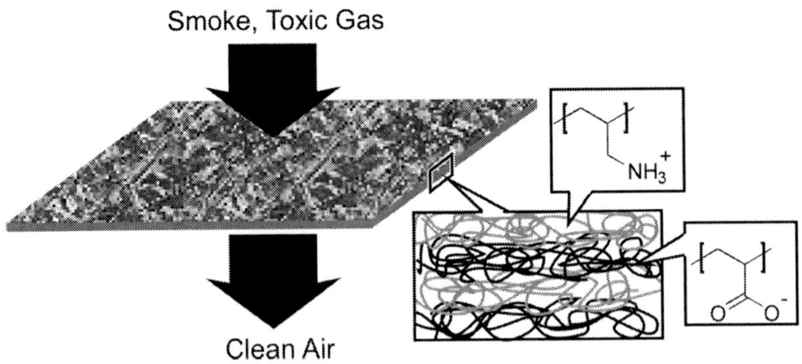

Figure 13. LbL assemblies of PAH and PAA on the surface of fiberglass cloths for air-cleaning.

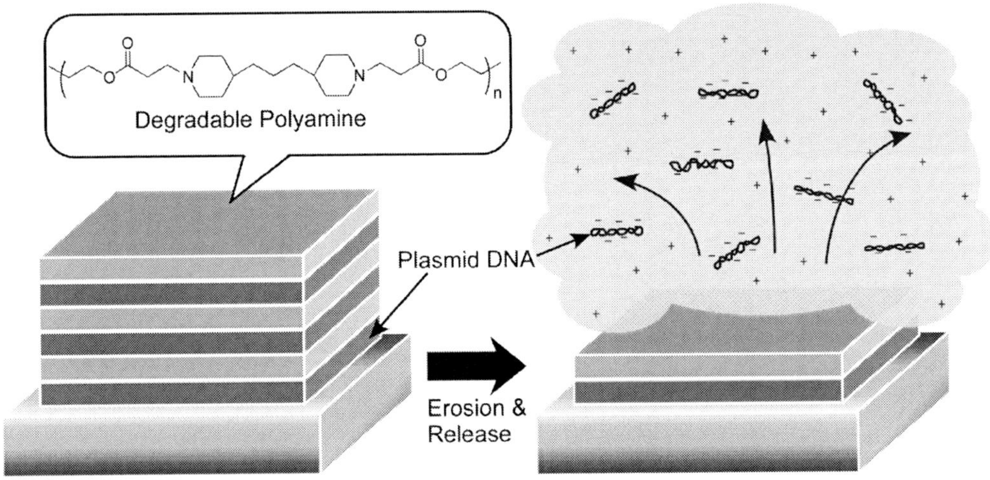

Figure 14. Controlled release of DNA based on the LbL film decomposition.

Ammonia gas was trapped by the PAA layer through physical electrostatic interaction while the PAH layer played an important role in removal of acetaldehyde gas based on chemical Schiff's base formation. This LbL filter may have potential for use as a high-performance filter for air cleaning.

Dynamic modes of controlled release of drugs based on regulated decomposition of an LbL film was demonstrated by Lynn and co-workers [49]. LbL films are usually assembled through non-covalent bonding and changes in ionic strength or environmental pH were known to trigger the deconstruction, or dissolution, of LbL multi-layered assemblies fabricated from strong and/or weak polyelectrolytes. They realized this concept for DNA delivery and subsequent gene expression (Figure 14). LbL assemblies up to 100 nm thick containing plasmid DNA and a synthetic degradable cationic polymer were fabricated on a surface. Plasmid DNA released from the LbL films was transcriptionally active and promoted the expression of high levels of enhanced green fluorescent protein in the cell since the plasmid used in this research encoded for enhanced green fluorescent protein. This approach could be very useful for the controlled administration of functional DNA fragments from the surfaces of biomedical materials and delivery devices.

Encapsulation and controlled delivery of DNA using LbL capsules have also been reported. For example, Lvov and coworkers reported an approach for DNA encapsulation using a biocompatible polyelectrolyte capsule (Figure 15) [50]. In their method, a water-insoluble DNA/spermidine complex was first precipitated onto $MnCO_3$ particles. The mixed component cores obtained were covered with an LbL assembly comprising biocompatible polyarginine and chondroitin sulfate. The $MnCO_3$ template particles were then dissolved in deaerated 0.01 M HCl, resulting in biocompatible capsules containing DNA/spermidine complex. Further treatment with 0.1 M HCl led to the decomposition of the DNA/spermidine complex, and only low molecular weight spermidine was released from the capsule interior, which results in a DNA-entrapped biocompatible capsule was complete.

As described previously for the example of the glucose sensor based on an LB film structure, ultrathin films including biomolecular components such as enzymes are useful nanostructured supermolecules for design of biosensors and bioreactors. However, the LB technique requires good packing of lipidic components for successful transfer onto a solid support. This characteristic is rather disadvantageous for material diffusion within the film structure. Therefore, LB structures are not appropriate media for sequential multi-enzyme reactions. Unlike the LB method, alternate LbL adsorption provides multilayer structures with a rather high rate of permeation of small molecules. A good demonstration of this advantage was given by Kunitake and coworkers who prepared a multienzyme reactor, as illustrated in Figure 16, where two enzymes, glucoamylase and glucose oxidase, were assembled in the same film [51]. The substrate of this reactor is starch, which is hydrolyzed to glucose by glucoamylase through hydrolysis of the glycoside bond. The glucose produced is converted into gluconolactone by glucose oxidase with H_2O_2 as a byproduct.

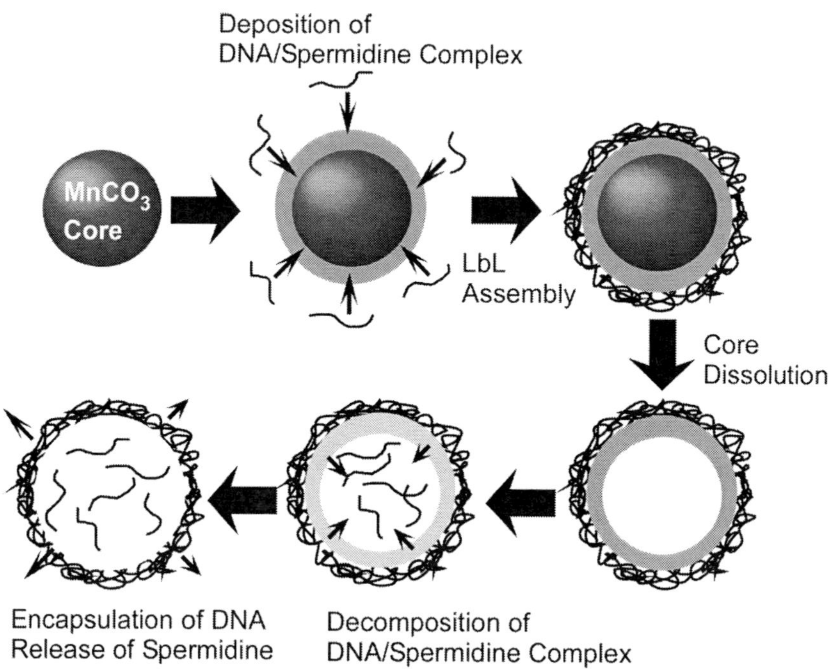

Figure 15. DNA encapsulation by a biocompatible polyelectrolyte capsule.

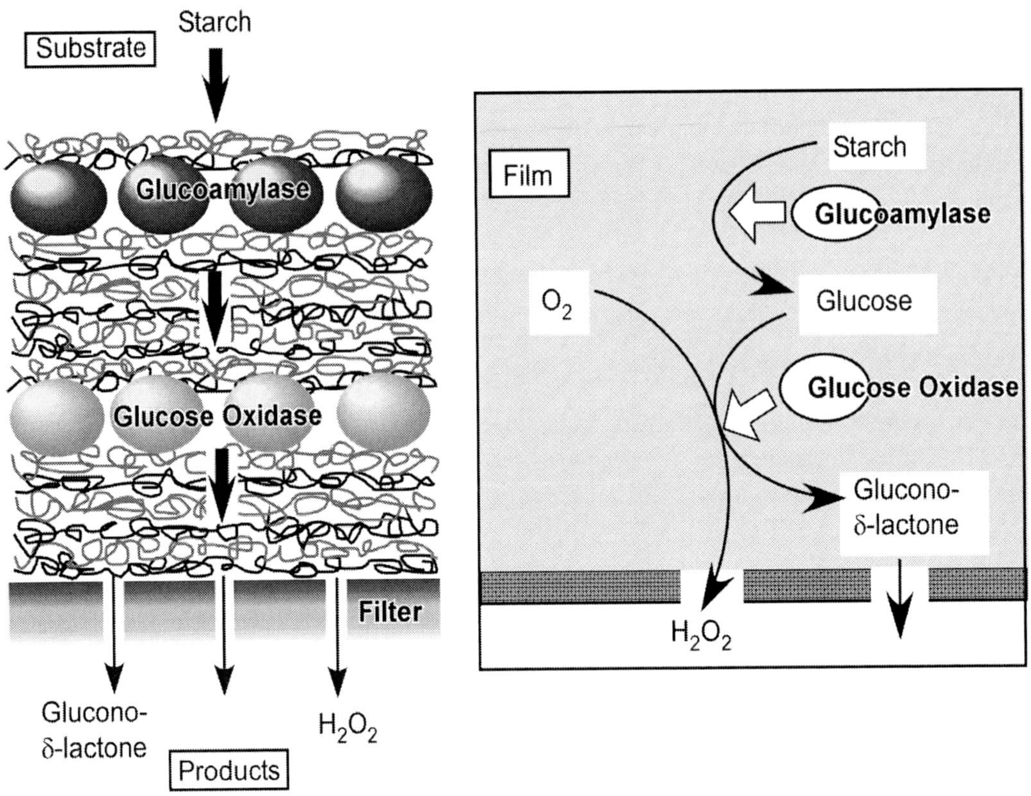

Figure 16. Multienzyme reactor with glucoamylase and glucose oxidase in LbL films.

A pronounced advantage of the LbL technique is freedom in film construction. Various film structures are easily prepared simply by changing the dipping number and sequence during LbL film preparation so that reactor structures can be easily optimized. Figure 17 summarizes relative enzyme activity depending on film design. The highest yield of products was obtained using film 1 in which the order of the enzyme layers agrees with the reaction sequence. Film 2, which also has an appropriate sequence of the enzyme layers, showed the second highest yield. The difference between film 1 and film 2 suggests the importance of the separation between the two enzyme layers, probably related to the inhibition of glucoamylase activity by gluconolactone. A larger separation between the two enzymes should reduce this inhibition. When the disposition of the two enzyme layers was reversed as seen in film 3, the reaction efficiency remains low. In that film structure, starch has to penetrate into the inner glucoamylase layer, and the resulting glucose has to diffuse to the outer glucose oxidase layer against the substrate flow. This film construction is apparently unfavorable for a sequential reaction. Film 4, with randomly assembled enzyme layers, displayed the lowest activity probably because of the inhibition of glucoamylase and/or unbalanced ratio of the two enzymes. This result supports strongly the superiority of controlled layering by using the LbL technique over random immobilization of the enzymes.

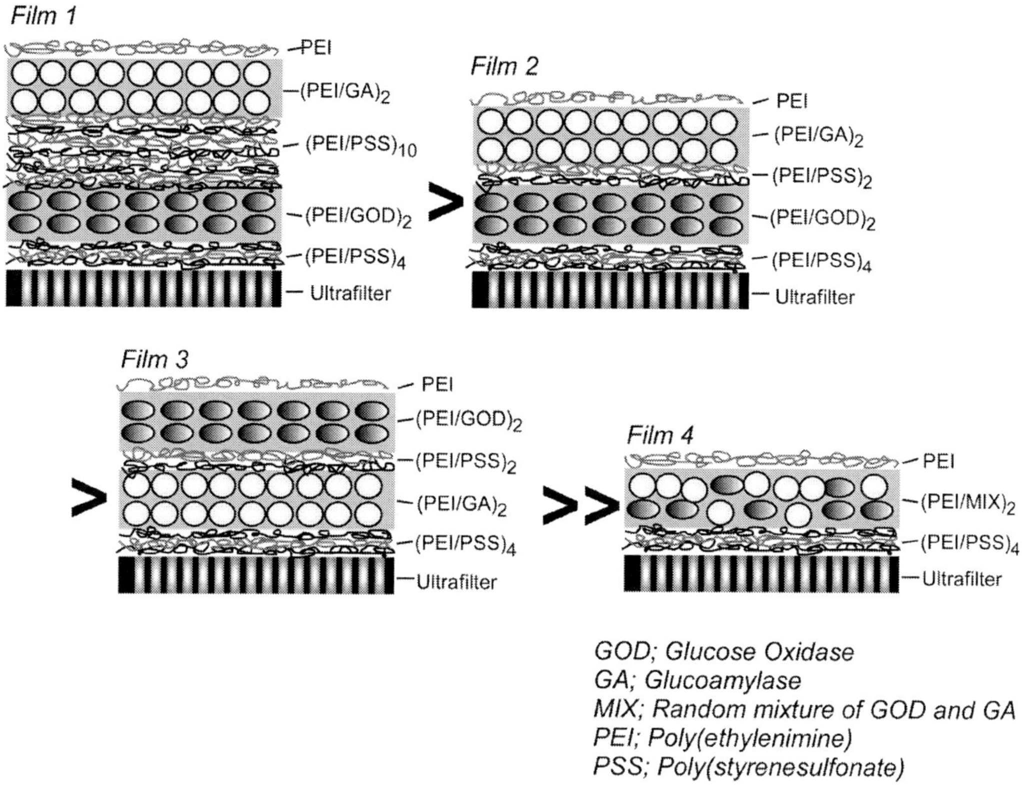

Figure 17. Order of enzyme activity among variously designed LbL films.

5. Mesoporous Materials

Nanostructured supermolecules can be used as templates for syntheses of stable nanostructured materials. A representative example is preparation of the mesoporous materials. For example mesoporous silica can be synthesized by structure transcription using micellar supramolecular assemblies (Figure 18). Kuroda and coworkers first reported the preparation of mesoporous silica with uniform pore size distribution from the layered polysilicate kanemite (FSM-16, Folded Sheet Materials) [52,53], followed by a significant breakthrough by Mobil scientists with disclosure of the M41S family of materials having large uniform pore structures, high specific surface areas and large specific pore volumes. The family includes hexagonal-MCM-41 [54,55], cubic-MCM-48 [56], and lamellar-MCM-50 [57]. HMS (Hexagonal mesoporous silica) [58], prepared using a neutral amine as template, and MSU-1 (Michigan State University) [59] using polyethylene oxide as a structure directing agent have also been prepared. Highly ordered large pore mesoporous silica, SBA-15 (Santa Barbara Amorphous) [60,61] with thicker pore walls and two-dimensional hexagonal structure, were prepared using amphiphilic triblock-copolymer of poly(ethylene oxide) and poly(propylene oxide) as the structure directing reagents in highly acidic media. MCF (meso cellular form) [62] materials can be prepared by using triblock-copolymers stabilized oil in water to result in aerogel-like structure.

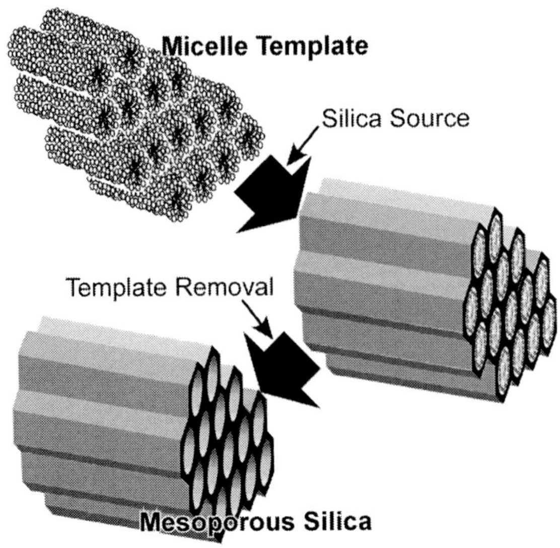

Figure 18. Outline of mesoporous silica synthesiS

Because mesoporous silica materials possess reliable pore geometries, huge surface area, and large pore volume, they can be used for storage and controlled release of drugs. Martens and coworkers reported on the enhanced release of the hydrophobic drug itraconazole from the ordered mesoporous SBA-15 silica material (Figure 19) [63]. In their report, rapid release kinetics from SBA-15 were explained in terms of a displacement desorption of itraconazole by the influx of water. Adsorbed itraconazole molecules can be desorbed from the surface by competitive adsorption with water molecules. They also expected that the presence of a sufficiently wide pore diameter could be the key for accelerating the release of the poorly soluble drug itraconazole. Suppression of the intermolecular itraconazole interactions apparent from its crystal structure by separating the drug molecules at the SBA-15 silica surface seems to be the key to circumvent slow dissolution kinetics. Therefore, there exists a minimal pore size at which the release rate of bulky hydrophobic itraconazole molecules is maximized.

Introduction of functional elements to mesoporous silica leads to controlled release properties regulated by external stimuli. For example, Fujiwara and coworkers reported pioneering work on development of photo-controlled regulation of drug storage and release (Figure 20) [64,65]. Mesoporous silicate MCM-41 was functionalized only at pore outlet with photoactive coumarin through grafting using a short reaction time. Iniitial incorporation of guest drugs such as cholestane into the mesopores of the modified silica occurred upon exposure to a solution of the guest. Subsequently, the attached coumarin moieties were dimerized upon irradiation of UV light (>310 nm), resulting in storage of the guest cholestane. With UV light irradiation at the different wavelength of 250 nm, the dimerized coumarin was efficiently cleaved and trapped cholestane was released. In this system, photochemical controls of uptake, storage, and release of the drugs in the mesopores of silica were realized.

Lin and coworkers proposed a colloid capping method for a mesoporous-silica-based controlled release delivery system (Figure 21) [66].

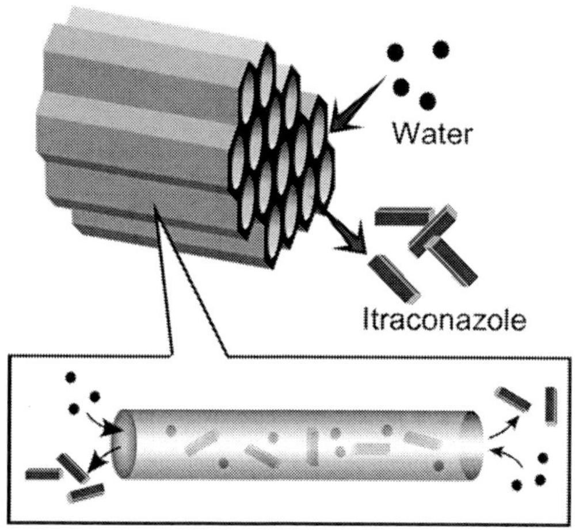

Figure 19. Enhanced release of the hydrophobic drug itraconazole from the ordered mesoporous SBA-15 silica.

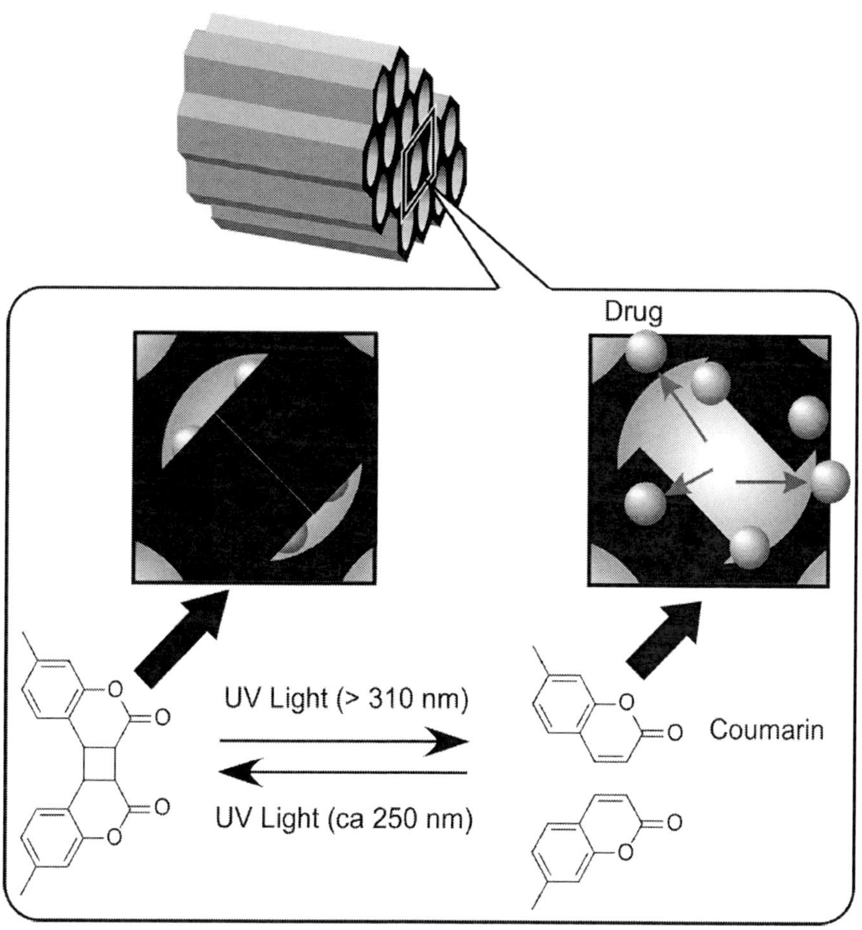

Figure 20. Photo-control regulation of drug storage and release from the coumarin-modified mesoporous silica.

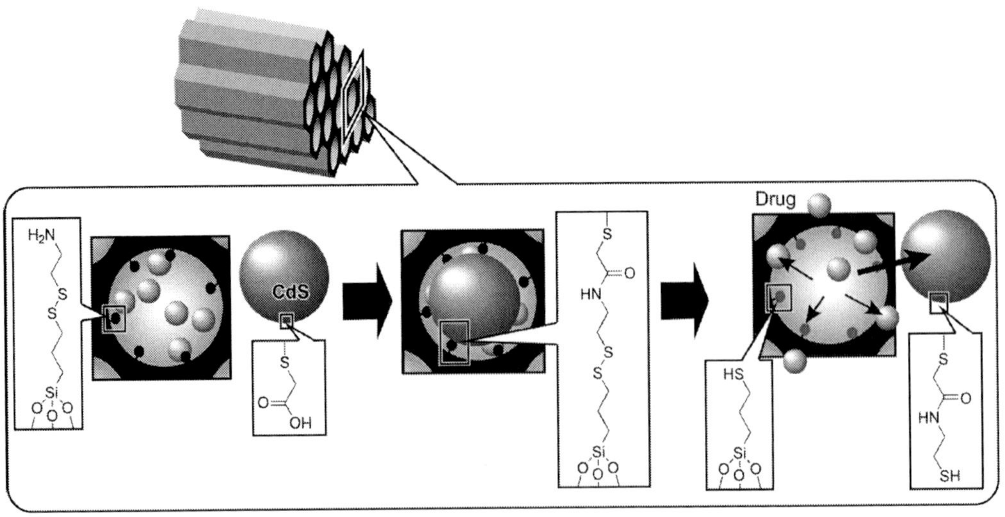

Figure 21. Colloid capping method for a mesoporous-silica-based controlled release.

A mesoporous silica nanospherical material was functionalized with 2-(propyldisulfanyl)ethylamine, and then various drug molecules such as vancomycin and adenosine triphosphate (ATP) were stored in the mesopores. The entrance of the mesopores of the drug-loaded mesoporous material was next capped by allowing the pore surface-bound 2-(propyldisulfanyl)ethylamine functional groups to covalently capture water-soluble mercaptoacetic acid-derivatized CdS nanocrystals. Disulfide linkages could be then chemically cleaved using various reducing agents, such as dithiothreitol or mercaptoethanol. They also demonstrated a biocompatibility with neuroglial cells for this stimuli-responsive system.

Through selective functionalization of external and internal surfaces of mesoporous silica, Lin and coworkers realized gate-controlled molecular sensing (Figure 22) [67]. Thiol-functionalized mesoporous silica MCM-41 nanospheres were synthesized using a co-condensation route. Epoxyhexyl groups were grafted selectively at the external surface of the mesoporous silica prior to removal of the surfactant template. The epoxy group was then converted to the dihydroxyhexyl group, and L-lactide was polymerized only at the external surface. *O*-phthalic hemithioacetal functional groups were immobilized within the mesopores as sensing elements because the *o*-phthalic hemithioacetal can emit fluorescence upon reaction with amines. In this system, the polylactate layer at the outside played the role of gatekeeper, resulting in selectivity in the detection of neurotransmitters such as dopamine, tyrosine, and glutaric acid.

The typical supramolecular element, rotaxane, has also been used for valve gating in controlled release from mesoporous materials. As illustrated in Figure 23, Stoddart and coworkers reported the tethering of pseudorotaxanes as gates at the entrances of the cylindrical pores in mesostructured silica creating nanovalves capable of trapping luminescent molecules and release them on demand [68]. In this design, a tethered 1,5-dioxynaphthalene-containing derivative acts as the gatepost, and a cyclic unit, which recognizes dioxynaphthalene units, serves as the gate. An external reducing reagent such as $NaCNBH_3$ liberated the rotaxane structure and opened the nanovalve.

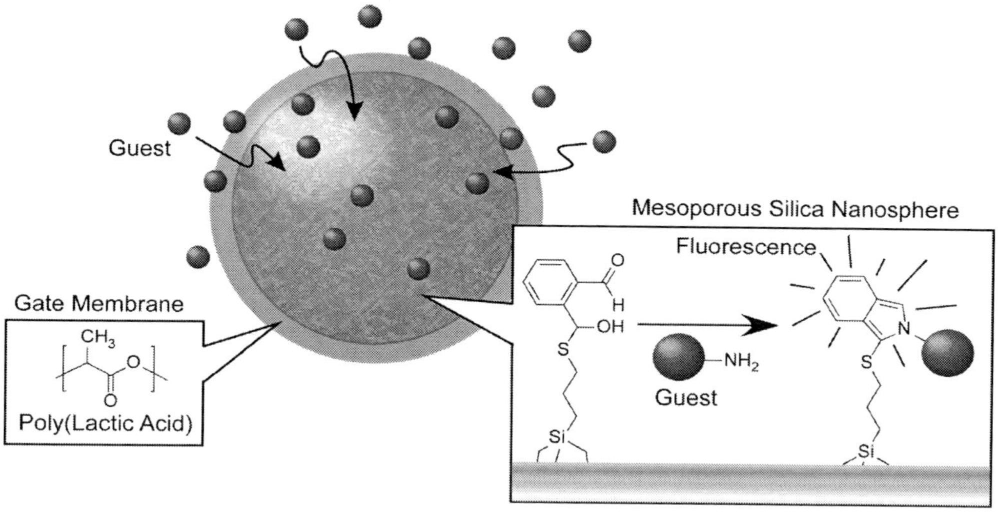

Figure 22. Gate-controlled molecular sensing system by polylactate-grafted mesoporous silica.

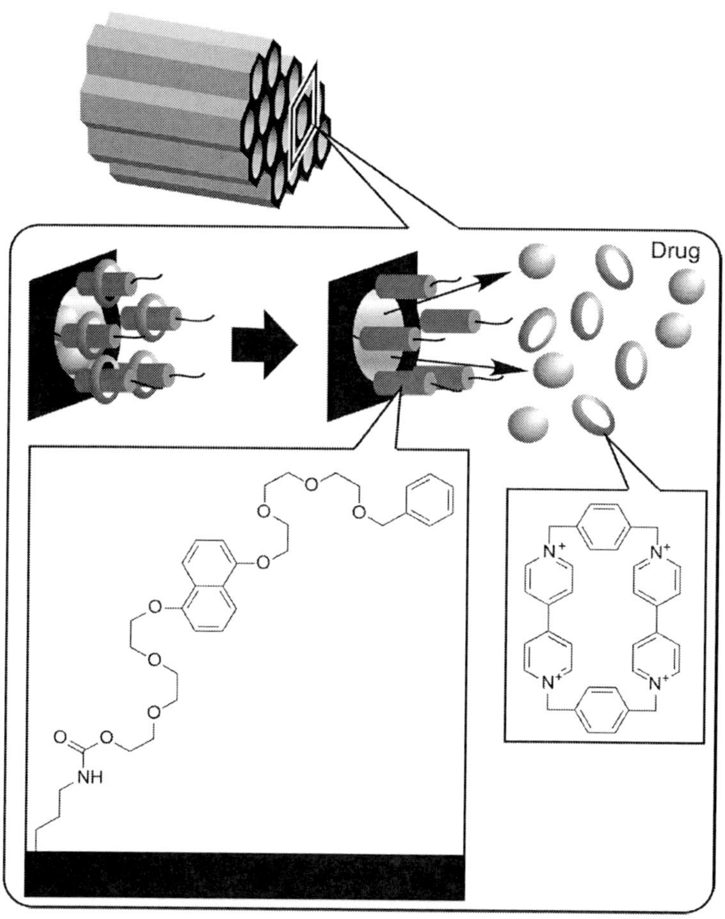

Figure 23. Pseudorotaxane gate for controlled release from mesoporous silica.

The same research group also realized a reversibly operating nanovalve that can be toggled on and off using redox chemistry on bistable[2]rotaxane molecules tethered at the openings of mesopores [69].

The above-mentioned examples are scientifically elegant, but are often accompanied by complicated synthetic processes. In order to eliminate this drawback, Choy and coworkers proposed a drug-polymer silica (DPS) - drug delivery system (DDS) (Figure 24) [70]. The DPS-DDS could be directly synthesized by incorporating the silica source tetraethyl orthosilicate with a C_2-ceramide (a water-insoluble model drug) and the mesoporous structural template triblock-copolymer. This one-pot process resulted in encapsulation of non-ionic drug molecules that could be doubly encased in soft (polymer) and hard (inorganic) capsules. The simplicity of the one-pot synthesis suggests potential for large-scale production. The therapeutic efficacy of DPS-DDS on melanoma cells was found to be higher than that of C_2-ceramide itself.

Martínez-Máñez and coworkers reported nucleotide-dependent controlled release using an MCM-41 mesoporous solid support functionalized at its external surface with polyamines for controlled entrapment and release of guest anions (Figure 25) [71]. At neutral and slightly basic pH, the attached polyamines were only partially protonated, and the effects of pore blockage were much less pronounced. The ability of the materials to deliver the $[Ru(bipy)_3]^{2+}$ guest at neutral pH was studied in the presence of fluoride, chloride, bromide, iodide, nitrate, phosphate, sulfate, acetate, and carbonate anions without notably affecting the delivery process. In contrast, upon addition of aqueous solutions of adenosine triphosphate (ATP) and adenosine diphosphate (ADP), the pores of the mesoporous materials were blocked. The chromogenic response of these materials was related to the ability of the protonated polyamines to strongly coordinate nucleotides. Pore blockage was observed for anions that form strong complexes with polyamines (ATP > ADP > guanosine monophosphate GMP).

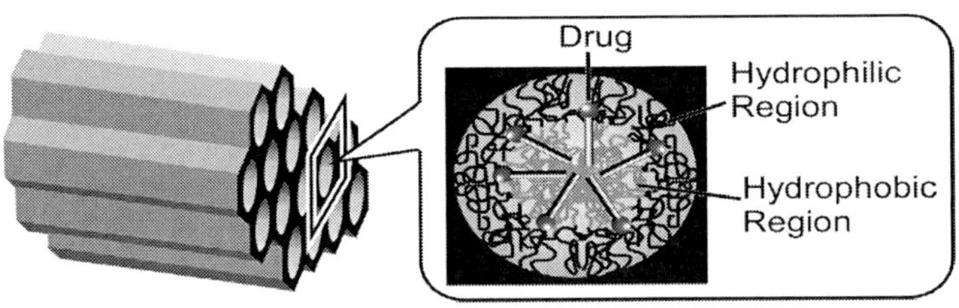

Figure 24. Drug delivery system of a drug-polymer silica.

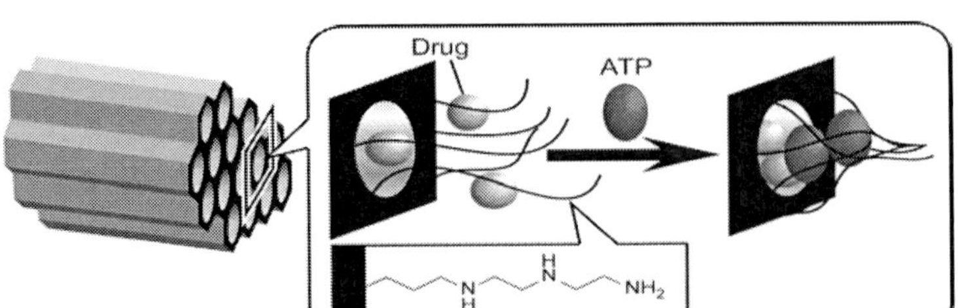

Figure 25. Nucleotide-dependent controlled release using an MCM-41 mesoporous silica functionalized polyamines.

Because solution pH is a generally applicable factor, pH-responsive controlled release could provide drug delivery systems applicable in a wide range of applications. Martínez-Máñez and coworkers demonstrated a controlled molecular gate using mesoporous functionalized materials with pH responsive groups at the pore outlets (Figure 26A) [72]. Oligoamine-type units, $-CH_2CH_2CH_2(NHCH_2CH_2)_2NH_2$, were introduced at the silica surface prior to template removal, leading to a preferential anchoring of the amino groups at the external surface. Removal of the template resulted in a locally functionalized mesoporous silica material. The pH-controlled "open-close" mechanism could arise from hydrogen-bonding interactions between amines (open-gate) and Coulombic repulsion between ammonium groups (closed-gate). Xiao and coworkers also reported a pH-responsive carrier system based on carboxylic acid modified SBA-15 silica rods, and complex formation with poly (dimethyldiallylammonium chloride) (PDDA) was used for storage and release of drug molecules from pore voids (Figure 26B) [73]. Polycations (PDDA) absorbed to carboxylate-functionalized SBA-15 at basic pH through electrostatic attraction, resulting in gate closure for storage of drugs in the mesopores. Transformation of ionized carboxylic acid species into protonated groups by lowering of the pH led to liberation of PDDA from the surface of modified SBA-15, leading to opening of the gates for release of drugs from the mesopores. Recently, Kawi and coworkers reported a similar system as a pH-sensitive drug delivery system [74]. Amine-functionalized mesoporous SBA-15 silica loaded with bovine serum albumin (BSA) was successfully encapsulated with a thin layer coating of poly(acrylic acid) PAA, with the entrapped BSA being released from the PAA-encapsulated SBA-15 at the higher pH value of 7.4 rather than at the lower pH value of 1.2.

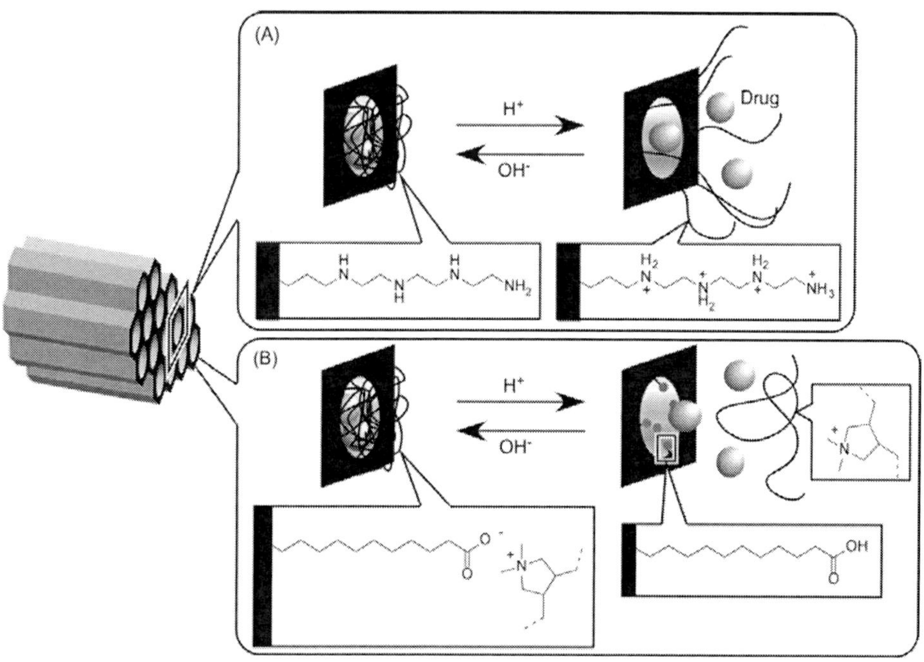

Figure 26. pH-Responsive controlled release from mesoporous silica; (A) by protonation of oligoamine chains; (B) through complex formation process between carboxylate group and poly (dimethyldiallylammonium chloride) (PDDA).

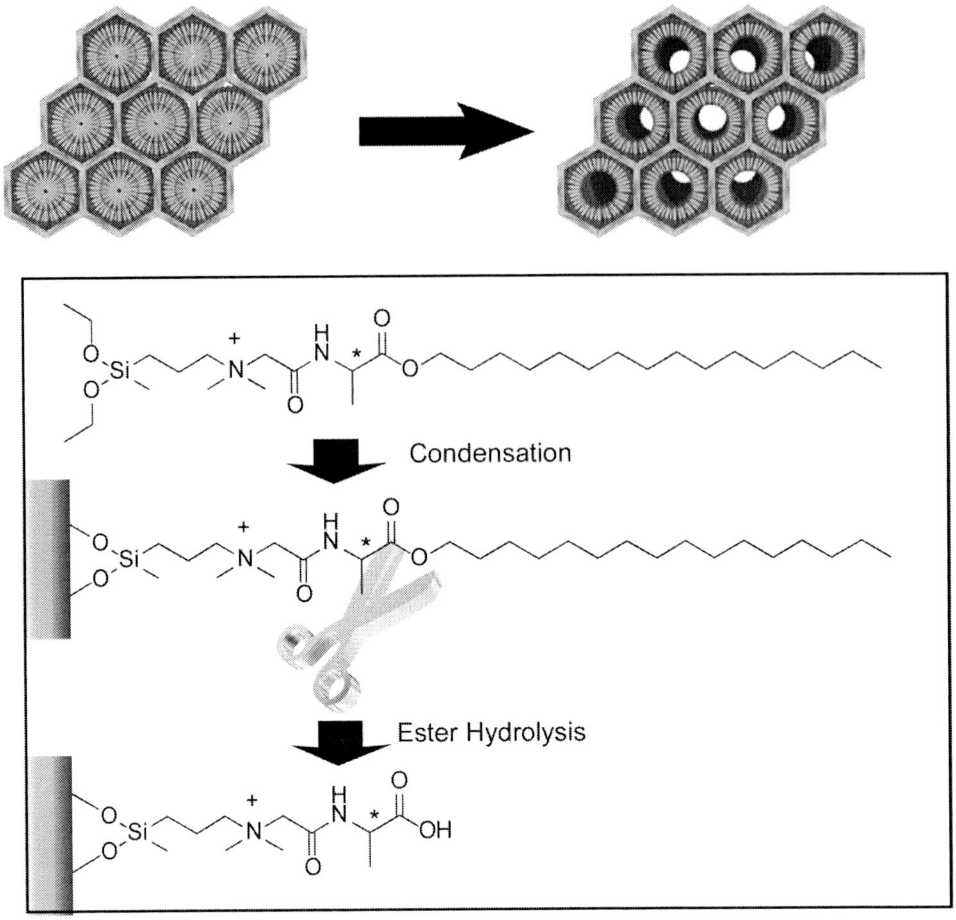

Figure 27. Lizard template method for dense functionalization of the mesopore inside. Reprinted with permission from Q. Zhang et al., *J. Am. Chem. Soc.* 2004, *126,* 988. © 2004, American Chemical Society.

As unique methods to prepare supramolecular hybrids from bio-components and mesoporous silicates, Ariga, Aida, and coworkers developed a new synthetic method that addresses the issues of both dense functionalization of the pore interior and increased accessibility for external guests (Figure 27) [75]. Since the amphiphile used had dialkoxysilane functionality as a part of the head group, the template amphiphile could be covalently attached to the silica framework upon sol-gel reaction with tetraethyl orthosilicate, resulting in mesoporous silica channels of filled with the organic group of the template. Cleavage and removal of the alkyl tail by selective hydrolysis of the ester at the C-terminal leaves open pores with a surface covalently functionalized with alanine residues. Here the template behaves like a 'lizard,' whose head bites the silica wall and whose tail can be cleaved. Temperature-programmed desorption (TPD) analysis with NH_3 as a basic guest confirmed exposure of the alanine C-terminal in the silicate channel. From the desorption profile, it was found that the hydrolyzed porous silica adsorbed an amount of NH_3 comparable with the quantity of immobilized alanine C-terminal.

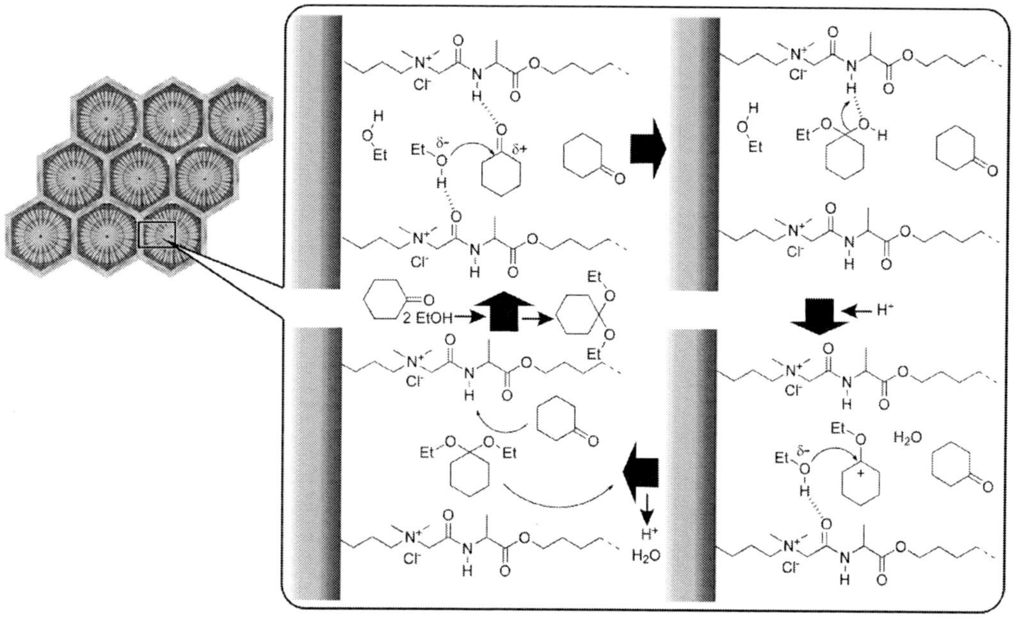

Figure 28. Mesoporous silica hybrid as a reactor for acetalization of a ketone.

It is anticipated that the lizard templating method will be used for the fabrication of mesoporous silica with a variety of organic functional groups. In particular, the immobilization of biological residues by this method is an attractive procedure because biologically important functionalities can be covalently confined in pores with defined dimensions. This should provide artificial active sites for biological functions.

Similarly prepared hybrid structures were used for reactor applications. Aida and coworkers illustrated the catalytic capability of unhydrolyzed materials on the acetalization of a ketone, such as cyclohexanone, in ethanol (EtOH) under mild conditions (Figure 28) [76]. Cyclohexanone and EtOH were incorporated simultaneously into the silica channels at the hydrophobic inner domain and hydrophilic (ionic) outer shell, respectively. This caused their activation through hydrogen-bonding interactions with the amide NH and carbonyl groups of the peptide functionalities, located at the core–shell interface. The acetalization should involve transient carbocationic intermediates, whose generation is favored by a highly polar environment containing concentrated ammonium salt functionalities. As shown above, in the amphiphilic core–shell architecture of the immobilized rod-like micelle, hydrophobic and hydrophilic reactants can be incorporated simultaneously into the silica channels and activated through hydrogen-bonding interactions with the peptidic functionalities located at the core–shell interface. This work should provide a novel and general strategy for rational molecular design of bio-inspired solid catalysts as a novel type of artificial enzyme.

Mesoporous silica structures have been used as hard templates for further material processing. For example, carbonization within mesoporous silica templates and selective edging of silica components can provide carbon materials with highly regular porous structures that are called mesoporous carbon. Mesoporous carbon materials structures were first reported by Ryoo using sucrose as a carbon source and mesoporous silicate templates such as MCM-48, SBA-1 and SBA-15 and were named CMK-x [77,78]. Hyeon et al.

reported, independently and somewhat later, a similar approach to the well-ordered mesoporous carbon materials designated SNU-x [79]. Vinu and coworkers recently reported synthesis of the novel nanocarbon, "carbon nanocage" [80,81], through template synthesis using a three-dimensional large cage-type face centered cubic mesoporous silica material (KIT-5) as inorganic templates (Figure 29A). High resolution transmission electron microscopy (HR-TEM) revealed the regular structure of carbon nanocage (Figure 29B). When compared with conventional mesoporous carbon, both the specific surface area and specific pore volume of carbon nanocage are much greater so that carbon nanocage possesses superior adsorption capability. For example, adsorption capacity of lysozyme on the carbon nanocage is much larger than that observed with mesoporous carbon CMK-3.

Carbon nanocage also shows powerful capabilities for separation of small molecules [82]. In Figure 30A, the superior adsorption capability of carbon nanocage was demonstrated by removal of a dyestuff (Alizarin Yellow). An aqueous solution was passed with application of a slight pressure through a bed of the respective carbon material deposited on top of a cotton plug in a pipette. When compared with the control test without carbon, the carbon nanocage materials completely removed the dye, while activated carbon powder and CMK-3 were not effective for removal of the dye under these conditions. A more practically meaningful application was tested using tea components.

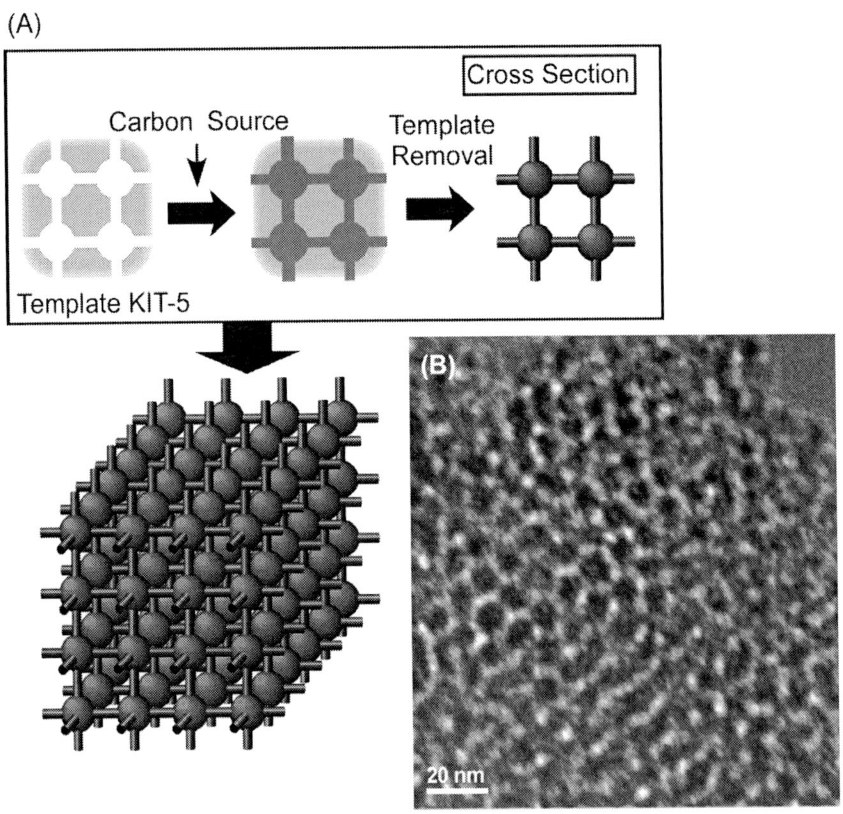

Figure 29. (A) Synthesis of carbon nanocage and (B) its TEM image. Reprinted with permission from K. Ariga et al., *J. Am. Chem. Soc.* 2007, *129,* 11022. © 2007, American Chemical Society.

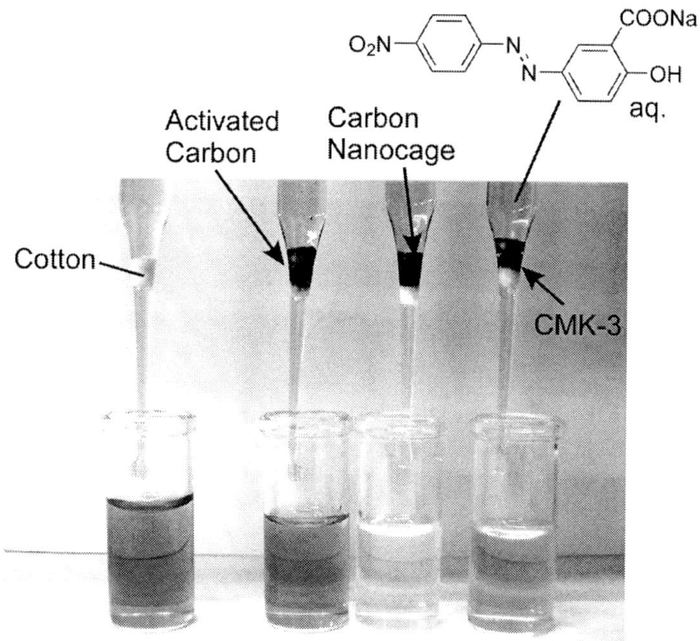

Figure 30. Adsorption experiment of Alizarin Yellow through carbon materials: from left, without carbon materials, through activated carbon, through carbon nanocage, through CMK-3. Reprinted with permission from K. Ariga et al., *J. Am. Chem. Soc.* 2007, *129,* 11022. © 2007, American Chemical Society.

As illustrated in Figure 31, selective separation of catechin and tannic acid by the carbon nanocage material in a one-pot process was realized. Adsorption ratios between these tea components at 1 g L^{-1} are also summarized in Figure 31, where complete inversion in the selectivity for catechin and tannic acid is demonstrated going from activated carbon, CMK-3 to carbon nanocage. Surprisingly, use of carbon nanocage as adsorbent provided a highly selective adsorption of tannic acid (ca. 95%) in a simple one-pot process. This result could originate from the very unique geometry of the carbon nanocage materials.

6. Future Perspectives

In this chapter, we have described briefly some types of biomimetic function, controlled release and materials conversion, which are operate within nanostructured supermolecules. The main topics were categorized as lipid thin films, layer-by-layer assemblies, and mesoporous materials. All of the methodologies provide excellent hybrid structures which display fine properties related to controlled release and materials conversion. In addition, these approaches could be useful for preparation of hybrid structures between supramolecular elements and rigid framework materials such as polymers and inorganic substances. Therefore, the nanostructured supermolecules obtained should be of great use for practical applications such as drug delivery, reactor applications, and sensor preparation.

However, currently available nanostructured supermolecules are at a primitive state of development relative to those observed in naturally occurring systems.

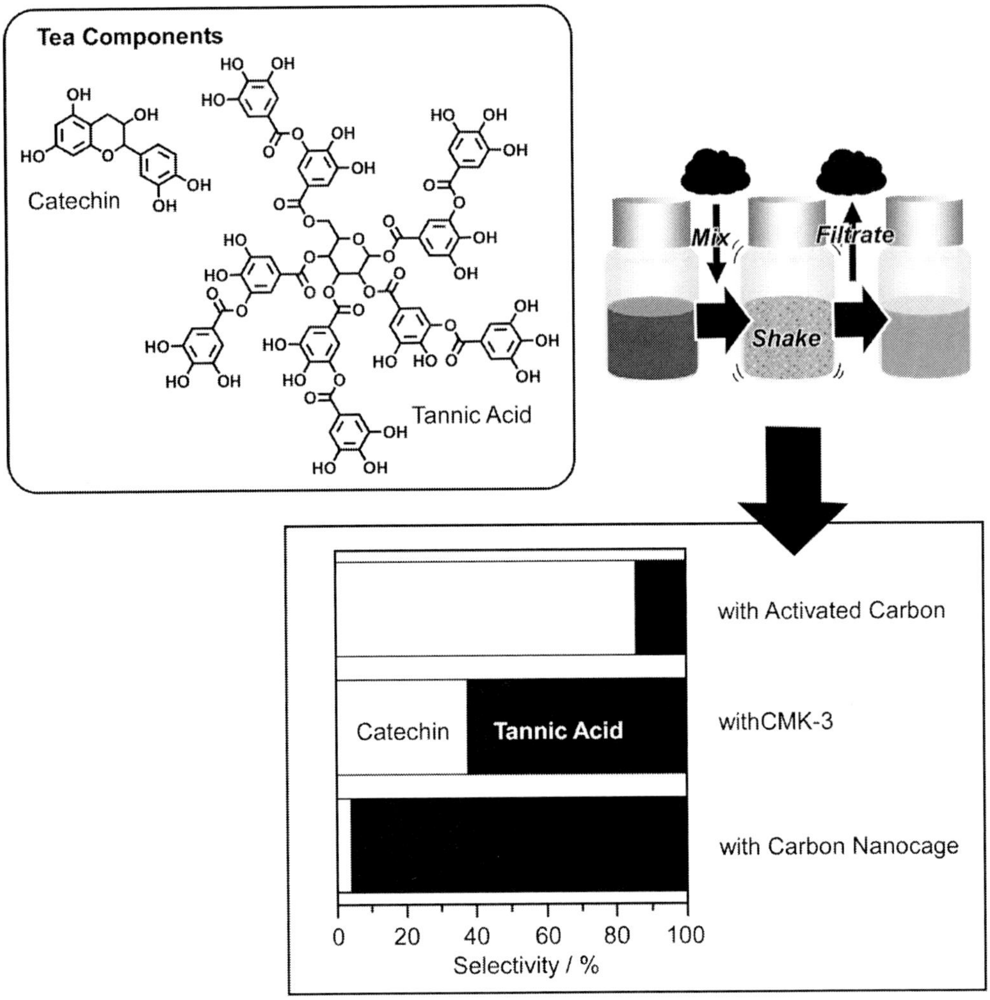

Figure 31. Selective separation of catechin and tannic acid upon adsorption. Reprinted with permission from K. Ariga et al., *J. Am. Chem. Soc.* 2007, *129,* 11022. © 2007, American Chemical Society.

Functional structures existing in biological systems are extremely complex and often adopt hierarchical organization, enabling them to perform highly sequential, parallel, and feed-back controlled processes. These characteristics are not available in the current simple supramolecular systems. Therefore, design and construction of supramolecular structures with hierarchical architectures could be an important key for the future direction of the corresponding research areas. Some attempts along this line have been already demonstrated. One recent approach for preparation of hierarchical thin film structures can be seen in the LbL assemblies of organic-inorganic hybrid vesicles (cerasome) reported by Katagiri et al [83,84]. The assemblies formed can be regarded as hierarchic multi-cellular mimics. Recently, Akagi at Kyoto University proposed the novel concept of "super-hierarchical structures" as a new research field [85]. Cumulative efforts should lead to establishment of methodologies for construction of much more complex nanostructured supermolecules that are more capable of mimicking sophisticated biological functions.

Acknowledgment

The researches described in this chapter were partially supported by World Premier International Research Center Initiative (WPI Initiative), MEXT, Japan, Grant-in-Aid for Scientific Research on Priority Areas "Chemistry of Coordination Space", MEXT, Japan, a Grant-in-Aid for Science Research in a Priority Area "Super-Hierarchical Structures", MEXT, Japan, and Grants-in-Aid for Scientific Research (B), JSPS, Japan.

Reference

[1] Litzinger, D. C.; Huan, G. L. *Biochem. Biophys. Acta* 1992, 1113, 201.
[2] Bellomo, E. G.; Wyrsta, M. D.; Pakstis, L.; Pochan, D. J.; Deming, T. J. *Nat. Mater.* 2004, 3, 244.
[3] Gregoriadis, G. *Trend Biotechnol.* 1995, 13, 527.
[4] Torchilin, V. P.; Rammohan, R.; Weissig, V.; Levchenko, T. S. *Proc. Natl. Acad. Sci., USA* 2001, 98, 8786.
[5] Rezler, E. M.; Khan, D. R.; Lauer-Fields, J.; Cudic, M.; Baronas-Lowell, D.; Fields, G. B. *J. Am. Chem. Soc.* 2007, 129, 4961.
[6] Okahata, Y. *Acc. Chem. Res.* 1986, 19, 57.
[7] Okahata, Y.; Lim, H. J.; Nakamura, G.; Hachiya, S. *J. Am. Chem. Soc.* 1983, 105, 4855.
[8] Okahata, Y.; Seki, T. *Chem. Lett.* 1984, 1251.
[9] Okahata, Y.; Seki, T. *J. Am. Chem. Soc.* 1984, 106, 8065.
[10] Okahata, Y.; Lim, H. J.; Nakamura, G. *Chem. Lett.* 1983, 755.
[11] Okahata, Y.; Lim, H. J. *J. Am. Chem. Soc.* 1984, 106, 4696.
[12] Okahata, Y.; Lim, H. J.; Hachiya, S. *Macromol. Chem. Rapid Commun.* 1983, 4, 303.
[13] Okahata, Y.; Noguchi, H. *Chem. Lett.* 1983, 1517.
[14] Okahata, Y.; Hachiya, S.; Ariga, K.; Seki, T. *J. Am. Chem. Soc.* 1986, 108, 2863.
[15] Okahata, Y.; Ariga, K.; Seki, T. *J. Am. Chem. Soc.* 1988, 110, 2495.
[16] Okahata, Y.; Ozaki, K.; Seki, T. *J. Chem. Soc., Chem. Commun.* 1984, 519.
[17] Okahata, Y.; Noguchi, H.; Seki, T. *Macromolecules* 1986, 19, 493.
[18] Okahata, Y.; Nakamura, G.; Noguchi, H. *J. Chem. Soc., Perkin Trans 2* 1987, 1317.
[19] Okahata, Y.; Ariga, K.; Seki, T. *J. Chem. Soc., Chem. Commun.* 1985, 920.
[20] Okahata, Y.; Ariga, K. *J. Org. Chem.* 1986, 51, 5064.
[21] Okahata, Y.; Ariga, K.; Seki, T. *J. Chem. Soc., Chem. Commun.* 1986, 73.
[22] Okahata, Y.; Ariga, K. *J. Chem. Soc., Perkin Trans.* 2 1987, 1003.
[23] Okahata, Y.; Ariga, K.; Shimizu, O. *Langmuir* 1986, 2, 538.
[24] Okahata, Y.; Ariga, K.; Nakahara, H.; Fukuda, K. *J. Chem. Soc., Chem. Commun.* 1986, 1069.
[25] Ariga, K.; Okahata, Y. *J. Am. Chem. Soc.* 1989, 111, 5618.
[26] Okahata, Y.; Truruta, T.; Ijiro, K.; Ariga, K. *Langmuir* 1988, 4, 1373.
[27] Okahata, Y.; Tsuruta, T.; Ijiro, K.; Ariga, K. *Thin Solid Films* 1989, 180, 65.

[28] Decher, G. Science 1997, 277,1232.

[29] Ariga, K.; Hill, J. P.; Ji, Q. *Phys. Chem. Chem. Phys.* 2007, 9, 2319.

[30] Lvov, Y.; Ariga, K.; Onda, M.; Ichinose, I.; Kunitake, T. *Colloid Surf. A- Physicochem. Asp.* 1999, 146, 337.

[31] Shiratori, S. S.; Rubner, M. F. *Macromolecules* 2000, 33, 4213.

[32] Lvov, Y.; Ariga, K.; Ichinose, I.; Kunitake, T. *J. Am. Chem. Soc.* 1995, 117, 6117.

[33] Onda, M.; Lvov, Y.; Ariga, K.; Kunitake, T. *Biotechnol. Bioeng.* 1996, 51, 163.

[34] Caruso, F.; Furlong, D. N.; Ariga, K.; Ichinose, I.; Kunitake, T. *Langmuir*1998, 14, 4559.

[35] Ariga, K.; Lvov, Y.; Kunitake, T. *J. Am. Chem. Soc.* 1997, 119, 2224.

[36] Lvov, Y.; Ariga, K.; Ichinose, I.; Kunitake, T. *Langmuir* 1996, 12, 3038.

[37] Lvov, Y.; Ariga, K.; Onda, M.; Ichinose, I.; Kunitake, T. *Langmuir* 1997, 13, 6195.

[38] Stockton, W.; Rubner, M. F. *Macromolecules* 1997, 30, 2717.

[39] Lee, H.; Kepley, L. J.; Hong, H. G.; Mallouk, T. E. *J. Am. Chem. Soc.* 1988, 110, 618.

[40] Shimazaki, Y.; Mitsuishi, M.; Ito, S.; Yamamoto, M. *Langmuir* 1998, 14, 2768.

[41] Serizawa, T.; Hamada, K.; Kitayama, T.; Fujimoto, N.; Hatada, K.; Akashi, M. *J. Am. Chem. Soc.* 2000, 122, 1891.

[42] Serizawa, T.; Hamada, K.; Akashi, M. *Nature* 2004, 429, 52.

[43] Ikeda, A.; Hatano, T.; Shinkai, S.; Akiyama, T.; Yamada, S. *J. Am. Chem. Soc.* 2001, 123, 4855.

[44] Lvov, Y.; Ariga, K.; Ichinose, I.; Kunitake, T. *J. Chem. Soc., Chem. Commun.* 1995, 2313.

[45] Caruso, F.; Caruso, R. A.; Möhwald, F. *Science* 1998, 282, 1111.

[46] Park, M.-K.; Deng, S.; Advincula, R. C. *J. Am. Chem. Soc.* 2004, 126, 13723.

[47] Lee, D.; Nolte, A. J.; Kunz, A. L.; Rubner, N. F.; Cohen, R. E. *J. Am. Chem. Soc.* 2006, 128, 8521.

[48] Shiratori, S.; Inami, T.; Kikuchi, M. *Thin Solid Films* 2001, 393, 243.

[49] Zhang, J.; Chua, L. S.; Lynn, D. M. *Langmuir* 2004, 20, 8015.

[50] Shchukin, D. G.; Patel, A. A.; Sukhorukov, G. B.; Lvov, Y. M. *J. Am. Chem. Soc.* 2004, 126, 3374.

[51] Onda, M.; Lvov, Y.; Ariga, K.; Kunitake, T. *J. Ferment. Bioeng.* 1996, 82, 502.

[52] Yanagisawa, T.; Shimizu, T.; Kuroda, K.; Kato, C. *Bull. Chem. Soc. Jpn.* 1990, 63, 988.

[53] Inagaki, S.; Fukushima, Y.; Kuroda, K. *J. Chem. Soc., Chem. Commun.* 1993, 680.

[54] Kresge, C. T.; Leonowicz, M. E.; Roth, W. J.; Vartuli, J. C.; Beck, J. S. *Nature* 1992, 359, 710.

[55] Beck, J. S.; Vartuli, J. C.; Roth, W. J.; Leonowicz, M. E.; Kresge, C. T.; Schmitt, K. D.; Chu, C. T. W.; Olson, D. H.; Sheppard, E. W.; McCullen, S. B.; Higgins, J. B.; Schlenker, J. L. *J. Am. Chem. Soc.* 1992, 114, 10834.

[56] Vartuli, J. C.; Schmitt, K. D.; Kresge,C. T.; Roth, W. J.; Leonowicz, M. E.; McCullen, S. B.; Hellring, S. D.; Beck, J. S.; Schlenker, J. L.; Olson, D. H.; Sheppard, E. W. *Chem. Mater.* 1994, 6, 2317.

[57] Dubois, M.; Gulik-Krzywicki, T.; Cabane, B. *Langmuir* 1993, 9, 673.

[58] Tanev, P. T.; Pinnavaia, T. J. *Science* 1995, 267, 865.

[59] Bagshaw, S. A.; Prouset, E.; Pinnavaia, T. J. *Science* 1995, 269, 1242
[60] Zhao, D.; Feng, J.; Huo, Q.; Melosh, N.; Fredickson, G. H.; Chmelka, B. F.; Stucky, G. D. *Science* 1998, 279, 548.
[61] Zhao, D.; Q. Huo, Q.; Feng, J.; Chmelka, B. F.; Stucky, G. D. *J. Am. Chem. Soc.* 1998, 120, 6024.
[62] Schmidt-Winkel, P.; Lukens, W. W.; Zhao, D.; Yang, P.; Chmelka, B. F.; Stucky, G. D. *J. Am. Chem. Soc.* 1999, 121, 254.
[63] Mellaerts, R.; Aerts, C. A.; Humbeeck, J. V.; Augustijns, P. Van den Mooter, G.; Martens, J. A. *Chem. Commun.* 2007, 1375.
[64] Mal, N. K.; Fujiwara, M.; Tanaka, Y. *Nature* 2003, 421, 350.
[65] Mal, N. K.; Fujiwara, M.; Tanaka, Y.; Taguchi, T.; Matsukata, M. *Chem. Mater.* 2003, 15, 3385.
[66] Lai, C.-Y.; Trewryn, B. G.; Jeftinija, D. M.; Jeftinija, K.; Xu, S.; Jeftinija, S.; Lin, V. S.-Y. *J. Am. Chem. Soc.* 2003, 125, 4451.
[67] Radu, D. R.; Lai, C.-Y.; Wiench, J. W.; Pruski, M.; Lin, V. S.-Y. *J. Am. Chem. Soc.* 2004, 126, 1640.
[68] Hernandez, R.; Tseng, H.-R.; Wong, J. W.; Stoddart, J. F.; Zink, J. I. *J. Am. Chem. Soc.* 2004, 126, 3370.
[69] Nguyen, T. D.; Tseng, H.-R.; Celestre, P. C.; Flood, A. H.; Liu, Y.; Stoddart, J. F.; Zink, J. I. *Proc. Natl. Acad. Sci. USA* 2005, 102, 10029.
[70] Son, Y.-H.; Park, M.; Choy, Y. B.; Choi, H. R.; Kim, D. S.; Park, K. C.; Choy, J.-H. *Chem. Commun.* 2007, 2799.
[71] Casasús, R.; Aznar, E.; Marcos, M. D.; Martíınez-Máñez, R.; Sancenón, F.; Amorós, P. *Angew. Chem. Int. Ed.* 2006, 45, 6661.
[72] Casasús, R.; Marcos, M. D.; Martíınez-Máñez, R.; Ros-Lis, R. V.; Soto, J.; Villaescusa, L. A.; Amor´os, P.; Beltrán, D.; Guillem, C.; Latorre, J. *J. Am. Chem. Soc.* 2004, 126, 8612.
[73] Yang, Q.; Wang, S.; Fan, P.; Wang, L.; Di, Y.; Lin, K.; Xiao, F.-S. *Chem. Mater.* 2005, 17, 5999.
[74] Song, S.-W.; Hidajat K.; Kawi, S. *Chem. Commun.* 2007, 4396.
[75] Zhang, Q.; Ariga, K.; Okabe, A.; Aida, T. *J. Am. Chem. Soc.* 2004, 126, 988.
[76] Otani, W.; Kinbara, K.; Zhang, Q.; Ariga, K.; Aida, T. *Chem. Eur. J.* 2007, 13, 1731.
[77] Ryoo, R.; Joo, S. H.; Jun, S. *J. Phys. Chem. B*, 1999, 103, 7743.
[78] Jun, S.; Joo, S. H.; Ryoo, R.; Kurk, M.; Jaroniec, M.; Liu, Z.; Ohsuna, T.; Terasaki, O. *J. Am. Chem. Soc.*, 2000, 122, 10712.
[79] Lee, J.; Yoon, S.; Hyeon, T.; Oh, S. M.; Kim, K. B. *Chem. Commun.*, 1999, 2177.
[80] Vinu, A.; Miyahara, M.; Sivamurugan, V.; Mori, T.; Ariga, K. *J. Mater. Chem.* 2005, 15, 5122.
[81] Vinu, A.; Mori, T.; Ariga, K. *Sci. Technol. Adv. Mater.*, 2006, 7, 753.
[82] Ariga, K.; Vinu, A.; Miyahara, M.; Hill, J. P.; Mori, T. *J. Am. Chem. Soc.*, 2007, 129, 11022.
[83] Katagiri, K.; Hamasaki, R.; Ariga, K.; Kikuchi, J. *J. Am. Chem. Soc.*, 2002, 124, 7892.
[84] Katagiri, K.; Hamasaki, R.; Ariga, K.; Kikuchi, J. *Langmuir*, 2002, 18, 6709.
[85] Akagi, K. *Polym. Int.* 2007, 56, 1192.

In: Biomimetic and Supramolecular Systems Research
Editor: Arturo H. Lima

ISBN: 978-1-60456-405-1

Chapter II

Supramolecular Systems Based on Dielectric Elastomer Actuators for Biomimetic and Biomedical Applications

Federico Carpi*[1] *and Danilo De Rossi
Interdepartmental Research Centre "E. Piaggio"
University of Pisa, School of Engineering
Via Diotisalvi, 2 - 56100 Pisa, Italy

Abstract

So-called ElectroActive Polymers (EAP) currently offer attractive properties as "smart" materials for supramolecular electromechanical transduction, suitable for actuation devices capable of mimicking functional characteristics of biological muscles and structures. Within the broad EAP family, dielectric elastomers represent one of the most performing classes of materials, able to show superior electromechanical properties. This chapter presents the ongoing research activities and the latest results achieved in our lab for the development of soft actuators made of dielectric elastomers, along with their possible uses for biomimetic and biomedical systems. Devices here described exploit the basic principle of operation of any dielectric elastomer actuator, i.e. an electrical squeezing of an elastomeric insulator, although specifically implemented by means of different structural configurations with tailored active features. In particular, new actuators shaped as contractile linear devices are emphasised, with reference to possible applications currently considered.

[1] E-mail: f.carpi@ing.unipi.it.

1. Introduction

The most used machines for the generation of mechanical work currently consist of thermochemical, electromagnetic and pneumatic systems. The physical principles exploited in such motors are completely different from those by which biological machines generate forces and accomplish work. Despite the advanced state of such conventional motor technologies, the development of actuators exhibiting biomimetic characteristics may provide major advances in several areas of technology, such as mechatronics, robotics and biomedical engineering.

In particular, actuators offering high power-to-weight ratios, high efficiency and large degree of compliance would result useful in several applications involving critical and delicate tasks. As an example, the development of 'artificial muscles' is certainly one of the most emblematic cases. Attempting to mimic muscular aspects in advanced robots has practical relevance in a number of significant ambits, particularly for those applications in which conventional robot control techniques score poorly with respect to human performance. This is the case, for example, of handling natural objects in a natural environment.

Human muscles are very unconventional actuators, from an engineering point of view. They are neither pure force generators (like dc electric motors), nor pure motion generators (like stepper motors); rather, they behave like springs, with tunable elastic parameters. This property turns out to be a key element while facing the complexity of motor problems, such as motor redundancy, trajectory formation, negotiation of impacts and motor learning. The built-in compliance of muscles is a winning feature for achieving versatility and robustness, although at the cost of precision. In a natural environment all sorts of uncertainties on interacting objects (on shape, location, surface and bulk properties) make pure mechanical precision unnecessary because what is of functional significance is a global characterization of the integrated sensory-motor system.

Aimed at reproducing the most salient functional features of biological muscles, biomimetic actuators necessarily should target their ideal performances to those of muscles, which also provide limiting design specifications: linear contractions, robustness, stability, long lifetime, built-in tunable compliance, compactness (low volume/power ratio, $< 10^{-8}$ m^3/W), high force density (skeletal muscles can generate stresses from 0.1 to 0.5 MPa), fast response time (mechanical tension of skeletal muscles becomes maximum within 30 ms after neural excitation for a single twitch), high power/weight ratio (typical output power densities of human muscles are of the order of 0.1 kW/kg) and high efficiency of energy conversion (biological muscles show values as high as 45-70 %). Some of these performances have been achieved or even exceeded with conventional motor systems, but most of them are still too ambitious to be approached with devices available today.

Aimed at the satisfaction of these and similar needs, in recent years new types of actuators have been proposed. They are based on so-called ElectroActive Polymers (EAPs) [1,2]. These materials exhibit interesting properties, such as sizable active strains and/or stresses in response to an electrical stimulus, low specific gravity, high grade of processability and down-scalability and, in most cases, low costs. EAPs can be classified in two major categories: 1) ionic EAPs (activated by an electrically-induced transport of ions or molecules), comprehending polyelectrolyte gels, ion polymer metal composites, conducting polymers and carbon nanotubes; 2) electronic EAPs (activated by an external electric field and by Coulombian forces), including piezoelectric polymers, electrostrictive polymers and dielectric elastomers.

Ionic EAPs can be advantageously activated by very low voltages (1-5 V), but can be operated only within a surrounding electrolyte medium and typically present long response times, low reliability and low durability.

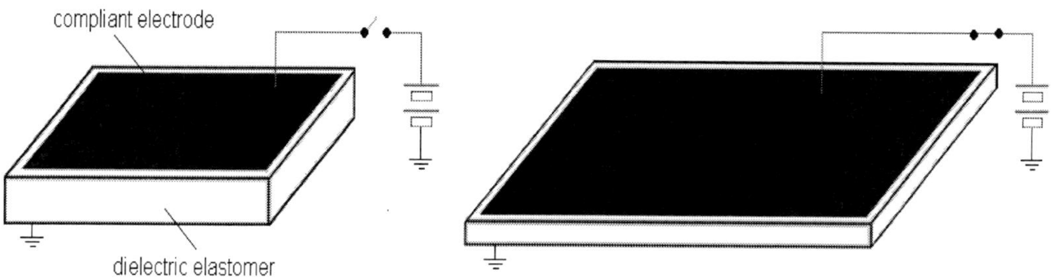

Figure 1. Working principle of a dielectric elastomer planar actuator. The electrode compliance favours the polymer squeezing from the rest state (left picture) to the electrically stressed one (right picture).

Electronic EAPs typically require high driving electric fields (order of 10-100 V/μm for electrostrictive polymers and dielectric elastomers) but posses interesting actuating performances, in terms of active strains and/or stresses, efficiency, response times, stability, reliability and durability.

Within the class of electronic EAPs, dielectric elastomer (DE) actuators represent a very attractive technology in order to develop biomimetic devices and systems. These actuation technology was pioneered by the Stanford Research Institute (SRI International, U.S.A.) [3-9]. The excellent electromechanical properties offered by this class of materials, in terms of large actuation strains and stresses, short response times, high coupling efficiencies, stability, reliability and durability, along with low costs, make it today one of the best choices for electromechanical polymer actuation [3-18]. This technology relies on electrically insulating dielectric polymers with a low elastic modulus, that are subjected to high electric fields applied via compliant electrodes. They can exhibit large deformations, mainly due to a Coulombian effect, that is by means of the electrostatic interactions among free charges on the electrodes. In particular, when a thin film of such materials is sandwiched between two compliant electrodes and a high voltage difference is applied between them, the polymer undergoes a deformation sustained by the electric field, consisting of a thickness squeezing with a related surface expansion, as sketched in Figure 1 [3,4].

The Coulombian force acting between the electrode free charges originates the so-called Maxwell stress. This kind of stress acts on any kind of dielectric material subjected to an applied electric field. However, it is evident that the corresponding deformations will be markedly emphasized by the polymer softness, as well as by the eventual compliance of the electrodes. These two key-features macroscopically distinguish actuating devices made of dielectric elastomers from those based on different electric-field-driven dielectrics, such as piezoelectric or electrostrictive materials.

Acrylic and silicone rubbers are the most representative materials of the dielectric elastomer class for actuation purposes. Such kinds of polymers comprehend representatives possessing very low elastic moduli and high dielectric strengths, so that to be able to show the highest actuating deformations among all EAP materials [3].

Acrylic polymers (or acrylates) can exhibit high-level actuating capabilities: thickness strains up to 60-70% at 400 V/μm, area strains up to 200% at 200 V/μm and corresponding stresses of some MPa have been reported [3]. Silicones are easily processable (by spin coating, casting, etc.) and permit the realisation of good rubber-like dielectrics. Dielectric breakdown at electric fields ranging from 100 to 350 V/μm have been reported for some silicones, enabling thickness strains up to 40-50% and area strains up to 100% with related stresses of 0.3-0.4 MPa [3].

Research activities of our group in this field move along three main directions: the first is material-oriented, in order to develop materials with improved actuation capabilities; the second is device-oriented, aimed at realising actuating configurations able to get the most of the material potentialities; the last is system-oriented, to pursue the ambitious aim of realising truly bioinspired systems.

The following sections separately summarize the work in progress in our lab in each of these fields. As a remark, we underline here that the extreme complexity of muscular structures makes today realistic only their functional biomimesis (in terms of resulting electromechanical properties), at least for what is concerned by engineering devices of practical utility.

2. Materials

The above-mentioned excellent electromechanical figures of merit possessed by DE actuators in several respects actually make them the most performing materials currently available for electromechanical transduction. Nevertheless, the price for achieving these high-level capabilities is represented by the high driving electric fields currently needed (order of 10-100 V/μm). For a definite polymer thickness, such field levels can be reached by applying high voltages, which may be disadvantageous in many applications. In order to reduce the driving electric fields, polymers with high dielectric constant would be advantageous.

Aimed at addressing this challenging and key issue, several strategies are currently being investigated by different research groups. In particular, a certain number of chemical and/or physical approaches could in principle be taken into account. The most significant of them are mentioned in Table. 1.

In particular, they can be classified in three major groups: random composites, field-structured composites and new synthetic polymers.

Table 1. Non exhaustive list of different methods suitable for improving the dielectric constant of elastomeric polymers

System	Processing	Filler types (examples)
Random composites	Mixing with organic and inorganic fillers	Ferroelectric/piezoelectric ceramics Conductive particles Organic monomers Organic polymers
Field-structured composites	Cross-linking in electric field of elastomers with fillers	Ferroelectric/piezoelectric ceramics
Synthesis of new polymers	Functionalization	--
	Copolymerisation (grafting and/or cross-linking)	--
	Polymer blending	--

The first approach is based on the dispersion into the polymer matrix of a filler, either solid (e.g. powder) or liquid. The second method exploits a composite approach as well, although the material is now cured in presence of an external electric field, in order to align dipoles. The third strategy deals with the synthesis of new materials with tailored characteristics. For a sake of shortness, this chapter can not discuss the very broad activities and results of ongoing studies in these fields. Nevertheless, the interested reader can find some useful information in [10]. Just a few basic considerations are here provided.

The first type of approach concerns the creation of random composites. This strategy for obtaining new dielectric elastomers would appear as the most simple and readily approachable, since it just requires a mixing of highly dielectric components with the dielectric elastomer during its processing. In fact, by filling an ordinary elastomer with a material having a greater dielectric permittivity, it is possible to obtain a resulting composite showing the fruitful combination of the advantageous high elasticity of the matrix and high permittivity of the filler. However, the composite approach typically introduces some basic problems. One of them concerns the mechanical properties of the obtained composite material. In fact, any attempt of increasing the material dielectric constant (which is actually readily achievable by means of highly polarizable fillers) should always preserve the mechanical compliance within an acceptable range, in order to make the material suitable for actuation. In fact, any advantageous increase of the dielectric constant frequently correspond to a disadvantageous mechanical stiffening. Such phenomena have, of course, antagonistic effects on the electromechanical strain response of the material. Another typical drawbacks of random composites consists of an observed typical decrease of the material dielectric strength(i.e. the maximum electric field that the material can sustain before breaking). This fact might be reasonably interpreted as mainly due to interfacial (Maxwell-Wagner) polarization phenomena. As an example, Figure 2 shows results from a silicone elastomer (having a relative permittivity of about 5) mixed with a titanium dioxide powder (having a relative permittivity of about 80) [19].

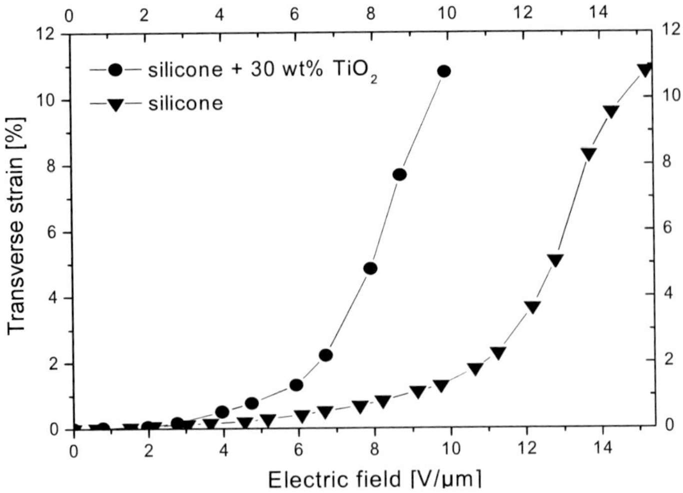

Figure 2. Strain-electric field curves of a planar actuator fabricated with a silicone elastomer in its pure form and loaded with a TiO_2 powder (adapted from [19]). Tests were performed by increasing the driving field, up to dielectric breaking of the material.

The data related to this example show that, although actual reductions of the driving fields are possible, usually they still are relatively poor. Accordingly, such a methodology may not be sufficient in order to approach the problem of achieving a drastic reduction of the driving voltages of DE actuators. For this reason, different types of solutions are also being investigated at present.

The dielectric constant of a random composite may result further increased by cross-linking the loaded elastomers under the action of a proper electric field, in order to obtain so-called field-structured composites. In fact, this may provide an alignment along a preferential direction of the intrinsic polar groups of both the material and the filler; this can happen especially if the latter exhibits ferroelectric properties. As a result, an increase of the global polarizability of the material may occur. Nevertheless, the actual electromechanical properties of field-structured systems may result sensibly affected by possible decreases of the breakdown strength. In fact, depending on the nature and geometry of the filler, field structuring may introduce additional pathways for breakdown currents. Such a method of material processing still poses many issues that need further basic investigations. Thus, deeper studies are currently necessary in order to assess the eventual suitability of this technique for effectively enhancing the actuation performances of elastomers.

An ideal approach to obtain elastomers with specific improved dielectric properties is represented by a challenging synthesis of new molecular architectures. For instance, they might be obtained either as by copolymerisation, by grafting of highly polarizable lateral chains to existing molecules or as blends of known polymers. In particular, blending different polymers can result in new materials with potentially attractive properties. As an example, in a recent work we described the eelctromechanical properties of a silicone/poly-hexyl-thiophene blend [20]. Interestingly, very low percentages of poly-hexyl-thiophene (1-6 wt%) yielded both an increase of the relative dielectric permittivity and an unexpected reduction of the tensile elastic modulus; both these factors synergically contributed to a remarkable increase of the electromechanical response, which reached a maximum at 1 wt% content of conjugated polymer, as presented in Figure 3. Interested readers can refer to [20] for details and discussions.

The most challenging synthesis of new highly-polarizable polymers may offer, of course, superior opportunities, although at the price of considerably greater development times and costs. This represents one of the major future challenges for achieving new generations of more efficient dielectric elastomers suitable for more performing actuators, possibly requiring lower driving voltages.

3. Devices

Many configurations for dielectric elastomer actuators have been proposed and demonstrated so far. They include planar devices, tubes, rolls, extenders, diaphragms, bimorph and unimorph benders [3-20]. Tube and, mainly, roll actuators offer interesting properties for linear actuation [4,7,14]. However, they present a mechanism of activation opposite to that of natural muscles: under electrical stimulation they elongate, instead of contract.

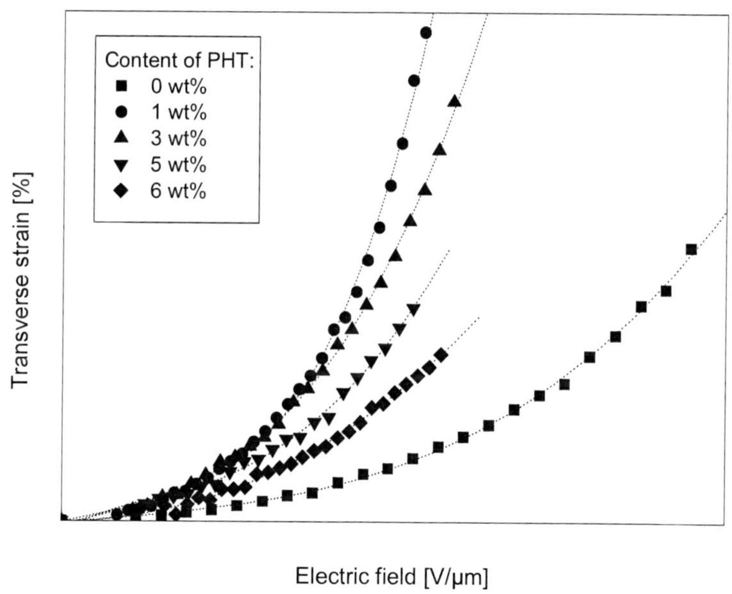

Figure 3. Comparison of the strain response exhibited by a pure silicone rubber and by its blends with different contents of poly-hexylt-hiophene (adapted from [20]).

In order to obtain linearly contracting devices, stack-like actuators have been proposed and demonstrated [15]. They consist of a multi-layer structure obtained by stacking many planar elementary devices connected in mechanical series and electrical parallel, as sketched in Figure 4a. Aimed at reducing the complexity of fabrication of this configuration, requiring layer-by-layer depositions of dielectric and conductive layers, we have proposed a couple of alternative solutions: the so-called *helical* and *folded* dielectric elastomer actuators, reported respectively in Figs. 4b and 4c. Pictures of samples of these two types of actuators are presented in Figure 5. Both these configurations are conceived to obtain new actuators able to undergo electrically-activated linear contractions, as separately described below.

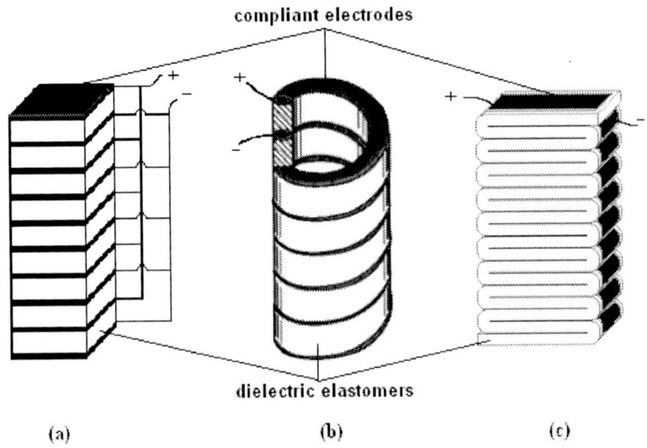

Figure 4. Configurations for DE linear contractile devices: stack (a), helical (b) and folded (c) actuators.

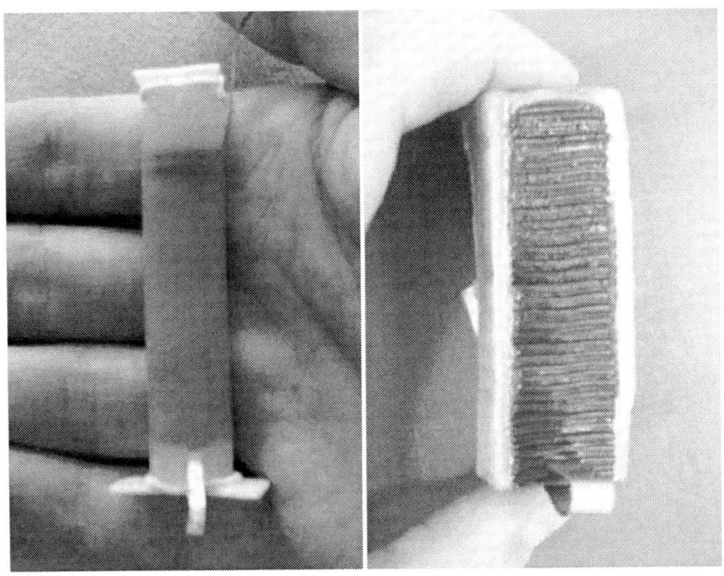

Figure 5. Prototype samples of helical (left picture) and folded (right picture) contractile actuators made of a silicone elastomer.

3.1. Helical Dielectric Elastomer Actuators

A helical dielectric elastomer actuator consists of a hollow cylinder of dielectric elastomer, having two helical compliant electrodes integrated within its wall (Figure 4b). By applying a voltage difference between them, the attraction among opposite charges causes an axial contraction of the actuator, as well as its radial expansion [21]. Axial contraction strains down to –5% in response to an electric field of about 14 V/μm were obtained from initial silicone-made prototypes [21]. Despite the potentialities of this actuating configuration, such relatively low performances achieved so far have to be merely ascribed to the poor custom fabrication technique that has been implemented at a prototype lab scale. In fact, it presents some drawbacks which actually complicate the manufacturing procedure and, in some cases, limit the reliability of samples. Although, several potential improvements are possible and have been already identified [21], the development of this type of actuator has been recently substituted by the fabrication of the folded configuration. In fact, at present it offers considerably better results, owing to its greater structural simplicity, as described below.

3.2. Folded Dielectric Elastomer Actuators

Folded dielectric elastomer actuators have been conceived to obtain a structure functionally equivalent to that of stack-like multilayer devices, but requiring an easier fabrication procedure. In particular, a folded actuator consists of a monolithic strip of electroded elastomer which is folded up, as sketched in Figure 4c [22]. The resulting compact structure is equivalent to a multilayer stack with interdigitated electrodes (Figure 4a). Accordingly, the application of a high voltage difference between the electrodes induces a contraction of the structure along its main axis and related lateral expansions.

With respect to the stack-like configuration, a folded device is advantageously not discontinuous and can be manufactured in one single phase, avoiding layer-by-layer multi-step procedures. Tests on silicone-made prototype samples of folded actuator have reported interesting contraction strains down to –15% with an electric field of about 16 V/μm [22]. Interested readers can refer to [22] for details and discussions.

Both helical and folded actuators may be useful in applications requiring spring-like linear contractions activated and modulated by an electrical signal. Accordingly, they may be regarded as possible candidates for future 'pseudo-muscular actuators' capable of replicating functional properties of natural muscles. They may find useful short-term applications as actuators for biomimetic and biomedical systems. In this respect, the following section briefly presents some applications which are currently under study or development in our lab.

4. Applications

4.1. Actuation of The Eyeballs of an Android Robotic Face

A research project called F.A.C.E. (Facial Automaton for Conveying Emotions) is currently under development in our lab [23]. It deals with the development of an android robotic face endowed with dynamic expressiveness and artificial vision. Aimed at realising an artificial attention system for such an automaton, a biomimetic actuation mechanism for its eyeballs is currently under development [24].

The artificial mechanism is inspired to the human ocular apparatus and is aimed at replicating actions exerted by the main ocular muscles. It is based on the use of contractile dielectric elastomer actuators as pseudomuscular devices capable of electrically driven linear contractions. In particular, the system under study consists of a sphere (eyeball) with a radius of 13 mm, which is hinged with negligible friction on a central vertical axis of rotation. A couple of linear contractile DE actuators in agonist-antagonist configuration is considered, in order to mimic the action of the lateral and medial rectus muscles. They are connected to the ball via inextensible tendon-like wires, which get into the sphere. In order to implement such a type of mechanism, a couple of folded actuators with circular cross-section has been considered, as sketched in Figure 6.

Folded actuators with a cylindrical structure can be easily obtained by opportunely shaping the starting strip to be folded [22]. A prototype sample of this type of device is presented in Figure 7, which also shows a couple of actuators used to implement the configuration sketched in Figure 6 for a simple test.

Figure 8 shows results of a preliminary test performed with rough actuator samples. Eyeball rotations up to 7° could be obtained in both directions [24]. The performances achieved in these first tests were rather limited, due to the poor quality of the fabricated actuator samples. Nevertheless, a theoretical analysis of this type of system suggested that eyeball rotations up to approximately ±25° can be theoretically foreseen with a couple of actuators having a rest length of roughly 54 mm and a Young's modulus of 100 kPa and being driven by an electrostatic pressure of 20 kPa (about three times higher than that applied in these tests) [24]. Such theoretical rotations are consistent with typical performances of the human ocular system.

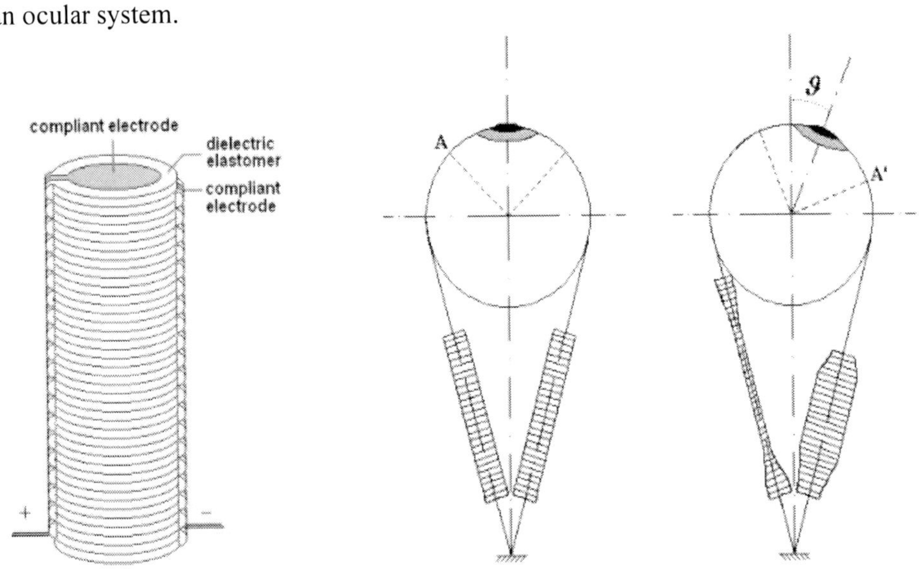

Figure 6. Schematic drawing of a cylindrical folded DE actuator and its proposed use for implementing an agonist-antagonist couple capable of driving an eyeball of the F.A.C.E. robot.

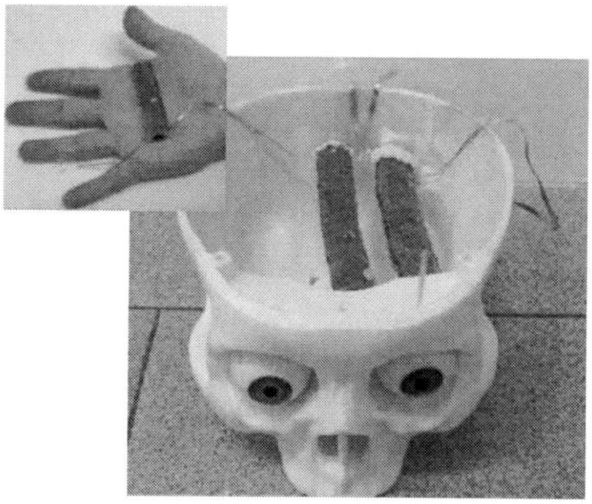

Figure 7. Prototype sample of a cylindrical folded DE actuator and resin skull equipped with an agonist-antagonist couple of these actuators, used to drive a plastic eyeball.

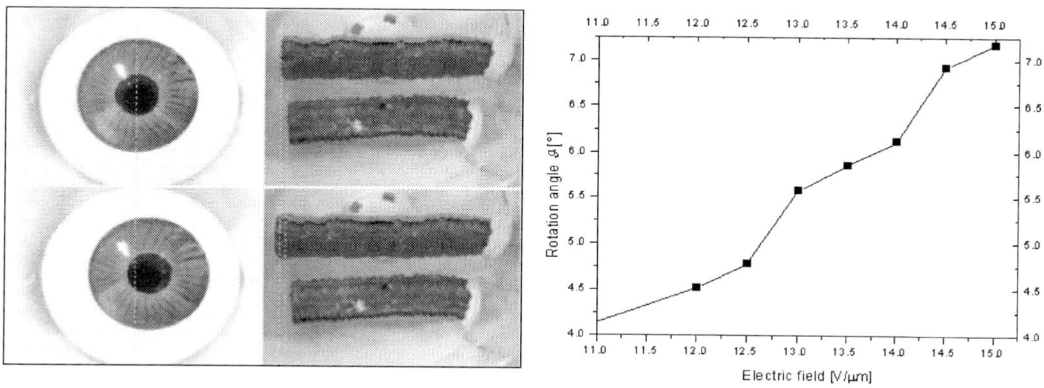

Figure 8. Example of an eyeball rotation induced by the activation of one actuator of the agonist-antagonist couple and plot of the eyeball rotation angle as a function of the electric field applied to the agonist actuator.

In fact, human eyes show rotations usually included within the range of ± 20° about 90% of time [25]. Moreover, most naturally occurring human saccades have magnitudes of 15° or less [26]. Accordingly, the potential features pointed out by the theoretical study make the system more attractive.

4.2. Active Hand Splints for Rehabilitation

As previously discussed, the major drawback that currently affects dielectric elastomer actuation consists of the high driving electric fields needed to drive the materials. This technology typically requires high voltages (although at low current), usually of some kV. This can be a disadvantage for certain types of applications and in some cases it can also completely prevent the possibility of use. This is the case, for instance, of devices implantable in a human body. Although the biomedical area could appear to be the most sensitive to this problem, it already offers possible fields of application. As an example, dielectric elastomer actuators may find valuable uses for wearable active orthoses.

With respect to this, we are currently developing a simple application of folded actuators for so-called *hand splints*: they consist of orthotic systems for hand rehabilitation. The dynamic versions of hand splints typically include elastic bands, which exert a passive elastic resistance to the voluntary elongation of one or more fingers, as shown in Figure 9a.

In order to provide such splints with the possibility of electrically modulating the compliance of the resistive elements, we are working on the substitution of the passive elastic bands with contractile DE actuators (Figure 9b). A prototype sample of such a type of system equipped with three folded actuators is presented in Figure 9c. The electrical activation of the contractile actuators is used to change the compliance of the system. This enables modulations of the force antagonist to the voluntary finger movement, according to programmable rehabilitation exercises. New generations of this type of system are currently under development and testing in our lab.

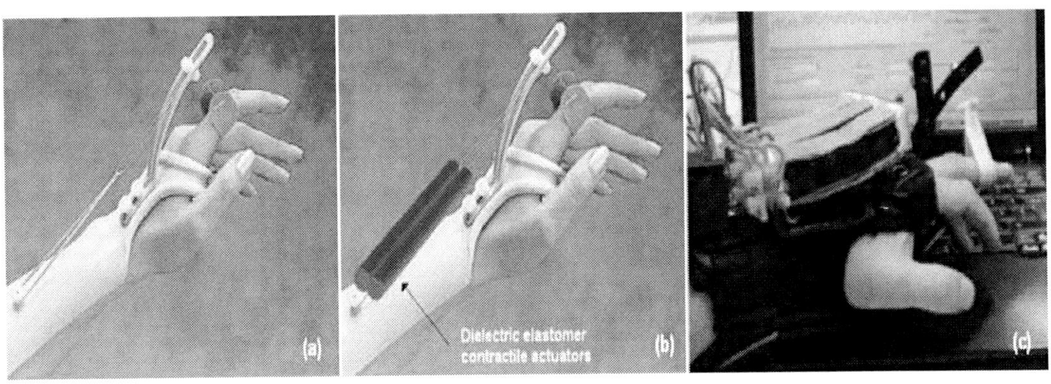

Figure 9. Dynamic hand splints: (a) passive system (state of the art); (b) concept for an active system; (c) first prototype implementation of an active system equipped with contractile folded DE actuators.

4.3. Man-Machine Interfaces With Pseudo-Muscular Actuation Systems

Man-machine interfaces represent an attractive domain for potential future applications of polymer actuators. By considering biomedical or biorobotic environments (as in the case of orthoses and biorobotic systems), it can be easily recognized that considerable advantages may derive from possible uses of human electrophysiological signals as control inputs for the actuators. The achievable combination of suitable mechanical features (elastic compliance), functional features (linear contractility) and control features (electrophysiological driving) may provide a major advantage in order to develop pseudo-muscular actuation systems, to be studied as possible candidates for future "artificial muscles".In particular, real-time activations of wearable actuators by means of electromyographic (EMG) signals may be useful for interfaces to be controllable by muscular activity, as suggested by the extremely simplified sketch reported in Figure 10. With respect to this, as an example Figure 11 shows results of a simple test of an EMG-based activation of a folded DE actuator.

4.4. Actuation of an Anthropomorphic Skeleton of Upper Limb

As mentioned above, contractile dielectric elastomer actuators may find possible applications as pseudo-muscular mechanisms. Within such a context, we are currently studying a system of actuation for a resin-made mould of the skeleton of a human upper limb [27]. In particular, we are studying the reproduction of the flexion/extension movements of the forearm around the elbow. To such an extent, several contractile units with parallel mechanical connection are considered, in order to mimic the action of the biceps muscle, as sketched in Figure 12.

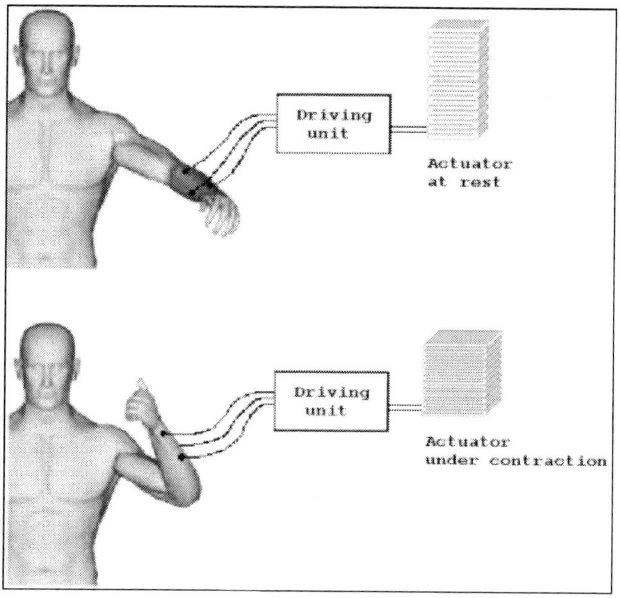

Figure 10. Electromyographic (EMG) control of a contractile DE actuator working as possible pseudo-muscular device.

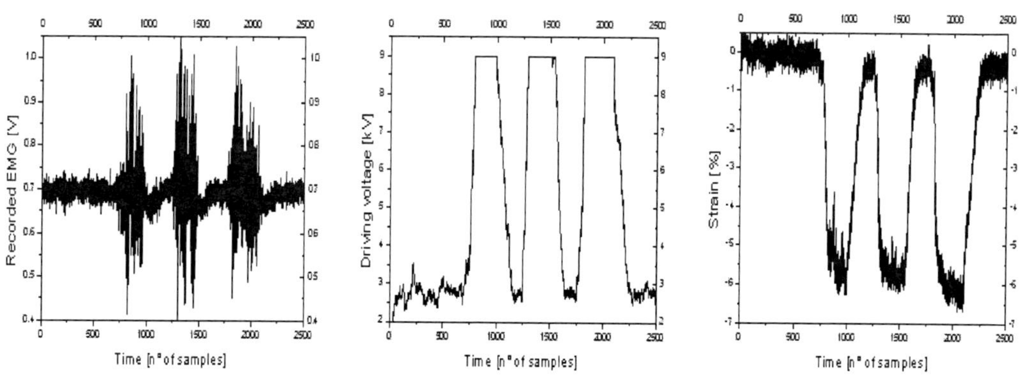

Figure 11. EMG-based real-time activations of a contractile folded actuator: EMG signal (left); elaborated driving signal (centre); actuator strain response (right).

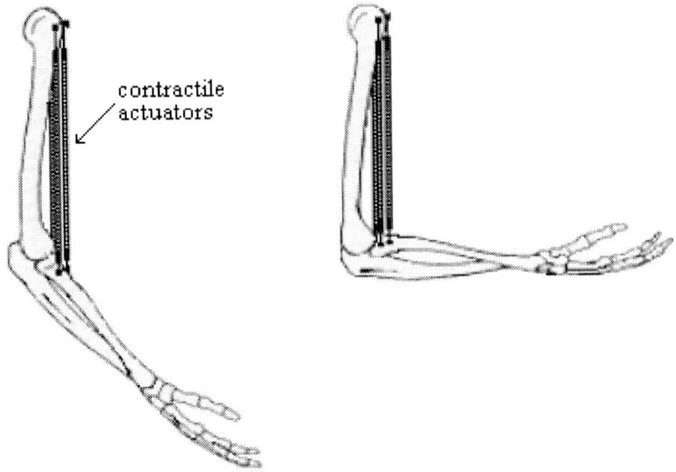

Figure 12. Schematic drawing of an upper limb skeleton equipped with contractile actuators, arranged so as to mimic the action of the natural biceps muscle.

No antagonist actuator has been considered so far, in order to simplify the system. The contraction of the actuators should enables the forearm rotation (against its weight) with respect to the humerus. So far, just a purely theoretical and simplified analysis of this type of system has been performed [27]. For this purpose, the system has been schematically modelled as composed of two rods, hinged with negligible friction. Results of these preliminary simplified investigations are encouraging [27] and promote further insights in the study of such an interesting application.

4.5. Actuation of a Robotic Gripper

A non-anthropomorphic system that we are currently developing consists of a robotic gripper equipped with contractile folded DE actuators, arranged according to the scheme reported in Figure 13.

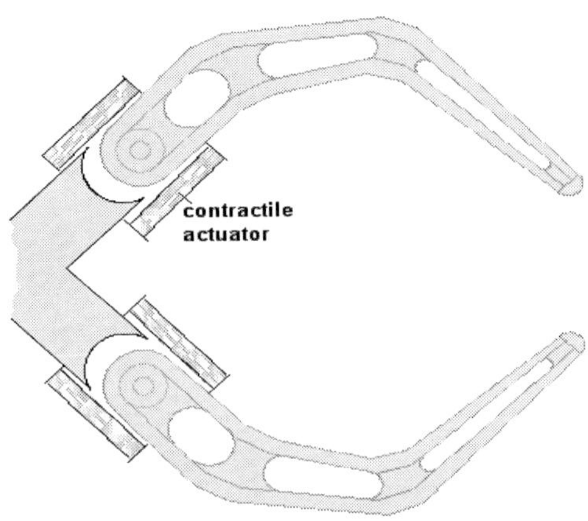

Figure 13. Concept of a robotic gripper equipped with contractile folded DE actuators.

Polymer contractile actuators represent possible new driving devices for gripper mechanisms, worthy of a specific investigation. In fact, the intrinsic compliance of the actuators would introduce new features in both the mechanical and in its control schemes, which may result particularly useful for certain types of applications. As an example, the manipulation of delicate objects is certainly a potential field in which such types of systems may gain a specific interest.

4.6.Biomimetic Robotic System for Explorations of Tubular Conduits

The previous sub-sections presented some applications currently being studied for folded-type contractile DE actuators. Nevertheless, the helical configuration previously described may result more useful for different types of applications. This is the case, for instance, of a robotic system that we have conceived for explorations of tubular cavities. A conceptual drawing for such a system based on helical DE actuators is presented in Figure 14

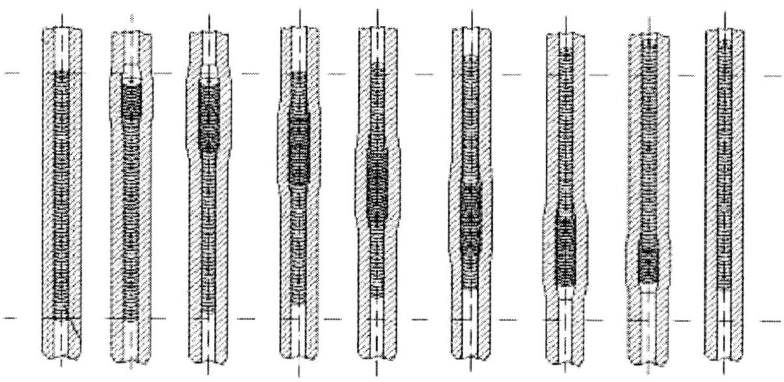

Figure 14. Conceptual drawing of a biomimetic robotic system for explorations of tubular cavities. In this example the system is made of six modules, consisting of helical DE actuators.

The proposed solution aims at obtaining a system capable of mimicking the peristaltic movement of earthworms. These animals implement forward motions by exploiting the alternating contraction of longitudinal and circular muscles of the numerous segments which constitute their body. Series of contractions periodically repeated consent the generation of a longitudinal wave which sequentially compresses and releases each segment. This mechanism provides longitudinal motion to the animal. A mimicry of such an earthworm locomotion can be thought to be achieved, in principle, by exploiting helical contractile DE actuators. The idea consists in the use of a mechanical series of such actuators, so that the resulting endoscope is made of a sequence of contractile modules, sequentially activated as shown in Figure 14. The activation of each module is responsible for its axial contraction and related radial expansion. The modules are sequentially activated starting from the head down to the tail. This can be expected to generate a forward movement of the system, as represented in Figure 14.

It is relevant to stress that such a solution based on the use of DE actuators currently is bound to the described need of high driving voltages. Accordingly, this fundamental issue prevents at present any possible investigation of this solution for intrabody applications and, in particular, for endoscopic explorations. Nevertheless, voltage reduction is a matter of developing new materials with improved electromechanical properties, as previously discussed. This is actually the great challenge for the new generations of DE actuators.

5. Conclusions

This chapter has highlighted some of the latest findings and research activities of our lab for the development of dielectric elastomer actuators suitable for biomimetic and biomedical applications. As it can be easily argued from results here presented, a great deal of work has still to be done and the solution of a large number of problems has to be given, in order to properly address several key issues. In particular, the main basic challenge is represented by the necessary development of new materials capable of being driven by lower electric fields. Fulfilment of this need would significantly contribute to enable the mimicking of the great performances of natural muscles, which represent an ideal model for our systems of actuation.

References

[1] Y. Bar-Cohen Editor, Electroactive polymer (EAP) Actuators as artificial muscles. Reality, Potential, and Challenges, 2nd ed., Washington: SPIE Press, 2004.

[2] J. D. W. Madden et al., "Artificial muscle technology: physical principles and naval prospects", *IEEE J. Oceanic Eng.*, vol. 29, no. 3, pp. 706-728, 2004.

[3] R. Pelrine, R. Kornbluh, Q. Pei and J. Joseph, "High-speed electrically actuated elastomers with strain greater than 100%", *Science*, vol. 287, pp. 836-839, 2000.

[4] R. E. Pelrine, R. D. Kornbluh and J. P. Joseph, "Electrostriction of polymer dielectrics with compliant electrodes as a means of actuation", *Sensor. Actuat. A-Phys*, vol. 64, pp. 77-85, 1998.

[5] R. Pelrine, R. Kornbluh and G. Kofod, "High-strain actuator materials based on dielectric elastomers", *Adv. Mater.*, vol. 12, no. 16, pp. 1223-1225, 2000.

[6] R. Pelrine, R. Kornbluh, J. Joseph, R. Heydt, Q. Pei, S. Chiba, "High-field deformation of elastomeric dielectrics for actuators", *Mater. Sci. Eng. C.*, vol. 11, pp. 89-100, 2000.

[7] Q. Pei, R. Pelrine, S. Stanford, R. Kornbluh, M. Rosenthal, "Electroelastomer rolls and their application for biomimetic walking robots", *Synth. Met.*, vol. 135-136, pp. 129-131, 2003.

[8] Q. Pei, M. Rosenthal, S. Stanford, H. Prahlad, R. Pelrine, "Multiple-degrees-of-freedom electroelastomer roll actuators", *Smart Mater. Struct.*, vol. 13, pp. N86–N92, 2004.

[9] S. Ashley, "Artificial muscles", *Scientific American*, pp. 52-59, October 2003.

[10] F. Carpi, D. De Rossi, R. Kornbluh, R. Pelrine and P Sommer-Larsen, Editors, *Dielectric elastomers as electromechanical transducers. Fundamentals, materials, devices, models and applications of an emerging electroactive polymer technology*, Elsevier Press: Amsterdam, 2008.

[11] Benslimane, P.; Gravesen, P.; Sommer-Larsen, P., Dielectric elastomer actuators with smart metallic compliant electrodes. In: Conference proceedings. Actuator 2002 - 8. International conference on new actuators, Bremen (DE), 10-12 Jun 2002. Borgmann, H. (ed.), (Messe Bremen GmbH, Bremen, 2002) p. 383-387.

[12] G. Kofod, P. Sommer-Larsen, R. Kornbluh and R. Pelrine, "Actuation response of polyacrylate dielectric elastomers", J. Intell. Mater. Syst. Struct., vol. 14, no. 12, pp. 787-793, 2003.

[13] F. Carpi, P. Chiarelli, A. Mazzoldi, D. De Rossi, "Electromechanical characterisation of dielectric elastomer planar actuators: comparative evaluation of different electrode materials and different counterloads", *Sensor. Actuat. A-Phys*, vol. 107, pp. 85-95, 2003.

[14] F. Carpi, D. De Rossi, "Dielectric elastomer cylindrical actuators: electromechanical modelling and experimental evaluation", *Mater. Sci. Eng. C*, vol. 24, pp. 555-562, 2004.

[15] Schlaak, H, Jungmann, M., Matysek, M., et al. (2005). Novel multilayer electrostatic solid-state actuators with elastic dielectric, In Smart Structures and Materials 2005: Electroactive Polymer Actuators and Devices (Y. Bar-Cohen ed.), Proceedings of SPIE 5759, pp. 121-133.

[16] C. Loewe, X. Q. Zhang, and G. Kovacs. Dielectric elastomers in actuator technology. Advanced Engineering Materials, 7(5):361–367, 2005.

[17] Lochmatter P., Kovacs G., Ermanni P. 2007, Design and characterization of a shell-like actuator based on soft dielectric EAP, Smart Mater. Struct., 16(4):1415-1422, 2007.

[18] Kovacs G., Lochmatter P., Wissler M. 2007 Arm Wrestling Robot Driven by Dielectric Elastomer Actuators Smart Materials and Structures 16 No 2 S306-S317.

[19] F. Carpi and D. De Rossi, "Improvement of electromechanical actuating performances of a silicone dielectric elastomer by dispersion of titanium dioxide powder", *IEEE T. Dielect. El. In.* vol. 12, no. 4, pp. 835-843, 2005.

[20] F. Carpi, G. Gallone, F. Galantini and D. De Rossi, "Silicone-Poly(hexyl-thiophene) blends as elastomers with enhanced electromechanical transduction properties", Advanced Functional Materials, Vol.18 (2), pp. 235 – 241, 2008.F. Carpi, A. Migliore, G. Serra, D. De Rossi "Helical dielectric elastomer actuators", *Smart Mater. and Struct.*, vol. 14, pp. 1210-1216, 2005.

[21] F. Carpi, C. Salaris and D. De Rossi, "Folded dielectric elastomer actuators", Smart Materials and Structures, Vol. 16, pp. S300-S305, 2007.

[22] G. Pioggia, A. Ahluwalia, F. Carpi, A. Marchetti, M. Ferro, W. Rocchia, D. De Rossi, "FACE: Facial Automaton for Conveying Emotions", *Applied Bionics and Biomechanics*, vol. 1, no. 2, pp. 91-100, 2004.

[23] F. Carpi and D. De Rossi, "Bioinspired actuation of the eyeballs of an android robotic face: concept and preliminary investigations", Bioinspiration and Biomimetics, Vol. 2, pp. S50–S63, 2007.

[24] Carpenter R H S 1998 Movements of the eyes, Pion Lt.

[25] Bahill A, Andler D and Stark L 1975 Most naturally occurring human saccades have magnitudes of 15 deg or less Investigative Ophthalmol. 14 468–469.

[26] F. Carpi, A. Migliore, D. De Rossi, "A new contractile linear actuator made of dielectric elastomers", In Smart Structures and Materials 2005: Electroactive Polymer Actuators and Devices, Y. Bar-Cohen Editor, Proceedings of SPIE, vol. 5759, pp. 64-74, 2005.

In: Biomimetic and Supramolecular Systems Research
Editor: Arturo H. Lima
ISBN: 978-1-60456-405-1

Chapter III

Supramolecular Techniques in MOS Systems for Definition of Nanoscale Features and for Bio-Sensors

R. A. Puglisi, S. Libertino and S. Lombardo
Consiglio Nazionale delle Ricerche (CNR), Istituto per la Microelettronica e Microsistemi (IMM), Catania, Italy

Abstract

Silicon microelectronics and MOS technology are among the most important fields of development in modern science and technology. On the other hand, the biotechnology community is feeling the need to implement fast, portable, low cost and user friendly biosensors to monitor different species in the everyday life (e.g. air and water pollution, biomedical applications, etc.). A promising approach to evolve biosensor technology appears to take advantage of the enormous potential offered by the silicon microelectronic integration technology.

In this chapter we discuss MOS based Si biosensors in which the sensitive elements (both proteins and DNA strands) are covalently immobilized on the transistor gate surface and are able to detect either organic or inorganic molecules.

To further improve both the biosensor responsivity and selectivity, the use of nanostructures is a viable tool. We describe how a supramolecular approach based on diblock copolymers self-assembling can be used for nanoscale definition of mono and two dimensional anchoring sites of the bio-elements, with control on size and position at nanometer level.

Introduction

The Metal-Oxide-Semiconductor (MOS) system based on Silicon as semiconductor and on Silica as dielectric is one of the key enabling technologies of these last fifty years. MOS is the work horse which has allowed the enormous success of microelectronics and in particular its pervasiveness in the consumer market. Complementary MOS (CMOS) technology has allowed the development of VLSI circuits with unprecedented performances at an exceptionally low cost for most of digital (microprocessors, microcontrollers, semiconductor memory devices, etc.), analog, mixed signal, and RF circuits. The basis for this impressive progress is the exceptional quality of the Si/SiO_2 system in terms of interface and bulk defects, low cost (silicon is the second element on the earth crust in terms of abundance) robustness to electrical and mechanical stress, scalability of the transistors, etc. These advantages resulted in the well known Moore's law, which is going on now from more than 40 years, and actually it has even accelerated the pace in the last about ten years, with a doubling of transistor density now every about two years. This pace evidently implies that soon or later some physical limit such as thickness of the gate dielectric, transistor channel length, noise margin, research and development cost, lithography limits, design issues, back-end materials limits, etc. will be hit. Nevertheless, the excellent level of confidence, competence and know-how in these technologies reached by the microelectronics community, and the size of the infrastructures for production and R&D imply an enormous effort worldwide to overtake those limits. For example, the limits in gate dielectric scaling which has slowed down the evolution in advanced CMOS in the last few years, will be overtaken by introducing a radical innovation in the MOS stack, i.e. the substitution of silica with high-k dielectrics and the introduction of metal gates. As Moore himself has put in evidence a few years ago, "no exponential is forever, but we can delay forever". This sentence implies a huge effort to find solutions to push CMOS beyond the limits, by using either radical innovations, the so called "revolutionary" approach, or by a more moderate mode, with the "device engineer" approach, "evolutionary", which implements smart solutions with a moderate but constant introduction of new materials, new architectures, but still maintaining the main CMOS concepts.

The self-assembly of molecules, quantum dots, or polymer layers can add other degrees of freedom to the Si/SiO_2 technology, to be exploited for numerous applications, ranging from traditional CMOS type concepts such as non volatile semiconductor memories, to novel sensor technologies.

In this chapter we discuss some basic aspects related to the self-assembly of quantum dots, co-polymers, and the grafting and self-organization of organic molecules on the Si/SiO_2 system. We also discuss some possible applications.

The first section of the paper deals with the self-organization of semiconductor nanocrystals formed by CVD on SiO_2 surfaces, and its exploitation for the realization of Flash memories, to push the scaling limits beyond those of the conventional NVM technology based on floating gates. The second section describes the self-organization of di-block co-polymers on SiO_2. This can be potentially used to improve the Si nanocrystal technology for NVMs, but also as self-organization approach for molecular sensor technologies. Finally, the third section briefly reviews the biosensor research field and describes the grafting of organic molecules including DNA segments, showing preliminary demonstration of the potential offered by the Si/SiO_2 system in molecular sensor technology.

Formation of Semiconductor Nanodots Controlled in Size and Position

Self-Assembling in Chemical Vapor Deposition of Si Dots

The quickening pace of MOSFET scaling coupled with sizes facing fundamental physical limits has implied a new approach, going from an evolutionary to a revolutionary method. This is accelerating the introduction of new technologies to extend bulk complementary MOS (CMOS) beyond the 65 nm node. These technologies include both new materials and advanced MOSFET structures. A particular challenge is represented by the scalability of nonvolatile memories [1], which imposes additional constraints on process integration and structure design, since NVM structures, that require relatively high voltages must be incorporated into CMOS technology that works at low voltage. Tipically, nonvolatility is achieved by storing and sensing the charge on a floating gate (see Figure 1). The dielectric between the floating gate and the control gate (the so called ONO, acronym of oxide-nitride-oxide stack) must scale with the tunnel dielectric to maintain adequate coupling to the control gate. The tunnel dielectric must be thin enough to allow charge transfer to the floating gate at reasonable voltage levels and thick enough to avoid charge loss when in read or off modes.

The aggressive scaling of the Flash devices seen so far [2] is pushing this technology towards a "wall" made of numerous technological key issues. Among these, the reliability constraints, in particular the requirement of 10-year non-volatility, is of primary importance for both high-density NVMs ("stand-alone") and for NVMs integrated with logic (the so-called embedded NVMs), widely used in portable consumer electronics (mobile phones, mp3 players, digital photography, usb drivers) automotive, and other digital applications. Even in present Flash memories, ensuring non-volatility for all bits after program/erase cycling is a challenge from the viewpoint of manufacturing and yield. The finite probability of having a cluster of two or three defects in the tunnel oxide, producing a huge local increase of the tunnel current (namely, stress induced leakage current, often indicated by its acronym, SILC), determines the anomalous failure bits, which are widely recognized as the most important concern in Flash devices. The amplitude of the SILC increases when the tunnel oxide is thinned, limiting the oxide scaling.

To extend Flash scalability, existing research efforts are exploring a variety of basic memory mechanisms. These mechanisms include charge isolated by surrounding dielectrics, charge held in place by Coulomb blockade potentials, chemical phenomena, magnetic phenomena, and material phase changes. Reviews on these approaches are reported in Refs. [2-5].

Among these, the use of discrete-trap storage nodes in conventional Flash memory technology has been invoked as one of the key items for Flash memories [5]. The basic idea of the "discrete-trap" mechanism is to replace the floating gate (FG) by many discrete trapping centers, as illustrated in Figure 2. The very special effect of the localized trapping is the high reliability associated with it. In fact, in Discrete-Trap memories a single leakage path due to a defect (intrinsic or SILC-related) in the oxide can only discharge a single storage node. This renders more reliable and robust with respect to SILC the memory operation.

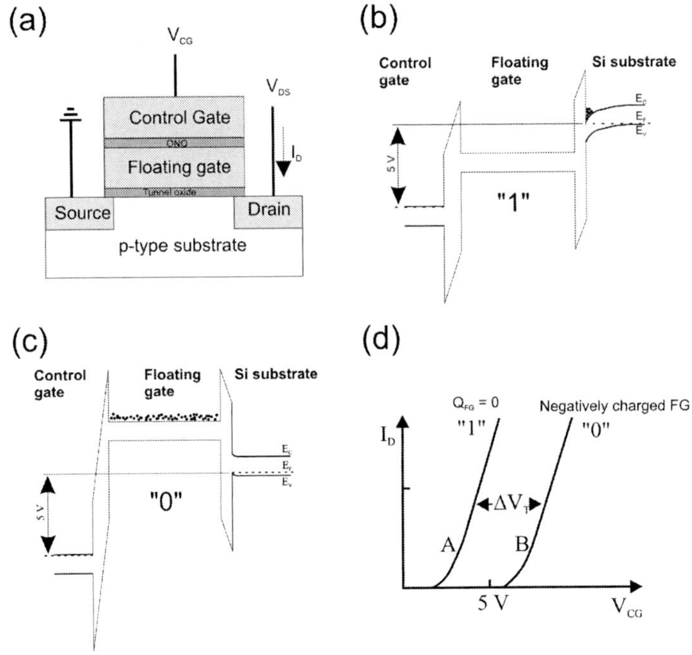

Figure 1. Schematic structure and working principle of a non volatile semiconductor memory cell. The sketch refers in particular to the most important NVM, i.e. the Flash. The name 'Flash' was suggested by the inventors because the erasure process of the memory content reminded the speedness of a flash of a camera. (a) shows the cell structure, i.e. a n-MOSFET with the gate dielectric containing a floating gate, electrically completely insulated from the environment. (b) and (c) represent the charge states of the floating gate (empty or filled), which in the simplest versions of NVMs are the two logic states of the bit. In this case a single cell is a bit. (d) shows the sensing of the floating gate charge state, performed by simply measuring the drain current under well defined bias conditions (VCG = 5 V in (d)).

Many different types of memory-cell structures, employing discrete-trap type storage nodes, have been demonstrated in the literature. In these devices the storage medium consists either of natural traps, normally in correspondence of a nitride layer [6-10], or semiconductor nanocrystals, in particular Si nanocrystals [11-14].

Several methods to synthesize the Si dots have been proposed and investigated: ion implantation [15], aerosol [16], and chemical vapor deposition (CVD) [17]. Specifically the use of CVD methods to synthesize the nanodots has demonstrated to be a convenient technique because of its immediate implementation in the ultralarge scale integration (ULSI) processing, a good control on the deposition parameters, and because of the possibility to obtain isolated storage nodes, immersed in a stoichiometric silicon dioxide matrix.

Typical CVD deposition of the Si dots are carried out on oxidized Si substrates, by the thermal decomposition of silane (SiH_4) or dichloro-silane (SiH_2Cl_2) at temperatures between 500 and 600 °C, at chamber pressures ranging between 10^{-3} - 10^{2} Torr, with SiH_4 as Si precursor and with H_2 as carrier gas. Figure 2 schematically shows, in the top part, the typical evolution of the Si-QDs density versus deposition time at fixed deposition temperature and SiH_4 partial pressure. It shows that the Si-QDs density increases continuously up to their impingement. Consequently, all along the process, new Si-QDs nucleate during the growth of the Si-QDs previously nucleated. Then, the Si-QDs nucleated at the end of the process might be smaller than the Si-QDs nucleated at the beginning of the process, as illustrated in the last schematic of the Figure. The concurrence of both nucleation and growth is then the main reason for the size dispersion of the CVD Si-QDs. To the purpose of the memory scaling it is important to quantify the contribution of each process, and its impact on the fluctuations in the quantity of silicon covering the substrate, i.e. the coverage ϑ. Recently, the Si-QDs nucleation kinetics have been followed by transmission electron microscopy since the very early stages of the nucleation process, with the scope to quantify the degree of coverage dispersion, and to understand if it is possible to control it [18, 19]. The bottom part of Figure 2 shows a sequence of Energy Filtered TEM micrographs of samples deposited for consecutive times, equal to 80s (a), 90s (b), 100s (c), 110s (d), 120s (e), and 140s (f). The white regions correspond to Si dots deposited on the oxidised surfaces. In the Figures are also reported the measured values of coverage ϑ. From the micrographs, as the deposition time increases, it is evident the increase in the amount of ϑ of the oxide substrate with Si, going from 6.5% to 100%. Moreover, in correspondence with the lower deposition times (micrographs (a) and (b)), it is evident a process of nucleation of the Si dots. The following period (micrographs (c), (d), and (e)) corresponds to a coalescence phase up to the complete coverage with Si (f). It is, however, particularly important to note in micrographs (c), (d), and (e) that, though a large fraction of the surface is covered with Si, in the uncovered regions new Si nuclei do continue to appear, thus indicating that the process is continuous.

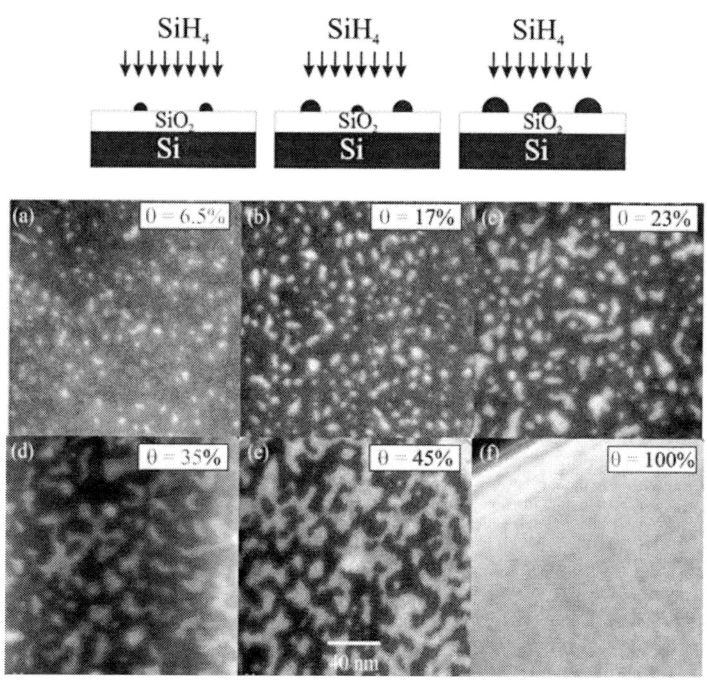

Figure 2. (Top) Schematic representation of the origin of the size dispersion obtained with a standard CVD process. (a) to (f) EFTEM micrographs of samples deposited at 550 °C for consecutive times, equal to 80 s (a), 90 s (b), 100 s (c), 110 s (d), 120s (e) and 140s (f) and relative values of fraction of the surface covered with silicon, ϑ.

Supported by this experimental evidence, the nucleation data have been analysed by adopting a model of continuos nucleation in the approximation of capillarity [20]. In this model the Frenkel-Zeldovich nucleation equation in the supercritical region is solved providing an analytical expression for the nuclei density and its time evolution. The model is based on the assumption that the nucleation is a random process, i.e. all the points of the substrate have same probability to be a nucleation site. The results of the fit procedure of the data relative to the dot radius distribution [19], indicate that the energy barrier to the nucleation is about 0.3 eV and the critical radius is 0.8 nm.

The same model has been used to fit the experimental data relative to the spatial distribution of the Si-QDs. Figure 3(b) reports in symbols the distribution of the edge distance between nearest-neighbor dots obtained by the analysis of the EFTEM images taken on the sample imaged in Figure 3 (a), and in solid line the results of the simulation based on the theoretical model.

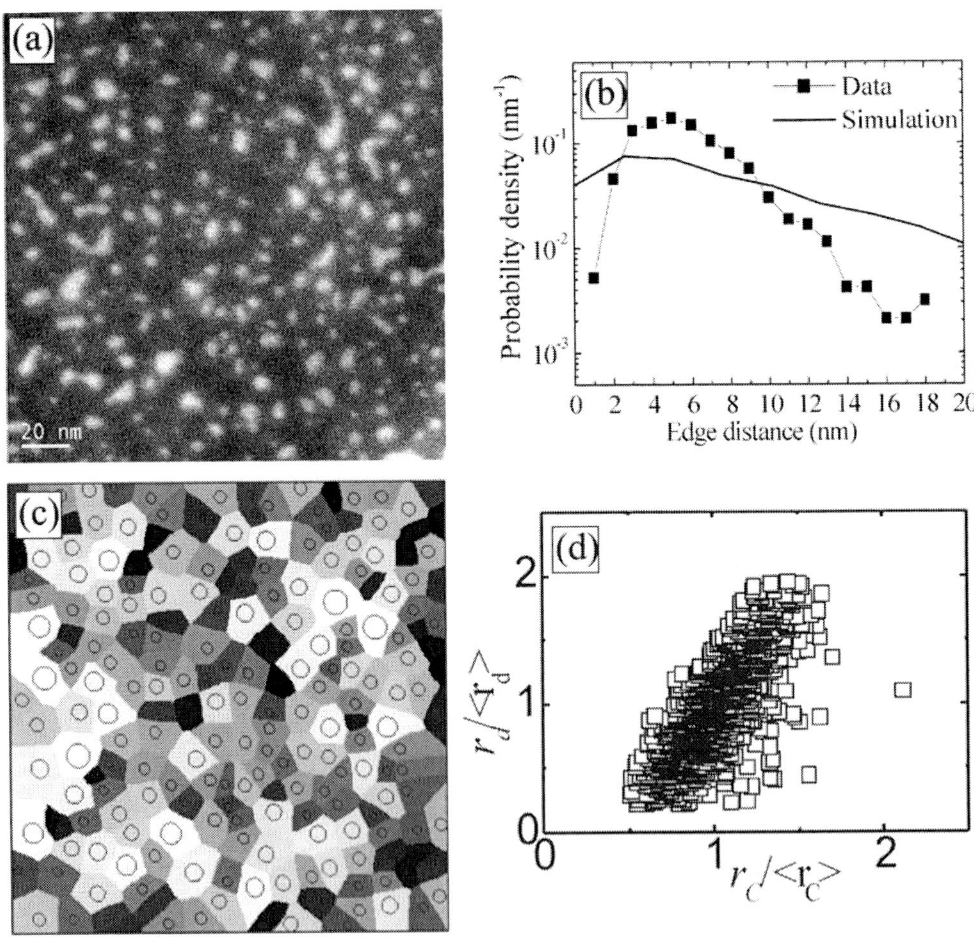

Figure 3. (a) EFTEM of a Si-QDs sample obtained by CVD at 550 °C 90 s, (b) Distribution of the nearest-neighbour edge distances, with the experimental data (symbols) and the simulation results (solid line), (c) capture zone relative to image in (a), and (d) scaled dot radius $r_d/<r_d>$ vs scaled capture radius $r_C/<r_C>$.

As it is possible to observe from Figure 3 (b), the simulation, which assume a random nucleation, provides a probability of having the nucleation of close dots much higher than the experimental observation. This suggests that the observed nucleation is not a random process and cannot be fully explained in terms of mean-field equations, but instead the local environment around each dot has to be taken into account. In particular the silicon adatoms, obtained from the dissociation of the silane, before aggregating to form a new nucleus, can diffuse toward previously nucleated clusters. So it is possible to define a capture zone centered around every nucleus where all the silicon adatoms, which fall within it, will preferentially contribute to the growth of this nucleus [for a complete review see, for example, Ref. 21]. In this condition the new nucleation event in the depletion region around each dot is strongly reduced and the nucleation sites inside are deactivated. As they spread with time, the capture zones may overlap adjacent active nucleation sites on which nuclei have not yet formed. One of the theoretical approaches present in literature simulates such a capture zone by using the Voronoi tessellation [22], by which for each dot it can be defined a boundary enclosing all the intermediate points lying closer to the border of this dot than to other points on the substrate. In this way the growth rate of each dot is limited by the diffusion of the deposited monomers, and depends on how much free substrate it has in its immediate neighborhood. This process plays a role in the range of coverage, ϑ, between about 10% – 30%, i.e. when the dots are sufficently close each other relatively to the diffusion length of the adatoms, but far from their impingement [23].

Figure 3(c) reports the capture zone pattern relative to the sample imaged in Figure 3(a). The circles representing the dots are superposed, as a guide to the eye, in order to easily compare the dot size to its capture size. By measuring the capture zone size calculated with this method a very good correlation with the respective radius size is found, as displayed by the graph reported in Figure 3(d). Specifically, it shows the scaled dot radius, r_d / $<r_d>$ as a function of the corresponding scaled capture radius, r_C / $<r_C>$. The sample presents a coverage ϑ of 14%. As it is possible to observe, the scaled dot radius increases with the scaled capture radius, and the data lie close to the graph bisectrix.

Since the scaled dot size is correlated to the scaled capture size, for the levels of coverage of interest for nanocrystal memories, the process of dot nucleation is a nonpurely random phenomenon, and the partial self-ordering in the Si-QDs is expected to produce a decrease in the fluctuations of ϑ compared to a pure random process. Figure 4 reports the simulated relative dispersion of ϑ, $\Sigma(\vartheta)$, defined as $\sigma(\vartheta)/\mu(\vartheta)$, where σ stands for standard deviation of ϑ, and μ for average of ϑ, versus gate area. The figure reports a comparison, under the same conditions of average dot radius, $<r_d>$ = 3 nm, and surface coverage, ϑ= 20%, of two cases: pure random nucleation (open triangles, well fitted by standard Poisson statistics [24], and nucleation with capture zone (circles). A typical result of the simulation process is also reported for the case of random nucleation events (Figure 4(b)) and for the capture zone mediated process (Figure 4 (c)). A decrease of the ϑ dispersion in the case of the capture zone is observed, and even for gate areas corresponding to NAND flash cells of the 20 nm technology node, $\Sigma(\vartheta)$ is expected to be about 20%, with dot densities close to about $1x10^{12}$ dots/cm^2 and ϑ of about 20%. Note, however, that if the capture zone effect is small (mainly at low ϑ), the estimates of Figure 4 may provide optimistic expectations [25, 26].

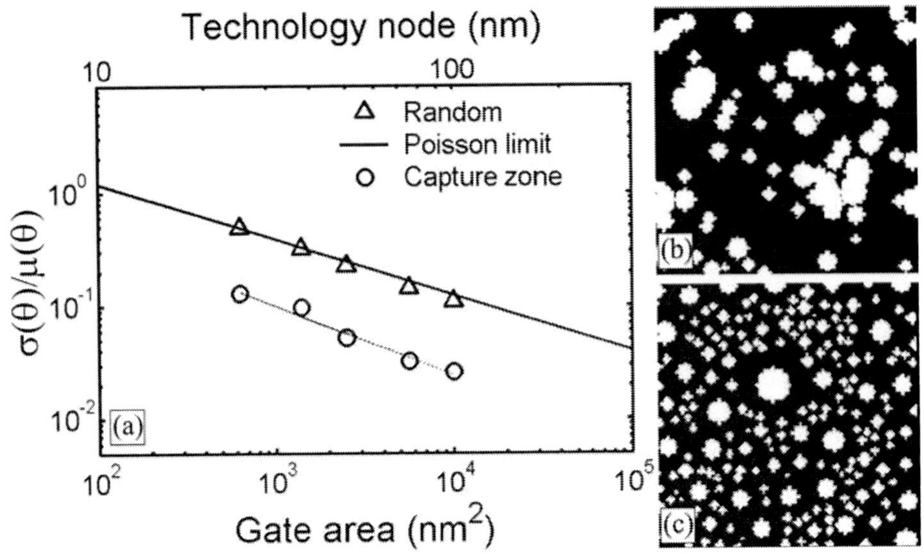

Figure 4. Relative standard deviation of ϑ vs gate area (a) for the case of purely random nucleation (b) and nucleation with capture zone (c). In both cases, same area coverage and average dot size are assumed.

In conclusion, CVD Si nanocrystals exhibit clear evidence of self-assembled nucleation, typical of supramolecular systems. The basic involved mechanism appears to be an adatom diffusion process toward previously nucleated Si clusters. Therefore, the ordering physics is not a direct inter-atom interaction, and the formed nanostructure is not a single molecule, but rather a cluster of atoms in the form of an amorphous solid. These circumstances are sources of fluctuations and disorder, so it appears more promising to improve the self-assembling by using directly molecules and their mutual interaction, closely following a supramolecular approach. In the next section we describe some preliminary experience in this area based on the di-block copolymer concept.

Nano-Templates Through Self-Assembling of Diblock-Copolymers

The above results show that through CVD processes, the deposition parameters controlling the adatom diffusion can be tailored to decrease the coverage fluctuations. To reduce to zero the coverage fluctuations, a simple, low-cost and VLSI compatible alternative resides on the self-assembly of block copolymers [27, 28]. In particular, the block copolymer system composed of polystyrene (PS) and polymethylmethacrylate (PMMA) can be an appropriate solution because it is compatible with standard semiconductor processing techniques, like the standard photoresists used in lithography. The self-assembled copolymers can be used as nano-templates for the production of nanoporous media, membranes, lithographic templates, and scaffolds for assemblies of electronic materials. The copolymers are realised on oxidized Si substrates with wet oxides of 10 nm of thickness (Figure 5(a)). We use a commercially available PS–PMMA diblock copolymer with average molecular weights M_n = 67000 g/mol or 131800 gr/mol, polydispersity index of 1.11, and PMMA weight fraction of 30%. The polymer is dissolved 1% by weight in toluene. Sample preparation starts with an initial surface pretreatment with a random copolymer, necessary to neutralise the substrate to the two block copolymers and to promote the proper orientation of the hexagonal template (Figure 5(b)). The polymer-coated substrates are then heated under vacuum at 140 °C for 48 hours, above the glass transition temperature of both PS and PMMA, to allow the terminal OH groups to diffuse to and react with the oxide substrate [29]. Unattached polymer chains are removed by rinsing in toluene the samples. The final thickness of the grafted layers as measured by ellipsometry is about 3 nm. A 30 nm thick film of copolymer is spin coated on the functionalised substrate, (Figure 5(c)), and annealed at temperatures ranging between 160 - 190 °C for times between 1 - 2 hours, to promote the microphase separation of the blocks and self-organization into the hexagonal array (Figure 5(d)). The template formation is completed by exposing the samples to ultraviolet light, which degrades the PMMA and cross links the PS (Figure 5(e)). The copolymer is developed to remove the PMMA cylinders and leave a porous PS template (Figure 5(f)).

Figure 6 (a) and (b) show respectively a tilted and a planar view SEM images showing the result of this fabrication procedure, relatively to the copolymer at 67000 gr/mol, annealed at 180 °C for 2 h, after development. In the SEM micrograph the dark spots correspond to the pores remaining after the PMMA removal, more evident in the tilted view. The image shows an ordered array of circular holes at the positions of the cylinders.

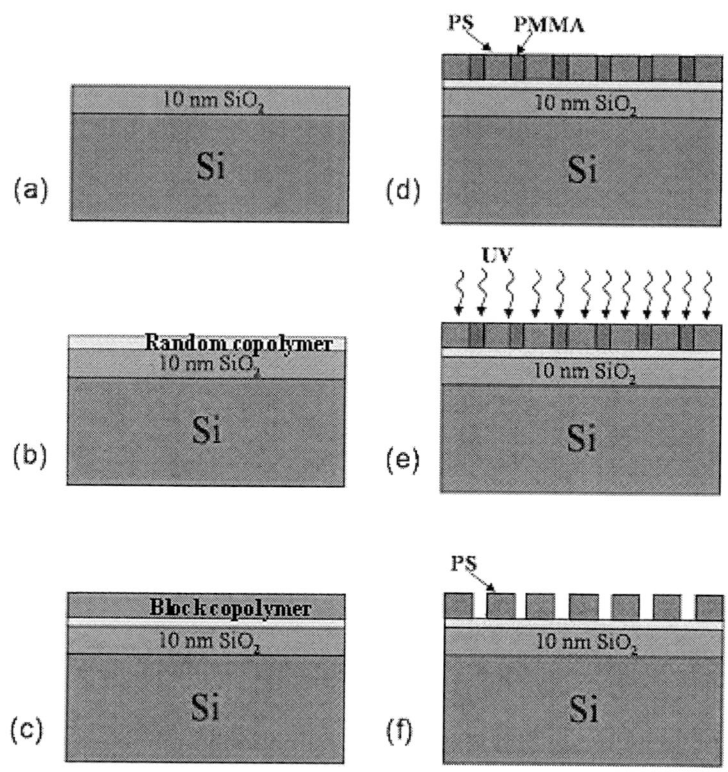

Figure 5. Schematic illustrating the several process steps to realise the P(S-b-MMA) self-assembled nanomask.

An important parameter to fabricate QDs with controlled spatial distribution is the distance between the pores. To check whether the annealing promoting the self-assembling process plays a role on the position of the pores, a statistical analysis of the inter-pore distance has been performed by measuring from the SEM images the pore position, and calculating the distance between the pores. The results are shown in Figure 7, which reports the probability distribution of the pore-to-pore distance for the 170 °C (circles), 180 °C (squares), 190 °C (triangles) annealing processes, for 2 hours. As it is possible to see the curves collapse and there is no evident difference between the three processes, and this suggests that the temperature does not play a role on the inter-pore distance.

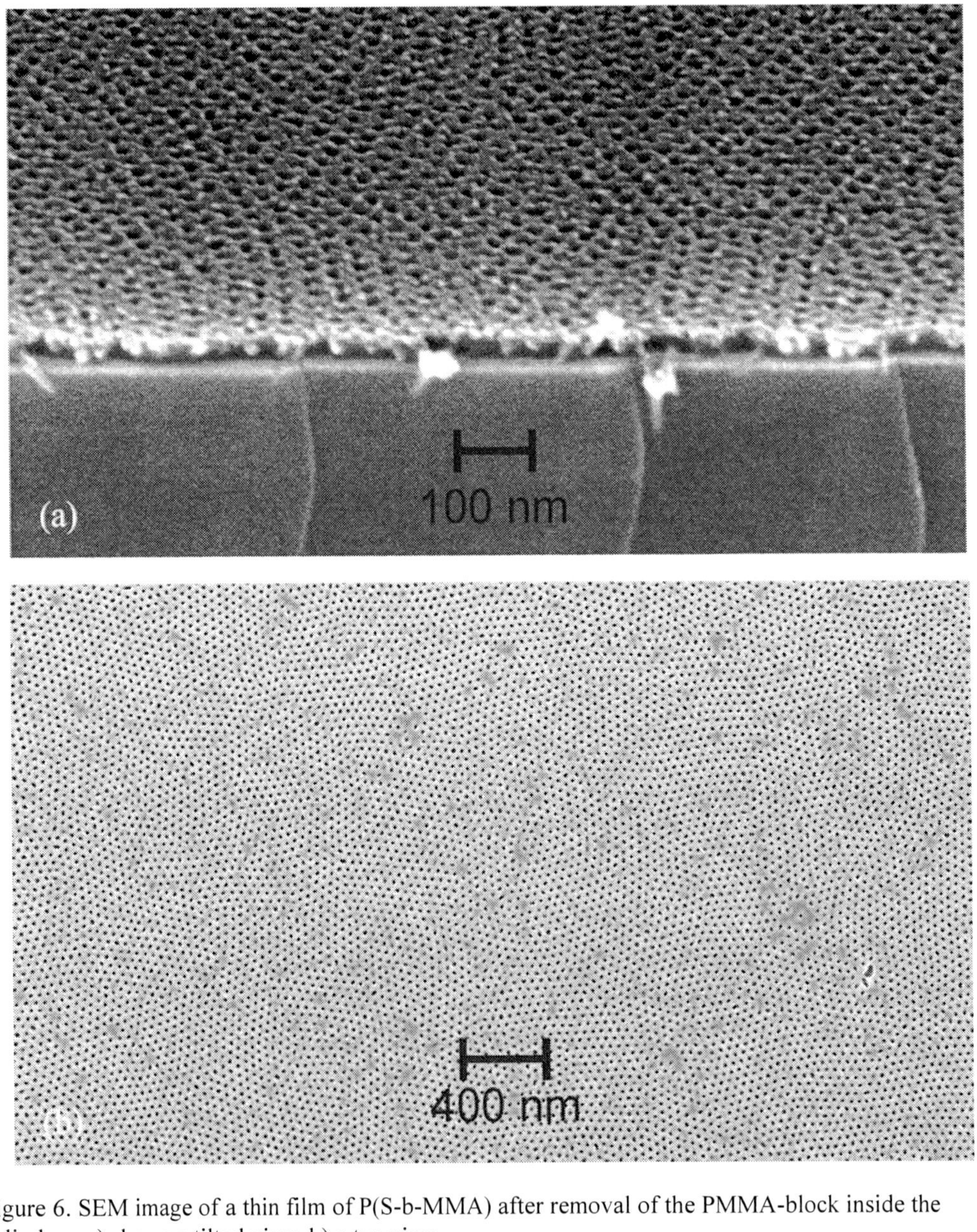

Figure 6. SEM image of a thin film of P(S-b-MMA) after removal of the PMMA-block inside the cylinders, a) shows a tilted view, b) a top view.

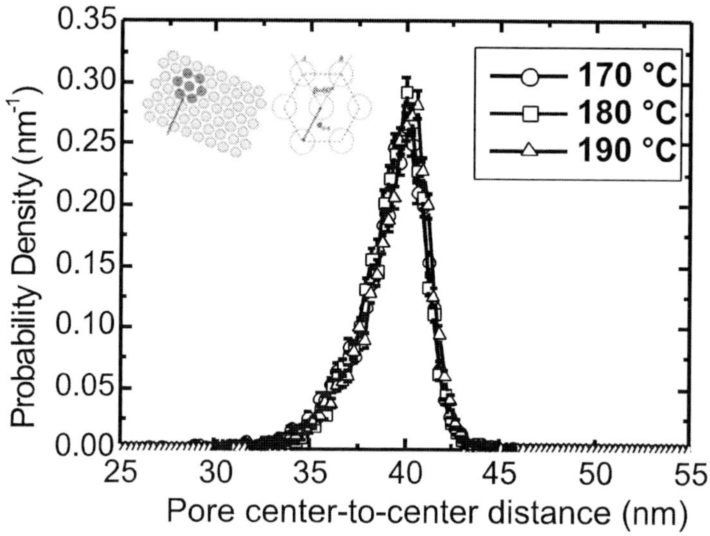

Figure 7. Probability density of the pore center-to-center distance for annealing of 2 hours at 170 °C (circles), 180 °C (squares) and 190 °C (triangles).

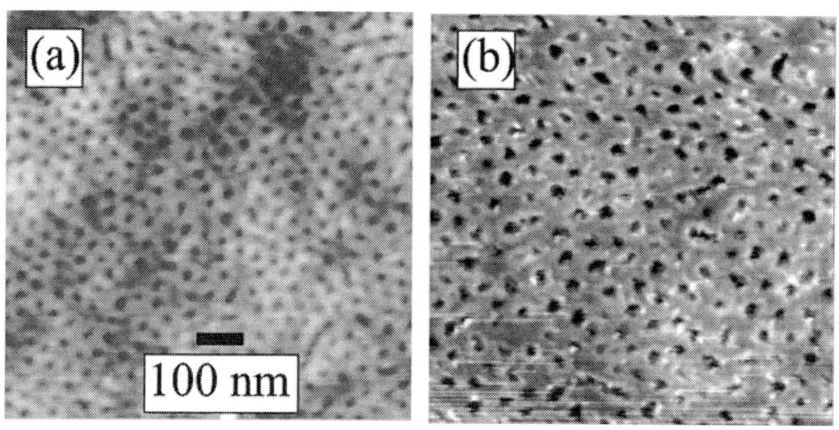

Figure 8. AFM micrographs illustrating the PS nanotemplate obtained by using PS-PMMA with molecular weight of M_n = 67000 gr/mol (a), and 131800 gr/mol (b). Both the films are annealed at the same temperature and time (180 °C, 2 h).

This result confirms previous similar results [30], which state that the pore distance is correlated to intrinsic characteristics of the copolymer itself, and it can be adjusted by changing the total copolymer molecular weight while keeping constant the molecular weight ratio between the blocks.

As a demonstration Figure 8 reports the Atomic Force Micrographs obtained on two samples relative to M_n = 67000 gr/mol (a) and 131800 gr/mol (b) copolymers. As it is possible to see, by increasing the copolymer molecular weight, both the distance and the size of the pores increase.

The size of the pores is expected to be 20 nm and 30 nm respectively, but due to the finite resolution of the AFM probe it was not possible to measure it. The pore-to-pore distance has been measured in these cases with the same procedure detailed above, and the results are reported in Figure 9. The curve relative to the low M_n (black circles) is centered at about 42 nm, while the high M_n case (grey squares) is centered at about 63 nm. As the graph shows, the curve relative to the high molecular weight case exibits a larger dispersion, around the average value, relatively to the low molecular weight case. This indicates that this sample presents a lower ordering degree, as it is evident also by the comparison between the two images in Figure 8. It is important to stress the fact the two samples are annealed by using the same thermal budget. To explain this result, it should be taken into account that the copolymer diffusion length, during the annealing, is:

$$\mathrm{L_D} = \sqrt{\mathrm{Dt}} \propto \sqrt{\mathrm{t}}\exp\left(\frac{-\mathrm{C}}{\mathrm{T}}\right) \tag{1}$$

where C is a constant which depends on the copolymer properties, T and t are the temperature and the time of the annealing process, and D the diffusion coefficient. For long polymeric chains D depends on the polymerization degree, N, i.e. on the number of monomers in the macromolecule:

$$D \propto \frac{1}{\mathrm{N}^2} \tag{2}$$

which substituted in Eq. (1) gives:

$$L_D = \sqrt{Dt} \propto \frac{1}{N}\sqrt{t}\exp\left(-\frac{C}{T}\right) \tag{3}$$

This equation indicates that for high values of N it is necessary to increase the thermal budget to have sufficient high polymer diffusion lengths, and consequently long range self-assembly.

One important aspect is that PS-PMMA copolymers self-assemble in hexagonal arrays forming *macrodomains* randomly oriented between each other. The macrodomain typical size is of about 200 nm. Figure 10 (a) shows an example relative to the PS-PMMA diblock copolymer with average molecular weight M_n = 67000 g/mol, annealed at 170 °C for 2 h, to promote the microphase separation of the blocks and self-organization into the hexagonal arrays.

The template presents the closepacked hexagonal array of pores arranged in the random macrodomains. The FFT associated to the whole image shows indeed that the film is "polycrystalline" (Figure 10 (b)). The black square in the Figure isolates one of these macrodomains, of about 200 nm in size. The six diffraction peaks in the associated FFT (Figure 10 (c)) indicate that the macrodomain is ordered in one single crystal with hexagonal symmetry. The randomic orientation of the macrodomains is the responsible of the large tails observed in the curves plotted in Figures 7 and 9, at small values of the inter-pore distances. This is in fact due to the shorter distance between the pores placed on the domain boundaries.

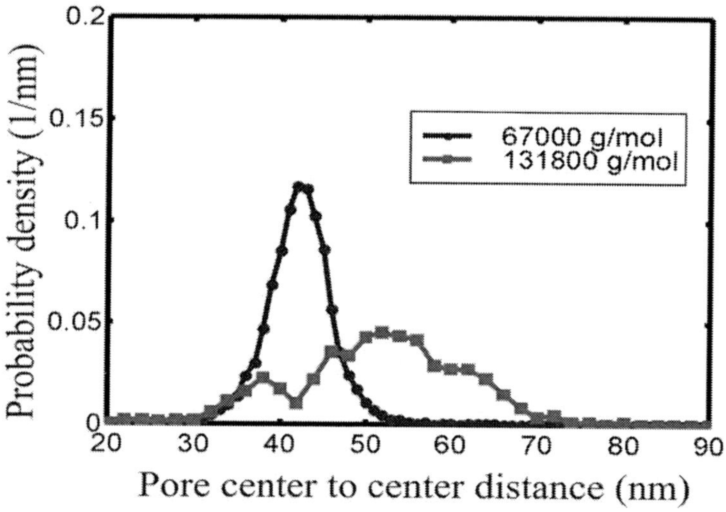

Figure 9. Probability density of the pore center-to-center distance for 67000 gr/mol (black circles) and 131800 gr/mol (grey squares) after annealing at 180 °C for 2 hours.

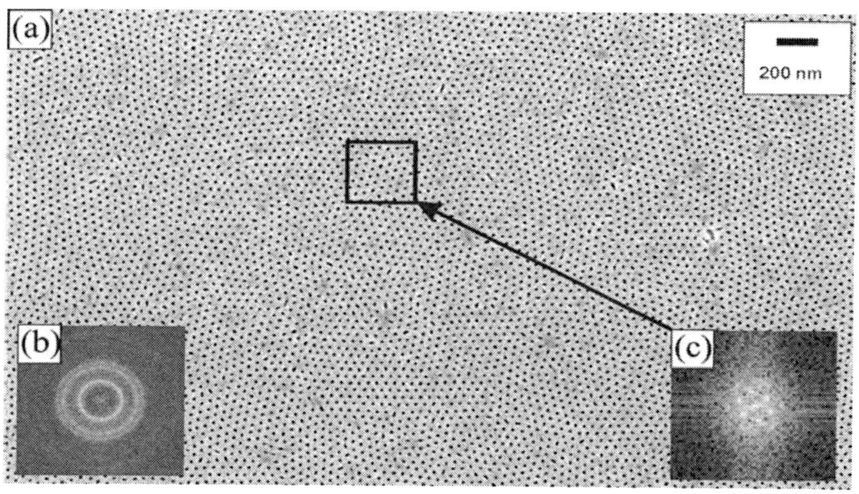

Figure 10. (a) SEM micrograph of the PS template obtained after annealing at 170 °C for 2 h. (b) FFT of the image (a). (c) FFT relative to the portion selected by the black square.

Some control has been gained in controlling the macrodomain size and orientation using external electrical fields [31], graphoepitaxy [32-36], temperature gradients, and solvent casting. Graphoepitaxy has been demonstrated among the others to be easy and efficient. It consists in registering the copolymer orientation into channels, pre-defined on the substrate, i.e. into structures confined along one axis.

To self-assemble the copolymer along more axes, i.e. on circular or triangular structures, or even on unpatterned substrates, it is possible to modify the parameters of the self-assembling process itself, without acting on the substrate geometry. The process step which plays the major role in the self-assembling of the copolymers macrodomains is the annealing process [37, 38].

To investigate the effect of the temperature on the macrodomain size, the SEM images are analysed by using the following method: first a Fast Fourier Transform (FFT) on the whole image is obtained. Then a mask is applied on the first diffraction ring to select the six spots relative to a specific local lattice orientation. Typically for each image and for each annealing condition, only three orientations are considered: 0°, 20° and 40°, since this already provides a statistics of about 300 macrodomains per condition. The remaining part of the FFT is then cancelled. In this way all the macrodomains oriented along the chosen direction are selected. The reduced FFT is then inverted and from the obtained image the density and the area of the macrodomains are measured. The radius of each macrodomain is taken as $r_m = \sqrt{\frac{A_m}{\pi}}$, where A_m is the area of the macrodomain. Figure 11 reports an example of this procedure. Figure 11 (a) is the SEM micrograph of a sample annealed at 170 °C for 2 h. The image is obtained by enhancing the contribution from the 1st ring of the diffraction pattern and by cancelling all the other contributions, which results in an enhancement of the macrodomains and in a removal of the image noise. Figure 11 (b) reports the FFT relative to the original image. Figure 11 (c) and (e) report the inverted images after the selection of the macrodomains oriented along 0° and 40° direction respectively, while Figures 11 (d) and (f) report the corresponding filtered FFTs.

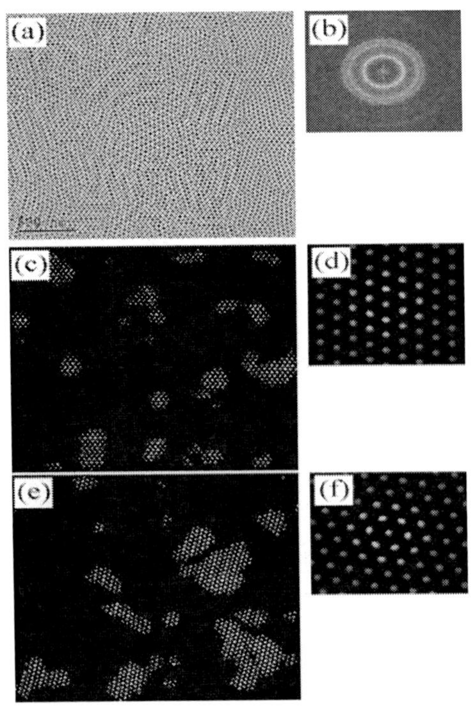

Figure 11. SEM image of sample annealed at 170 °C for 2 h (a) and relative FFT (b). Same image as in (a), but after the selection of the macrodomains oriented along 0° orientation (c), and the 40° orientation (e), and realtive FFTs ((d) and (f) respectively).

This procedure is systematically applied to a number of samples, annealed at different temperatures, and the results are shown in Figure 12. It shows the images relative to the 0° orientation, for annealing at 160 °C (a) and 190 °C (b) for 2 h, and the calculated domain size distributions (c). The graph shows that, by increasing the annealing temperature, the size distribution shifts toward large sizes. In particular, although the average size does not change appreciably (going from 230 nm to 290 nm), the maximum macrodomain size increases from about 380 nm at 160 °C to about 980 nm at 190 °C. Moreover, during the high temperature annealing the small domains tend to disappear, as evidenced by the curve at small sizes.

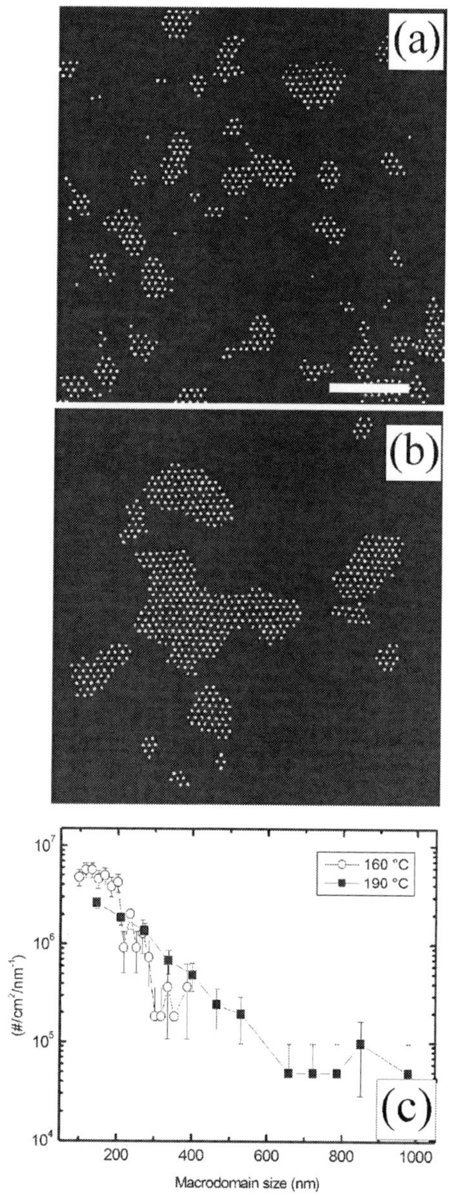

Figure 12. Macrodomains oriented along the 0° in the samples annealed at 160 °C (a) and 190 °C (b), and relative size distributions. The bar in Figure (a) is equivalent to 500 nm.

Indeed all these results are evident by observing the Figures 12 (a) and (b). They illustrate that the density of the domains decreases, the smaller domains tend to disappear while the large ones increase in size. This is probably related to a process of secondary crystallization, where smaller grains collapse into the larger grains, and the process is repeated with larger sizes. In fact, the large domain present in the center of Figure 12 (b), with a size much larger than the average, represents the typical equivalent of the secondary grain.

Based on these results, it can be concluded that it is possible to tailor the macrodomain size by changing the annealing conditions, and without changing the substrate topography. In particular, it is possible to obtain very large macrodomains at expenses of smaller domains.

Biosensors

The ability to detect biomolecular interactions is of extreme importance in medical, pharmaceutical and biotechnological research and development. To this purpose biosensors have been developed [39-41]. A biosensor is a device that transforms biochemical information (presence and/or concentration of a specific analyte), into an analytically useful signal. It can be schematically represented as two basic components connected in series: a biological recognition system (bio-receptor, usually based on supramolecular chemistry principles) and a physico-chemical transducer, as shown in Figure 13. The system may be completed by a signal amplifier and a microelectronic circuit to elaborate the signal. Usually the biosensor and the signal processing set-up are not integrated.

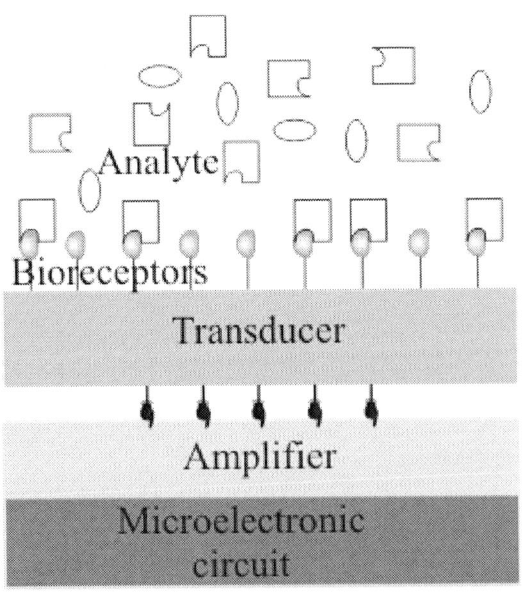

Figure 13. Schematic representation of a biosensor consisting of bio-receptor, transducer, amplifier and microelectronic circuit for signal processing.

The main purpose of the biological recognition system is to provide the sensor with a high degree of selectivity for the analyte to be measured. The high selectivity is the major driving force behind the significant increase of interest and research in biosensors. Different types of biologically sensitive materials can be applied as recognition elements. They can be enzymes, antibodies, antigens [42], proteins [43], nucleic acids [44,45] or even living biological systems (e.g., cells, plants, organs or whole organisms) [46].

The biosensor working principle is straightforward: as a result of the specific biological interaction, there is a change in one or more physico-chemical parameters. Examples of parameters are: pH; ion-concentration change, also due to an enzymatic reactions; adsorption of charged macromolecules such as proteins or DNA strands; affinity binding of molecules, examples are the antigen-antibody reaction or the DNA hybridization. Once the reaction occurs, these quantities are converted into a quantifiable signal by the transducer, amplified, processed and displayed in a suitable form.

Many techniques have been developed in recent decades for the detection of such interactions [47]. At present, the use of optical measurements, e.g. for DNA recognition, is predominant [48,49]. Known probe molecules (e.g. single strand DNA, ssDNA) are immobilized at selected locations and the target molecules are labelled with fluorophores. Interaction between probe and target (e.g. DNA hybridization) is then indicated by the presence of fluorescence at the probe's location. However, the need to use conventional organic fluorescent dyes, which suffer from weak signal intensity, photo-bleaching and self-quenching, limits the applications of fluorescence-based biosensors. Furthermore, optical methods are expensive and difficult to implement in portable instrumentation. Due to this limitation, in vivo, in situ and real time measurements are very difficult.

To overcome the above mentioned limitations label-free detection platforms have been proposed. The most intriguing approach is provided by electrochemical measurements even if they often require a pretreatment of the reference electrode surface [50,51]. They work well with electric control systems including signal amplification and filtering, since they provide an electrical signal as output, but they are not easy to integrate with the circuit on chip simultaneously. To solve these problems, (FET)-type microbiosensors have also been considered [52-58].

In this chapter, a brief overview of electrochemical biosensors [59-64] and some interesting solutions to the main issues related to their use will be presented. Interesting reviews on this subject have been published in the last ten years [65-68].

Electrical Sensing

Over the last few years there has been a growing interest in creating microbiosensors, fabricated in Si-compatible technologies, to be integrated within microelectronic circuits. The reason is that silicon-based devices would provide a lot of potential advantages such as small size and weight, fast response, high reliability, low output impedance, the possibility of automatic packaging at wafer level, on-chip integration of biosensor arrays and a signal processing scheme with the future prospect of low-cost mass production of portable microanalysis systems.

Electrical sensing is considered one of the main goals to achieve in the next generation of biosensors since it could promote their integration within complex electrical circuits, miniaturization and mass production. Due to their simple principle of measurement and integrable signal processing on chip, biosensors that are based on electrochemical transducer principles are the most common sensor devices investigated so far [59-68]. The increasing interest for biosensors led the research to find innovative devices that could satisfy both the needs of mass production and accurate analysis. For these reasons, many research groups are using Si compatible technology as an innovative platform to create biosensors. Many goals will be achieved using Si-based materials: *i*) the use of a low cost and mature technology, fundamental for mass production; *ii*) the possibility to shrink the devices, implying reduced molecular diffusion path, faster kinetics and an improvement of the analytical performance of the device [69,70]; *iii*) the possibility to create microstructured devices achieving complex functions, e.g. micro-total-analysis-systems; *iv*) the integration on the same chip of the electronics and/or photo-electronics needed for detection. Moreover, miniaturization opens the field of *in vivo* physiological monitoring; it implies lower reagent consumption, hence minimized sample volumes, lower energy consumption, and less space requirement (sensor portability). It should be mentioned that conventional biosensors need extensive packaging, complex electronic interfaces and regular maintenance or reactivation.

According to a working group of the International Union of Pure and Applied Chemistry (IUPAC) 'An electrochemical biosensor is a self-contained integrated device, which is capable of providing specific quantitative or semi-quantitative analytical information using a biological recognition element (biochemical receptor) which is retained in direct spatial contact with an electrochemical transduction element' [71].

Biosensors may be classified according to the biological specificity-conferring mechanism or, alternatively, to the mode of physico-chemical signal transduction. According to the official nomenclature, electrochemical transducers include amperometric, potentiometric, conductometric, impedimetric, ion charge and field effect [71]. Biosensors may be further classified according to the analytes or reactions that they monitor, e.g. glucose biosensors. Biosensors may operate a direct monitoring of analyte concentration or of reactions producing or consuming such analytes; alternatively, an indirect monitoring of inhibitor or activator of the biological recognition element (biochemical receptor) may be achieved.

Electrochemical biosensors can be easily divided in two main groups depending on the signal to detect: capacitive or in current. A schematic representation of the different kinds of electrochemical bio-sensors is shown in Figure 14.

Capacitive biosensors can be further divided in two main categories: interdigitated electrode affinity biosensors (Figure 14(a)) and electrolyte-insulator-semiconductor (EIS) biosensors (Figure 14(c)), while the typical configuration of a ion-sensitive field-effect transistor (ISFET) biosensor is shown in Figure 14(e). The typical responses measured for the three types of biosensor are also schematically shown in Figure 14(b), (d), (f).

All the device structures proposed are very sensitive to any kind of electrical interaction occurring at or nearby the gate insulator/electrolyte interface. Nearly each (bio-)chemical reaction leading to chemical or electrical changes at this interface can be measured by means of an EIS or ISFET coupled with a bioreceptor. The basic mechanisms to be used in an EIS or ISFET biosensor can be summarized as follows: *i*) potential changes that are caused by a catalytic reaction product (e.g., between an enzyme and its substrate); *ii*) potential changes that are caused by surface polarisation effects or the change of dipole moments (e.g., antigen–antibody affinity reactions); change of the electric field associated with affinity binding of biomolecules (DNA hybridisation); *iii*) potential changes that are coming from living biological systems as a result of more sophisticated (bio-)chemical processes (e.g., action potential of nerve cells). Biosensors can also be classified according to the biorecognition element that is used for detection. As an example, here report some of the nomenclatures [66]: enzyme-modified FET (EnFET); immunologically modified FET (ImmunoFET); DNA-modified FET (DNA-FET or GenFET: gene-modified FET); cell-based FET (or cell-potential FET (CPFET)); and 'beetle/chip' FET.

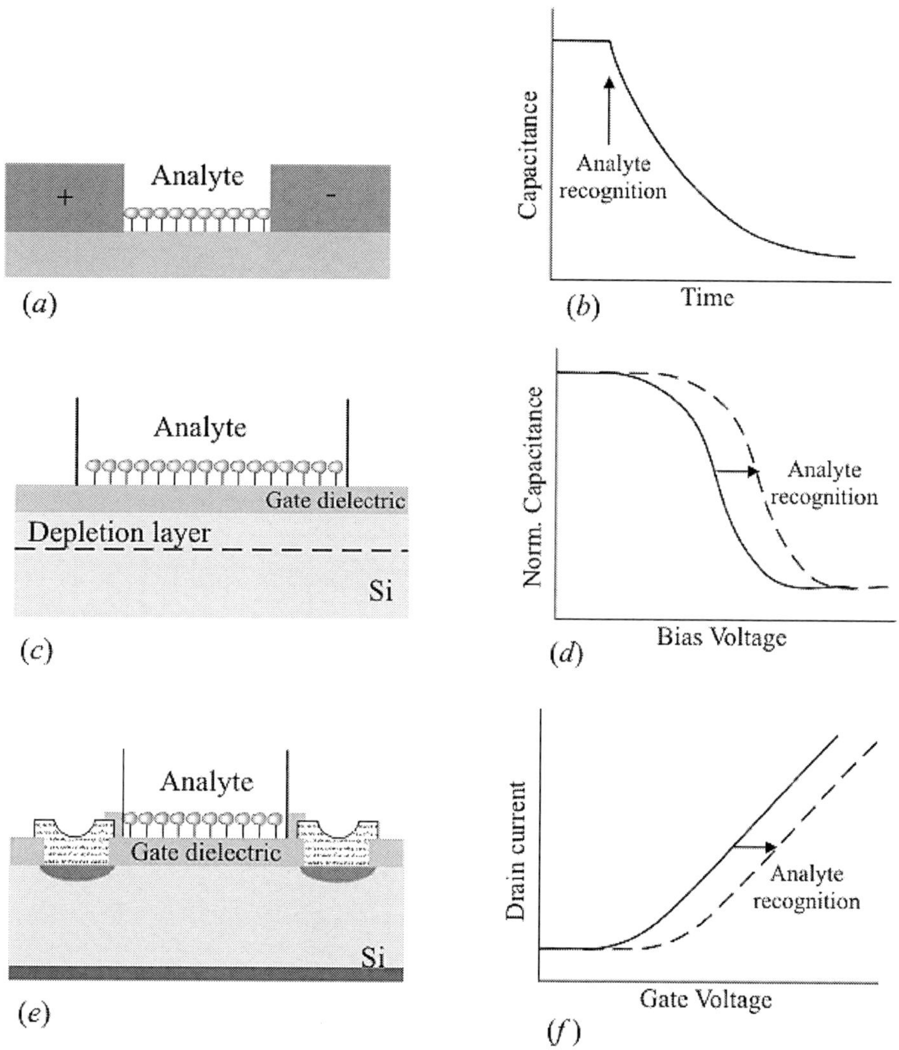

Figure 14. Schematic representation of the structure (a), (c), (e) and of the typical response (b), (d), (f) of: (a-b) interdigitated electrode affinity, (c-d) capacitive, (e-f) ISFET (ion-sensitive field-effect transistor) biosensors.

A particular attention must be devoted to DNA recognition for the importance it has assumed in the last years. DNA molecules are charged macromolecules and the direct hybridisation event is an affinity binding process. Nevertheless, the fabrication of a DNA-sensor must take into account the unique structure of DNA molecules. DNA is a poly-anion with negative charges along its phosphate backbone. It can be considered as a circular cylinder (with a diameter of about 1.5–2 nm) with electrostatic charges evenly distributed about the cylindrical surface [72]; the length of a DNA probe depends on the number of nucleotides and the length of a nucleotide (or base) is about ~ 0.34 nm [44]. To selectively recognise a unique human DNA sequence, DNA probes must be at least 16 bases long [73]. The negative charge of DNA macromolecule can allow immobilisation of ssDNA by electrostatic attraction between the positively charged transducer surface and the negatively charged sugar–phosphate backbone with the bases oriented towards the solution [74]. Finally, changing the surface electrostatics (*i.e.*, negative or positive surface potential) the DNA immobilisation and hybridisation process can be dramatically affected with the possibility of a DNA hybridisation control by regulation of the surface charge (potential).

Capacitive Biosensors

Capacitance measurements can be made in two principally different ways. In the first approach two parallel metal plates are used and the recognition element is located in between them. The capacitor has a dielectric formed by both the recognition element and the solution in which the device is immersed and can be measured with a high precision. The capacitance is described by the typical equation:

$$C = \varepsilon\, \varepsilon_0\, A/d \tag{4}$$

where, ε is the dielectric constant of the material between the plates, ε_0 is the dielectric constant for vacuum, A is the area and d the distance between the plates. The capacitance will increase when the distance between the two plates decreases. Hence, the device sensitivity strongly depends from the fabrication technology. The most sensitive device will be the one having the shortest distance between the plates, technologically reproducible. When the analyte will interact with the receptor, e.g. binding it, a variation in the dielectric medium (ε) is detected. Capacitive affinity biosensors can be constructed by immobilizing recognition elements in thin layers on an electrode or between two electrodes and measuring changes in the dielectric properties when an analyte binds. Changes may also be induced when a large receptor molecule is displaced by a smaller analyte [67]. Capacitance measurements can also detect if a protein on a surface changes its conformation after binding of an analyte, as demonstrated for heavy metal ions [75]. The capacitance versus time measurements provide the required information (Figure 14(b)). This approach was used to fabricate antibody-based biosensors for human IgG [76] with a high sensitivity. A limitation associated with this approach is that this method is very sensitive to changes in the bulk solution. This last problem can be overcome by using a reference cell, without the recognition element present.

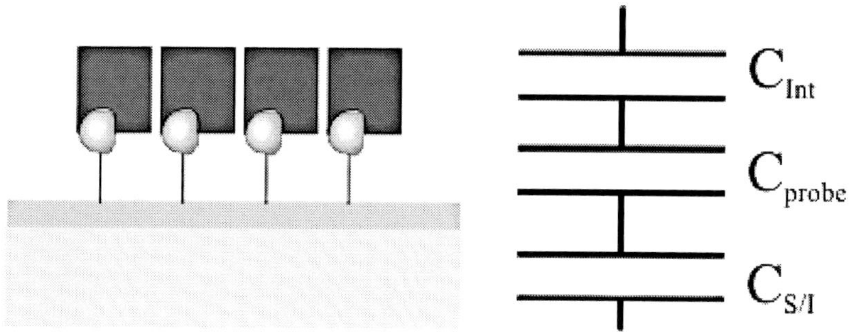

Figure 15. Schematic representation a capacitive EIS structure. On the right the equivalent capacitance are indicated.

The second group of capacitive biosensors (Figure 14(c)) is given by EIS (Electrolyte-insulator–semiconductor) structures. The capacitance measured gives information about the insulator–solution interface, hence, a chemical modification of this boundary should give rise to a change in capacitance, the size of which will depend on the nature and coverage of the modification material. In some cases the insulator is covered with a metal. In that case, the capacitance gives information on the metal-solution interface. The capacitance can be described as the built-up of several capacitors in series, as schematically represented on the right of Figure 15.

The first capacitor, $C_{S/I}$, represents the semiconductor/insulator capacitance. The second capacitor, C_{probe}, is the capacitance due to the anchoring groups (see later), the recognition element and any contribution from the Stern layer, which consists of hydrophobically bound water molecules between the recognition and the diffuse layers [77]. Finally, the third capacitor is given by the concentration dependent diffuse layer, C_{Int}. It extends from the outer device surface into the bulk of the solution. This last capacitance was first described by Gouy and Chapman at metal surfaces in solutions [78]. The recognition event will give a change in the capacitance of the recognition layer C_{probe}. The total capacitance C_{Tot} can then be described by the following equation:

$$1/C_{Tot} = 1/C_{S/I} + 1/C_{probe} + 1/C_{Int} \tag{5}$$

The equation clearly states that the lowest capacitance will dominate the total capacitance. It is therefore important to design the device in order to have the capacitance in the insulating layer as high as possible, otherwise it may dominate and analyte changes might not be detectable. Nevertheless, the oxide can not be too thick, otherwise the device sensitivity will strongly reduce. Another important issue is the device surface functionalization. In fact, if the bio-recognition elements do not cover completely the electrodes the influence of holes can not be neglected. The capacitances will be in parallel with resistors, thereby decreasing the range and overall sensitivity of the measurements. Since the solution is conductive, there will be a short-circuiting of the capacitances of the layers on the electrode, due to the solution presence in the holes. A procedure for filling the holes in the biological layer will therefore improve the quality of the sensors [79]. Capacitance changes can be detected either as a change in the dielectric constant or as a change of the thickness of a layer immobilized on the transducer [80]. A specific drawback for capacitive biosensors compared with other types of biosensors is that immobilization of the bio-recognition layer is more critical. If it is not sufficiently insulated, ions can move through the layer causing short-circuiting of the system, leading to a decrease or absence of the signal [66,67]. Moreover, if the distance between the probe molecules is larger that the Debye length in the electrolyte [81], the effectiveness of the method is also reduced. The density of probes can be controlled by applying a voltage to the gate metal or the back contact, while the Debye length can be controlled by changing the ionic concentration of the solution.

Ion Sensitive Field Effect Transistors

The working principle of ion sensitive field effect transistors (ISFET) the same of the MOSFET devices, see e.g. ref. [82] for an exhaustive explanation of MOSFET theory. In particular, in a MOSFET structure (see Figure 14(e)), the current in the transfer characteristic (see Figure 14(f)) depends on the flat band voltage (V_{FB}) and, indirectly, by the gate oxide capacitance. The drain current I_D is a function of the input voltage V_{GS}, according to the following equation:

$$I_D = C_{ox}\mu\frac{W}{L}\left[(V_{GS}-V_T)V_{DS}-\frac{1}{2}V_{DS}^2\right] \tag{6}$$

where Cox is the oxide capacitance per unit area, μ is the carrier mobility in the channel, W and L are the channel geometrical parameters (width and length, respectively) and V_T is the threshold voltage, and V_{GS} is the only variable of the equation. In fact, V_T in a "normal" MOSFET is a constant, which manifests itself only as a certain threshold voltage, hence its name. In the ISFET case, however, the interfacial potential in the input circuit in described in terms of V_T. Therefore, it is fundamental report the threshold voltage equation for the ISFET immersed in an electrochemical cell:

$$V_T = E_{ref} - \Psi + \chi^{sol} + \left(-\frac{\Phi_{Si}}{q} - \frac{Q_{ox}+Q_{ss}+Q_B}{C_{ox}} + 2\phi_f\right) \tag{7}$$

where E_{ref} is the reference electrode potential (constant), the term $\Psi + \chi^{sol}$ represents the potential at the solution/oxide interface. In fact, Ψ is the chemical input parameter, that is a function of the solution pH and χsol is the surface dipole potential of the solvent. and thus having a constant value. The terms in brackets are the parameters determining the traditional MOSFET V_T, with the exception of the metal work-function that is inglobated in the E_{ref} term. The first is the silicon work function (Φ_{Si}), the second is due to accumulated charges: in the oxide (Q_{ox}), at the oxide–silicon interface (Q_{ss}) and in the Si depletion region (Q_B), whereas the last term (ϕ_f) determines the onset of inversion depending on the doping level of the silicon [82,83]. The biological material presence affects the potential at the solution/oxide interface, thus modifying V_T. If also the recognition event modifies this potential, it can be detected using an ISFET.

In literature, two different kinds of ISFET devices can be found: in the first kind the biological probe is anchored to the gate oxide, without the upper metallization. The device is often incorporated in an electrochemical cell and the gate voltage is provided by a reference electrode, whose function is to keep the all solution to a known potential (E_{ref}) [54,62,83]. A modification of the layer due to the recognition event is recorded by the device as a charge variation on the gate. As a result, a shift in the Drain current is obtained (see Figure 14(e)). As an example the results by Sakata and co-workers [62] are shown in Figure 16. They used an ISFET to detect the DNA hybridization in an electrochemical cell. The ISFET did not have the gate metal and the gate oxide was given by a double layer of SiO_2/Si_3N_4 oxides, 157 nm thick (17 nm of SiO_2 and 140 nm of Si_3N_4). In Figure 16(a) is shown the device transfer characteristic in the all measured range, while in Figure 16(b) is reported a zoom to show the shift due to the DNA presence. The electrical signal of immobilization of oligonucleotide probes, hybridization with target DNA on the genetic field effect transistor (FET) is evident. The amount of the threshold voltage V_T shift was determined at a constant drain current of 600 μA. A first shift of 78 mV was detected as a result of the single strand DNA immobilization on the gate oxide, while a further shift of 11mV was detected due to the DNA hybridization with the complementary strand.

The results clearly indicate the device sensitivity to DNA immobilization and hybridization.

Nevertheless there are many drawbacks that prevent the commercialization of these devices. As an example, in this kind of structure there is a dependence of the sensor response to the on buffer capacity, there is a restricted dynamic measurement range and a non-linearity in the response.

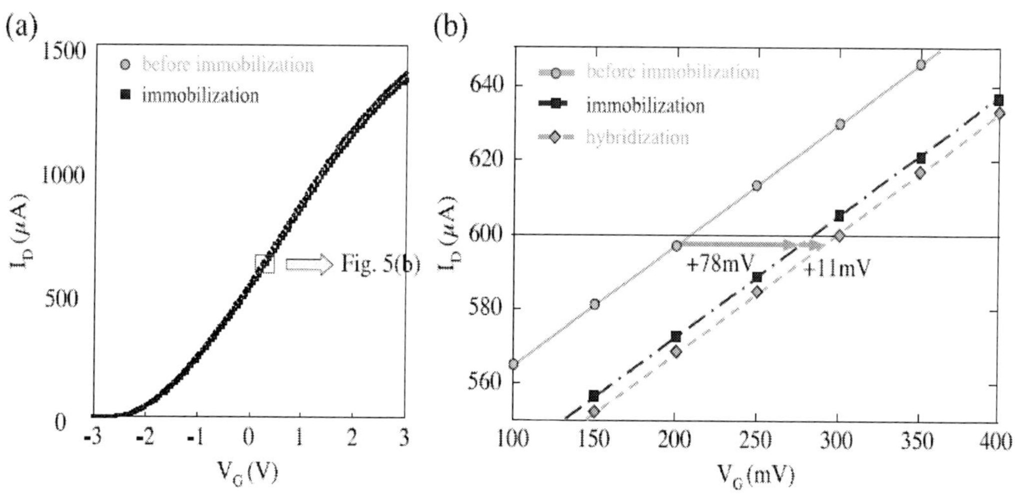

Figure 16. Electrical signal of immobilization of oligonucleotide probes, hybridization with target DNA on the genetic field effect transistor (FET). The amount of the threshold voltage V_T shift was determined at a constant drain current of 600 μA (from ref. 62, Copyright Wiley-VCH Verlag GmbH and Co. KGaA. Reproduced with permission).

Other drawbacks are: the relatively slow response and recovery times; the operating and storage stability; the reproducibility; the dependence of the sensor signal on the biological probe immobilization method [67-68]. Finally, the so-called counter-ion screening effect as well as the non-ideality of the molecular layer reduces the possibility to actually fabricate these devices for a direct electrostatic detection of charged macromolecules. This subject has been critically discussed in case of immunospecies, protein molecules and DNA in [68,84,85]. The estimated hybridization signals are in the range of several mV [86]. An enhancement of the sensor signal can be obtained when operating the devices in a low ionic-strength solution (<1 mM) and utilizing tightly packed probe ssDNA (>10^{13} ssDNA/cm^2) [68].

Probe Immobilization

The immobilization of the biological probe onto the transducer surface plays an important role in the overall performances of biosensors. Three main issue must be considered: *i*) the sensing surface must be biocompatible; *ii*) the immobilization protocol must not degrade the inorganic part of the sensor (for Si-based devices, VLSI compatibility); *iii*) the biological molecule must be anchored to the solid surface avoiding its denaturizing or the loss of its activity. Moreover, the immobilization step should lead to a well defined probe orientation, readily accessible to the target [87,88]. The environment of the immobilized probes at the solid surface depends upon the mode of immobilization and can differ from that experienced in the bulk solution. These issues have been the target of many works in literature [70,89-94]. Silanized SiO_2 semiconductor surfaces on various conducting supports like gold or silicon have been used [59,60,80,95-97]. Gold electrodes can be functionalized through self-assembled monolayers of thiols, sulfides or disulfides [98]. Moreover, if silicon dioxide is used, the thermal processing used (wet or dry) may produce quite different results [94]. The most used approach is the formation of covalent bonds between the molecule and the solid surface, often using linker molecules [70,79,89,99-101]. The immobilization procedure must be optimized to obtain the maximum surface coverage and to prevent the biological molecule denaturation and/or the loss of its specific property, e.g. for an enzyme its enzymatic activity. As an example, the immobilization protocol we optimized is described hereafter. The biological molecules immobilization procedure, reported schematically in Figure 17, lies in four steps. The first is oxide activation leaving OH terminated groups on the surface. This step is fundamental when the oxidation process is performed in dry ambient, in fact, the surface must be pretreated before silanization in order to maximize the OH groups available to bond the silane groups during the second step [94]. The second step is the silanization that was carried using 3-aminopropyltriethoxysilane (APTES) that terminated the surface with NH_2 groups. The last two steps are glutaraldehyde (GA) immobilization (surface terminates with –(C=O)-H groups) and amino terminated biological molecules bonding. The power of this method is that, once optimized, it may be used for any biological molecule having amino groups free for bonding.

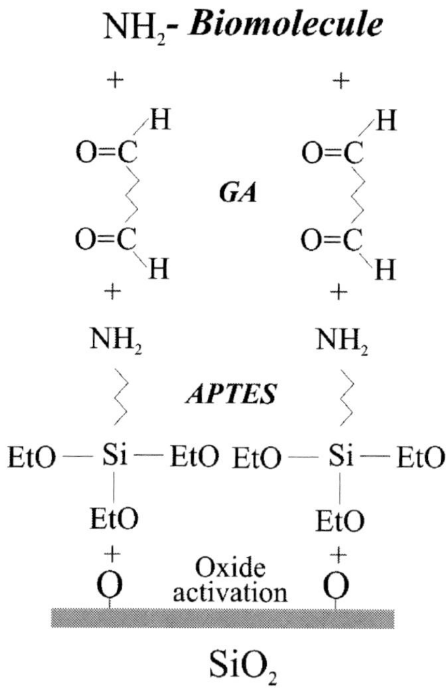

Figure 17. Schematic structure of the probe immobilization protocol. The four immobilization steps, oxide activation, silanization, linker molecule deposition and biological molecule immobilization are indicated as: oxide activation, APTES, GA, biomolecule, respectively.

In particular, in this chapter the results obtained for and enzyme, the Glucose oxidase (GOx) from *Aspergillus Niger*, and short synthetic oligonucleotides, immobilized on a silicon oxide surface are presented. GOx is 160 kDa homodimeric globular protein, with a tightly bound (K_a = $1x10^{-10}$) flavin adenine dinucleotide (FAD) per monomer. The overall dimer dimensions are $6.0\times5.2\times7.7$ nm^3 measured by X-ray crystallography [102]. GOx, as all peptides and proteins, is a polymer of α-amino acids; it includes 580 amino acids, the FAD cofactor, six N-acetylglucosamine residues, three mannose residues and 152 solvent molecule acids [102]. As it is well known, the general chemical structure of an α -amino acid (excluding proline) is R-CH-NH_2–(COOH). The enzyme catalyzes the oxidation of β-D-glucose to D-glucono-1,5-lactone and hydrogen peroxide, using molecular oxygen as electron acceptor [103]. This enzyme is usually employed when the glucose concentration in the blood must be measured, hence GOx based micro-biosensors [99] would have immediate applications in monitoring diabetes [104].

The importance to text DNA strands has been widely discussed in the previous section. The single strand DNA (ssDNA) used were 20 bases long functionalized at the 5' end with a NH_2-C_6 group. The sequence used was ATGCATGCATCGTACGTACG. DNA hybridization was monitored utilizing its complementary strand (ssDNA-c), having the sequence ATGCATGCATCGTACGTACG.

The details of the immobilization procedure for both GOx [100,101,105] and DNA [106] probes are fully reported elsewhere. After immobilization, the samples were either dried under a gentle nitrogen flow and stored at RT or stored in PBS at 4°C, until measurement, usually performed within two days from preparation. As reference, various samples were prepared: some were simultaneously treated with only water, while other samples were prepared by applying only the biological molecule immobilization step (only DNA and only GOx).

Since the probes immobilization strongly affects the overall device performance, a careful inspection of the immobilized biological layer must be performed. The surface can be characterized using various techniques, but our choice was to use techniques normally employed in microelectronics, in order to maintain a full compatibility with the microelectronic process flow. Both X-ray Photoelectron Spectroscopy (XPS) and Atomic force Microscopy (AFM) were used to test the sample surface after each step of the immobilization protocol.

In Figure 18, the AFM analyses for some intermediate steps of the immobilization process are reported. The samples that underwent only oxide activation did not show any macroscopic modifications during the first immobilization step, as previously mentioned.

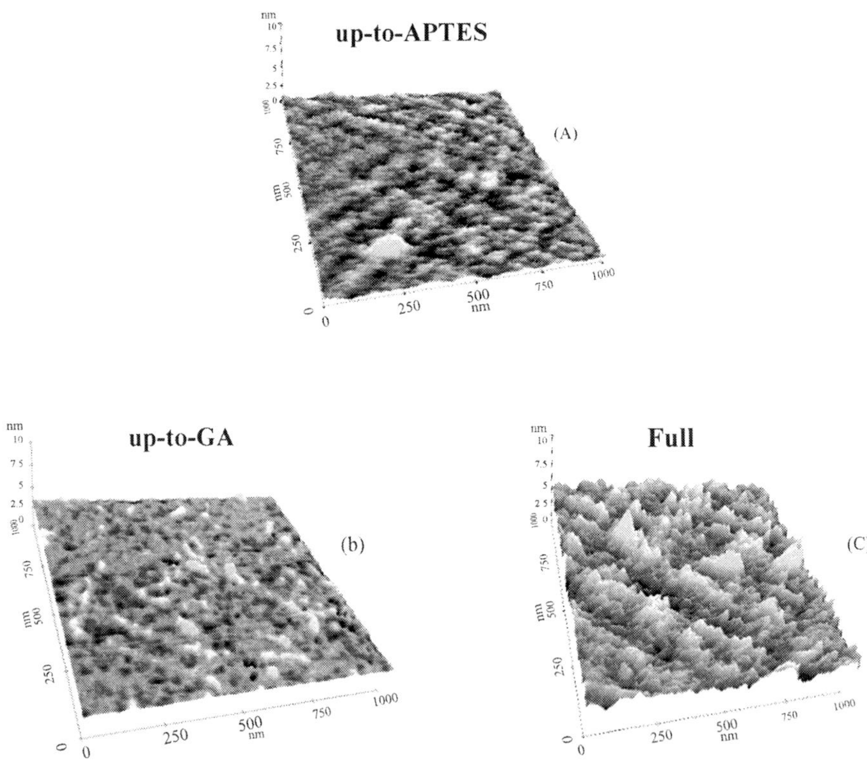

Figure 18. AFM images of: (a) up-to-APTES, (b) up-to-GA and (c) Fully processed (with GOx) samples. The vertical scale is 10 nm while the horizontal scales are both 1μm. Higher peaks appear lighter in a grey color scale.

The second step, silanization, allowed us to functionalize the sample surface with amino groups that easily bond the linker molecule, glutaraldehyde, if silane polymerization did not occur. AFM measurements of this sample, shown in Figure 18(a), indicated a surface RMS value of 0.22 nm. We used the same vertical scale in all three-dimensional images to allow a direct comparison of the samples. It should be noted that the APTES layer, if the surface is properly functionalized, has a physical thickness of 0.8±0.1 nm [107]. The AFM results suggested a good sample coverage. The highest peaks observed in Figure 18 are due to APTES polymerization. Samples that did not undergo oxide activation showed more polymerized sites [105] suggesting a lower number of active sites available for GA immobilization, and a much higher RMS value. The measurements showed that the protocol used produced a uniform surface silanization, as confirmed by the contact angle measurements reported in ref. [105].

During immobilization, one glutaraldehyde (GA) end bonds the amino groups of the silane, while the other end can bond the protein amino groups. When the GA was deposited on the sample surface only small modifications of the surface morphology occurred (Figure 18(b)). In fact, the surface RMS, as measured by AFM, was 0.19 nm, barely distinguishable from the data extracted from the previous sample measurements. The data clearly indicated that the surface did not have relevant modifications, detectable by AFM. Finally, AFM measurement carried out on a sample that underwent the full immobilization process with GOx are shown in Figure 18(c). A clear modification of the surface occurred due to enzyme immobilization, with the appearance of peaks having the same height and well distributed on the sample surface. The RMS value for this sample was 0.59 nm. These qualitative observations of the surface evolution following the four steps of the immobilization protocol were quantitatively demonstrated extracting the power spectral density (PSD) from the AFM images [105]. A complete explanation of the observed morphological modifications, and their absence after the first three immobilization steps, was provided by XPS analysis. As an example, the XPS Si2*p* spectral region for some immobilization steps are summarized in Figure 19. Also XPS data showed differences between the fully processed samples and those stopped before the final step.

The reference sample (treated with water alone, Figure 19 top spectrum) showed two components with binding energies of 99.7 eV and 104 eV, assigned to Si° and SiO_2 respectively [108,109]. The SiO_2 component was centered at 104 eV instead of 103.4 eV, as expected for silicon oxide, due to differential charging between the SiO_2 layer and the Si substrate [110]. The SiO_2 layer thickness (d=6.5 nm) was calculated from the intensities of the Si2*p* peak components assigned to the Si° and SiO_2 following the procedure reported in ref. [100].

The progressive depositions of APTES, GA, and GOx produced extra layers on the SiO_2 surface whose thickness increases with the process step. The final film deposited on the Si substrate, given by both the SiO_2 and the organic layers (APTES+GA+GOx) had a thickness that completely shielded the Si2*p* substrate signal, as it can be seen from the spectrum of the fully processed (with GOx) sample of Figure 19 (bottom). The sample fully processed with DNA produces the same Si2*p* signal already observed for the GOx and shown in Figure 19. The observation of the up-to-GA sample (center) clearly shows that such signal was already fully shielded after GA deposition (SiO_2+APTES+GA, see insert), as demonstrated by the presence of the SiO_2 layer component and the absence of any signal in the region where the Si° should be.

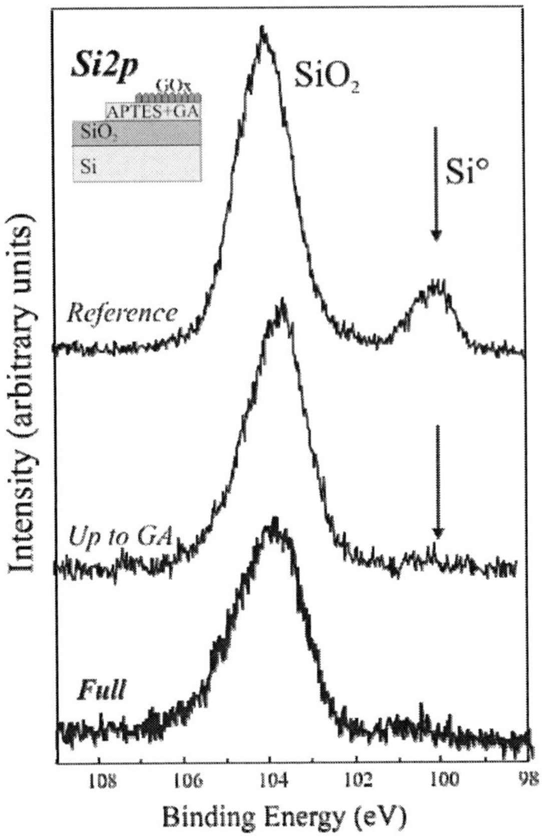

Figure 19. XPS Si2*p* spectral regions of the samples after the various protocol steps. From the top: reference, up-to-GA, fully processed (with GOx). The insert shows the progressive surface coverage during the immobilization steps.

Two conditions must be fulfilled to have a complete substrate shielding: *i)* a film thickness above the photoelectrons escape depth, *ii)* the uniformity of the layer. According to literature data [107], since APTES and GA are small molecules, uniform thin layers on the SiO_2 surface can be obtained if their immobilization protocol is optimized. In particular, the physical thickness of an APTES monolayer is 0.8±0.1 nm, while the total thickness of the APTES+GA layer is 1.5±0.1 nm. As a result, when the third step of the immobilization protocol was carried out, the film on the Si substrate had a total thickness of 8.0 nm (6.5 nm of SiO_2 + 1.5 nm of organic compounds). The progressive layer deposition is summarized in the insert of Figure 19, for the GOx immobilization. The Si substrate signal intensity is attenuated by two layers (SiO_2 + organic), based on the calculations reported in ref. [100]. Our results confirmed that the immobilization protocol was correctly optimized to provide a uniform GA layer ready for GOx deposition, as demonstrated by XPS data. In fact, the up-to-GA sample did not exhibit the Si2*p* substrate signal to confirm the uniform GA film deposition. Since APTES tends to polymerize a uniform GA distribution is not easily obtained. The final issue to be addressed is the layer uniformity. In fact, it could be pointed out that a thick, but non uniform, layer could produce the same results.

The inspection of the C1*s* peak, reported in Figure 20 allowed us to have conclusive proof of the biological probe deposition on the sample surface. The up-to-APTES sample (top spectrum of Figure 20) shows a peak at 284.8 eV assigned to C–C and C–H bonds, as expected if C is present on the sample surface. It should be mentioned that the C peak was also detected on the reference sample (not shown) it is due to the presence of adventitious contaminations on the sample surface. Also after the third step of the immobilization protocol was finished (up-to-GA sample), only the C-C, C-H bonds are observed. It should be mentioned that the GA molecule has two aldehyde groups, separated by a flexible chain of three methylene bridges (CHO-$(CH_2)_3$-CHO). The aldehyde groups contribution (286.5-287 eV of binding energy) produced an asymmetry in the high binding energy side of the spectrum. The simulation of these bonds is superimposed to the measured spectrum as a dashed line in Figure 20 and clearly evidences there are no additional signals.

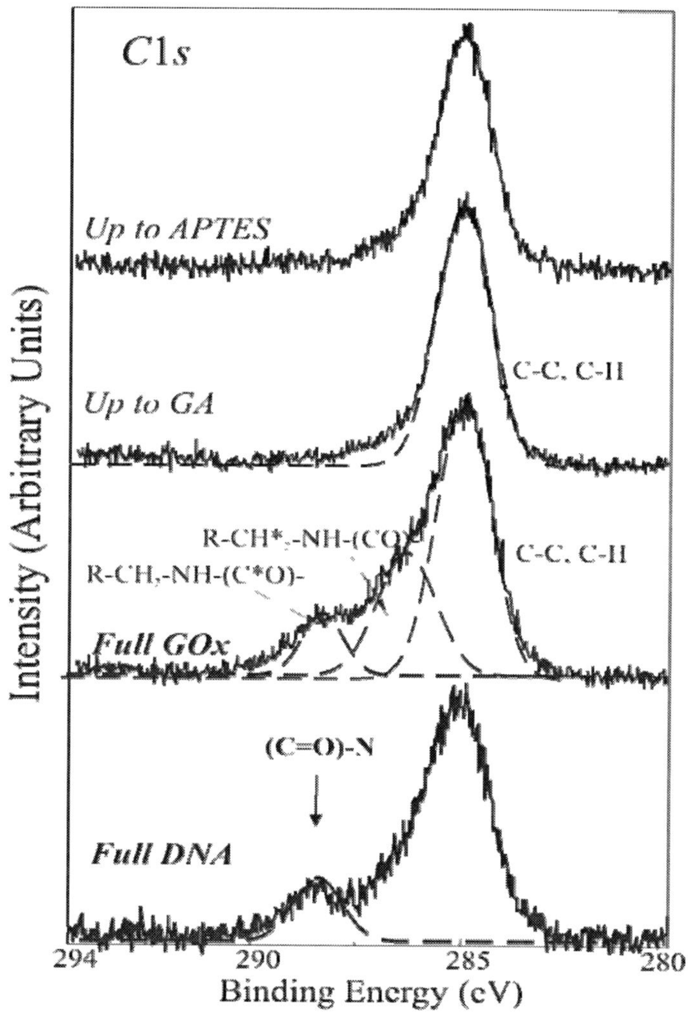

Figure 20. XPS C1*s* spectral regions of the samples after the various protocol steps. From the top: up-to-APTES, up-to-GA, fully processed with GOx and fully processed with DNA. The dashed lines are the simulated signals, their labels are also reported in figure.

Dramatically different is C1*s* peak for the fully processed samples, both with GOx (one above the bottom) and with ssDNA (bottom spectrum in Figure 20). The fully processed sample with GOx exhibits the C1*s* peak is centered at 284.8–285 eV and shows at least two additional components at 286.3 eV assigned to R–CH_2*–NH–(CO)–, and 288.3 eV, assigned to R–CH_2–NH–(C*O)– chemical groups respectively, characteristic of the proteins, as expected if the enzyme is deposited on the sample [101,108,109]. The three different components were simulated and the results are plotted as dashed lines in the same figure. Finally, the sample fully processed with ssDNA clearly showed the C1*s* peak at 285 eV, but it also exhibited another component at about 288 eV, assigned to –(C=O)-N chemical groups. They are characteristic of the DNA nitrogen bases and their presence is expected only when the DNA is deposited on the sample [106,108,109] as demonstrated by their absence in the up-to-GA sample. These results confirmed the DNA presence on the sample. It should be mentioned that the component of C1*s* at 288 eV we assigned to the nitrogen bases of DNA is specific of DNA presence in our samples. In fact it is not detected in any spectra of the other samples (e.g. the up to GA sample) and is detected in a reference sample where only DNA was deposited, without previous surface functionalization (not shown). The XPS data fully confirm that the biological probes were correctly immobilized on the sample surface.

As previously mentioned, one of the main issues in biological probes immobilization is that they must maintain their specific biological function, for en enzyme, its enzymatic activity. Glucose oxidase enzyme catalyzes the oxidation of β-D-glucose to δ-gluconolactone by a reaction that can be summarized in two steps: i) glucose oxidation with the enzyme reduction, ii) re-oxidation of the enzyme with consumption of molecular oxide (O_2) and production of hydrogen oxide (H_2O_2) [102]. In the presence of peroxidase (POD), if p-hydroxybenzoic acid and 4-aminoantipyrine are present in the solution, hydrogen peroxide participates in a reaction and a quinoneimine dye complex is formed, which is measured at 510 nm using a spectrophotometric assay. The GOx enzymatic activity is shown in Figure 21 for a sample immobilizes using the full protocol and a sample with GOx immobilized without the sample functionalization (only-GOx). An increase in the enzymatic activity was observed when the immobilization protocol is carried out (Full), while in the only-GOx sample the enzymatic activity was lower than the value measured for the sample that underwent chemical immobilization. [105].

The results so far presented are intended as an example of sensitive layer characterization for the fabrication of a biosensor.

Electrical Characterization

The immobilization protocol described in the previous section can be used for the fabrication of MOS-like, biosensor structures [106]. In particular, thin oxides (15 nm thick) were thermally grown on p-type epitaxial Si layers. On some samples the immobilization protocol described in the previous section was applied and the resulting devices compared with reference structures. Both the reference and the fully processed samples were electrically characterized, using a PBS drop (0.1 M, pH 6.5) as gate electrode and placing the probe directly in the drop. The first test was carried out on reference samples (SiO_2/Si). In particular, we tested the measurement reproducibility carrying out CV measurement with a repetition frequency of 10 Hz. The results are shown in Figure 22, where the normalized capacitance as a function of the applied voltage is reported. In particular the figure reports all the curves acquired during 100s of measurements. All curves are overlapped, clearly indicating the reproducibility of the data. The same results were obtained for different samples (not shown).

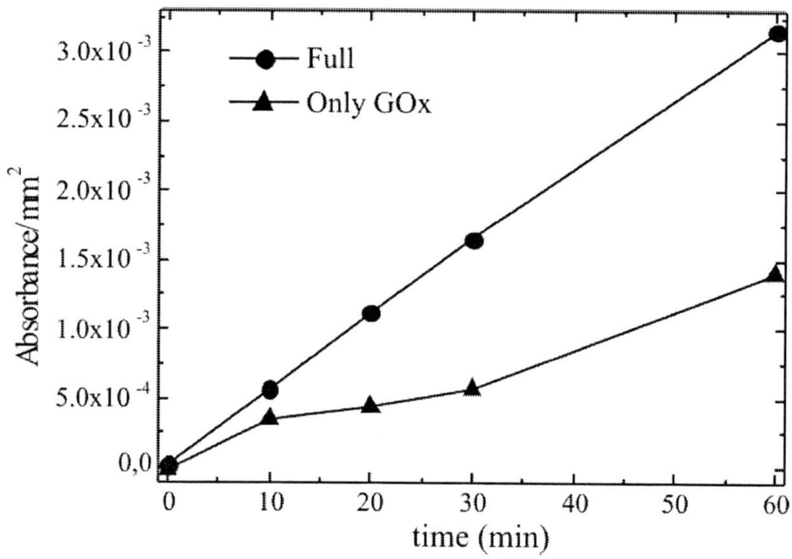

Figure 21. Glucose oxidase activity as a function of time for a sample fully processed (circles) and with only GOx deposited (triangles).

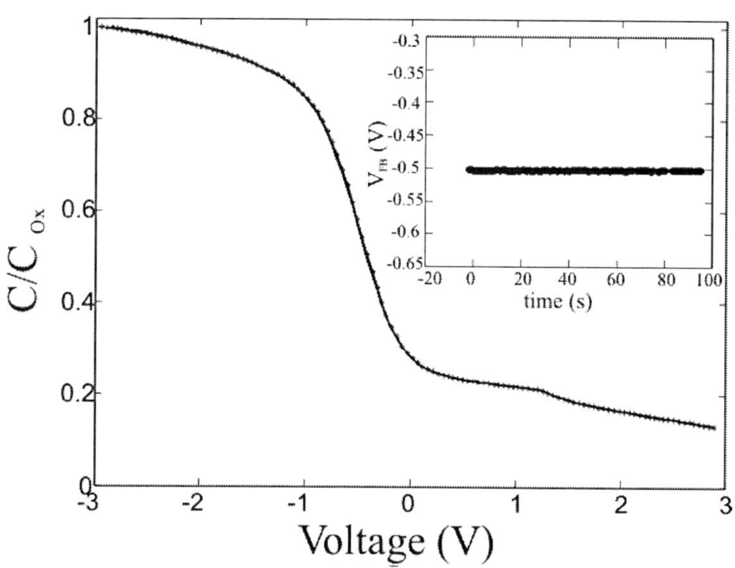

Figure 22. Normalized Capacitance Voltage curves of the reference sample. The measurements were repeated with a frequency of 10 Hz in dark for 100s. In the insert the flat band voltage (V_{FB}) as a function of time for those measurements is reported.

The measurements reported in Figure 22 provide interesting information of oxide aging. The capacitance was normalized to the oxide capacitance (C_{ox}), measured in the device accumulation region (negative biases), while it reduced increasing the voltage in the depletion region [82]. If oxide aging was present, a shift in the flat band voltage (V_{FB}) had to be observed as a function of time, having a direction depending on the trapped charges. The V_{FB} values as a function of time, as extracted from C-V measurements, are reported in the insert of Figure 22. For this set of measurements a constant value of -0.51±0.01 V was detected for the V_{FB} during the entire measurement time. The results clearly demonstrated that there is no oxide aging during the measurement time, hence any shift in V_{FB} must be attributed to the organic layer deposited on the oxide surface. The V_{FB} value used for the comparison is the average on three sets of measurements (each one including 60 curves) and it is -0.50±0.03 V.

Once verified the reference samples provide a reproducible signal, we tested the dependence of measurement from the solution parameters. In fact, it has been already discussed, that the solution parameters, pH and ionic concentration, can strongly affect the measurement (see eq. 5 and 7) thus reducing the sensor performances [52,66,67,83].

The CV measurements were carried out on the reference samples as a function of the solution pH. An example is shown in Figure 23. The curves perfectly overlap (within the experimental errors) regardless of the buffer solution pH, from 4.5 (dashed line) up to pH 8 (squares). It should be mentioned we tested from pH 3 to pH 9, even if the data are not shown, and obtained the same results. It is known that SiO_2 exhibits a pH dependence in the observed range of ~ 30 mV/pH [62]. Since no shift was observed, within the experimental errors, in the measured curves it is reasonable to assume that the SiO_2 surface is perfectly passivated. In fact, it is known that the pH sensitivity depends on the density of OH groups at the SiO_2/solution surface [94].

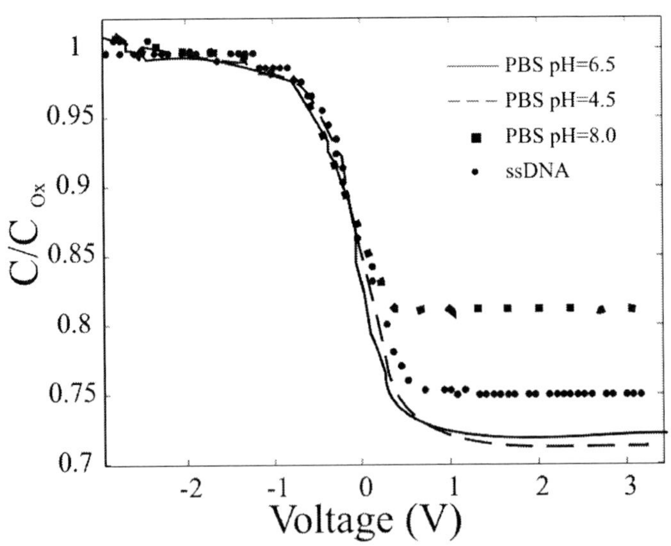

Figure 23. Normalized Capacitance Voltage curves of the reference sample as a function of the electrolyte pH (6.5 solid line, 4.5 dashed line, 8 squares) and in presence of ssDNA at pH of 6.5 (circles).

The oxidation process used to prepare the sample used in this work was in a dry ambient, hence the number of OH groups is strongly reduced. Moreover, since no permanent V_{FB} shift was observed as a function of time, also stressing the samples (+8V for 50 ms before each measurement for a total time of 200s, not shown) we may conclude that there was no dielectric aging during the measurements. Finally, a PBS solution with the complementary oligonucleotide was spotted on the reference sample and the C-V measured to verify if the presence of the additional negative charge in solution cold cause a V_{FB} shift. The data, also shown in Figure 23 as circles, demonstrate that the C-V was not affected by both pH and DNA presence in the PBS solution.

Once verified the measurement reproducibility and that the samples are not sensitive to the solution parameters, GOx immobilized samples were tested, using the procedure previously reported. Also in this case no dielectric (SiO_2 + organic) aging was detected (not shown), but a clear shift of V_{FB} towards negative biases was monitored, as shown in Figure 24, where the reference and the GOx fully immobilized sample C-V are compared.

The V_{FB} approaches to -1.25±0.04V (average value), hence, a shift of 0.74 V, well above the measurements indetermination (±0.05), was detected. This result confirmed that GOx immobilized on the dielectric surface introduced a positive charge as already observed by Wang and co-workers [58]. The most important conclusion that can be drawn from these data is that this device is very sensitive to the presence of organic layers, hence it is a promising candidate for the fabrication of fully electric (MOS-based) glucose sensors.

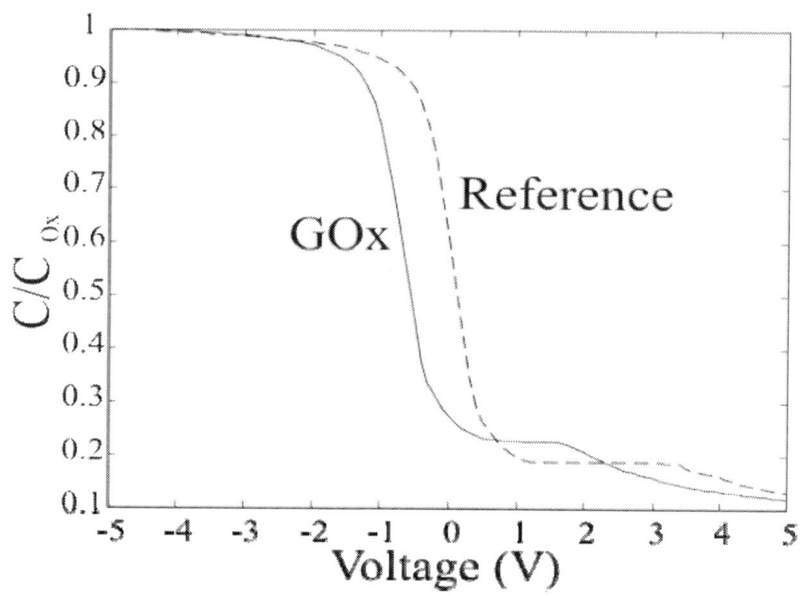

Figure 24. Normalized Capacitance Voltage curves of the reference sample (dashed line) and of the GOx fully processed sample (solid line).

Quite different is the V_{FB} behaviour when ssDNA is immobilized on the SiO_2 surface.

Since DNA molecules have negative charges derived from the phosphate groups in aqueous solutions, a negative charge is trapped on the sample surface. The negative charge presence on the oxide causes a V_{FB} shift, with respect to the reference sample [58, 111].

Conclusion

In conclusion, a partial self-ordering of CVD Si nanodots is observed in the neighborhood of pre-existing dots. This is attributed to the silicon adatoms capture mechanism during the deposition. The calculation of the capture zone in case of continuous nucleation is performed. A correlation between the final dot size and the available capture zone is found and this effect is shown to increase with increasing the substrate coverage. The data on the dot size distribution are fitted by using a scaling model based on the modification of the Voronoi tessellation. From the fit to the data a percentage of 30% of nucleation sites of the substrate is found to be deactivated by the nearest neighbor exclusion mechanism. In the nanocrystal memory, the presence of a denuded zone surrounding each dot leads to less dispersion from bit to bit of the surface coverage, compared to the assumption of random nucleation. This implies favourable projections concerning the ability of the nanocrystal memory concept to meet the scaling targets of the future nodes. Experimental data of threshold voltage window dispersion fully support such picture. To further decrease the coverage dispersion, and finally bring it to zero, we investigate a low cost, simple and VLSI compatible approach for the formation of nanodots controlled in size and position. This approach is based on the self-assembling of polystyrene/poly-methylmethacrylate block copolymer films. We follow the copolymer self-assembling process on flat substrates, and in particular the evolution of the hexagonal macrodomains in terms of density, size and relative orientation, during the annealing promoting the self-assembling. The high statistics is acquired through an original method based on the elaboration of the electronic micrographs in the reciprocal space. The results indicate that while the annealing conditions do not play a role in the pore to pore distance distribution, thus confirming previous results, they influence the macrodomain density and size distributions, inducing an onset of secondary grain growth crystallization behavior in this system. This indicates that it is possible to tailor the long range order of the block copolymer, without the need of any substrate prepatterning.

We covalently immobilized two biological molecules, an enzyme, the glucose oxidase (GOx), and amino modified single strand DNA (ssDNA) fragments on a SiO_2 surface, using a four steps protocol: oxide activation (carried out to increase the OH bonds), silanization, GA deposition and biological molecule bonding. The correct immobilization was demonstrated using XPS and AFM measurements.

The electrical characterization of MOS-like capacitors having the biological molecules directly immobilized on SiO_2, was carried out directly dropping a PBS solution on the SiO_2+organic layer. The dielectric did not experience aging in the measurement time regardless of the immobilized biological species. The MOS-like structures tested are sensitive to the biological species immobilized on the dielectric. In fact, GOx immobilization produced a negative shift (~ 0.7 V), typical of positive charges presence, of the V_{FB}, while DNA immobilization produced a shift in the opposite direction. Also ssDNA immobilization causes a V_{FB} shift, further increased if the hybridization process takes place.

Acknowledgments

The authors acknowledge A. Spada and N. Parasole of CNR – IMM for the expert technical assistance. This work was partially funded by the regional project "POR Sicilia 2000-2006, Mis. 3.15"

References

[1] Cappelletti, P.; Golla, C.; Olivo, P.; Zanoni, E. Flash Memories; Flash Memories (Computers); Springer: Berlin, DE, 1999; pp 1-556.

[2] International Technology Roadmap for Semiconductors (ITRS); *Process Integration, Devices, and Structures and Emerging Research Devices*, 2006 Edition.

[3] Wade, W.; and Lammers, D. (July 16 2001). Intel grooms pair of substitutes for Flash. EE Times. http://www.eetimes.com/.

[4] Yoshikawa, K. In *Proceedings of 30th European Solid-State Device Research Conference*; Frontier Group; Ed.; ESSDERC 2000: Cork, IR, 2000; pp 72.

[5] De Salvo, B.; Lombardo, S. In *Proceedings of the Workshop "Non volatile memories with discrete storage nodes", Special issue of Solid-State Electronics*; Elsevier: Oxford, UK, 2004; pp 48.

[6] Minami, S.; Ujiie, K.; Terasawa, M.; Komori, K.; Furusawa, K.; Kamigaki, Y. IEICE *Trans. Electron* 1994, Vol. E-776C, pp 1260-9.

[7] Eitan, B.; Pavan, P.; Bloom, I.; Aloni, E.; Frommer, A.; Finzi, D. *IEEE Electron Devices Letters* 2000, Vol. 21, pp 543-5.

[8] van Schaijk, R.; van Duuren, M.; Mei, W. Y.; van der Jeugd, K.; Rothschild, A.; Demand, M.; *Microelectron. Eng.* 2004, Vol. 72, pp 395.

[9] Lee, C. H. ; Kang, C.; Sim, J.; Lee, J. S.; Kim, J.; Shin, Y.; Park, K. T.; Jeon, S.; Sel, J.; Jeong, Y.; Choi, B.; Kim, V.; Jung, W.; Hyun, C. I.; Choi, J.; Kim, K. in *Non-Volatile Semiconductor Memory Workshop 21st*, 2006, pp 54-55.

[10] Park, Y.; Choi, J.; Kang C.; Lee, C.; Shin, Y.; Choi, B.; Kim, J.; Jeon, S.; Sel, J.; Park, J.; Choi, K.; Yoo, T.; Sim, J.; Kim, K. In *Techological Digest of IEDM*, 2006, pp 29-32.

[11] Tiwari, S.; Rana, F.; Chan, K.; Hanafi, H.; Chan, W.; Buchanan, D. *Proceedings of IEEE International Electron Devices Meeting* 1995, pp 521.

[12] De Salvo, B.; Gerardi, C.; Lombardo, S.; Baron, T.; Perniola, L.; Mariolle, D.; Mur, P.; Toffoli, A.; Gely, M.; Semeria, M. N.; Deleonibus, S.; Ammendola, G.; Ancarani, V.; Melanotte, M.; Bez, R.; Baldi, L.; Corso, D.; Crupi, I.; Puglisi, R. A.; Nicotra, G.; Rimini, E.; Mazen, F.; Ghibaudo, G.; Pananakakis, G.; Monzio Compagnoni, C.; Ielmini, D.; Spinelli, A.; Lacaita, A.; Wan, Y. M.; VanderJeugd, K. IEDM Technical Digest. *IEEE International*, 2003, 26.1.1 - 26.1.4.

[13] Lombardo, S.; De Salvo, B.; Gerardi, C.; Baron, T. *Microelectronic Engineering* 2004, 72, 388–394.

[14] Steimle, R. F.; Muralidhar, R.; Rao, R.; Sadd, M.; Swift, C. T.; Yater, J.; Hradsky, B.; Straub, S.; Gasquet, H.; Vishnobhotla, L.; Prinz, E. J.; Merchant, T.; Acred, B.; Chang, K.; White, B. E. Jr. *Microelectronics reliability* 2007, 47, 585-892.

[15] Gebel, T.; Rebohle, L.; Zhao, J.; Borchert, D.; Fro¨b, H.; Borany, J. V.; Skorupa, W.; *Mater. Res. Soc. Symp. Proc.* 2001 638, F18.1.

[16] De Blauwe, J.; Ostraat, M.; Green, M. L.; Weber, G.; Sorsch, T.; Kerber, A.; Klemens, F.; Cirelli, R.; Ferry, E.; Grazul, J. L.; Baumann, F.; Kim, Y.; Mansfield, W.; Bude, J.; Lee, J. T. C.; Hillenius, S. J.; Flagan, R. C.; Atwater, H. A. Proceedings of the *International Electron Devices Meeting, Technical Digest*; IEEE: Piscataway, NJ, 2000, pp 683.

[17] Baron, T.; Martin, F.; Mur, P.; Wyon, C.; Dupuy, M. J. Cryst. *Growth* 2000, 209, 1004.

[18] Nicotra, G.; Lombardo, S.; Puglisi, R.; Spinella, C.; Ammendola, G.; Gerardi, C. Inst. Phys. Conf. Ser. , *Microsc. Semicond. Mater. Conf. Cambridge* 2003, 180, 119-122.

[19] Nicotra, G.; Puglisi, R. A.; Lombardo, S.; Spinella, C.; Vulpio, M.; Ammendola, G.; Bileci, M.; Gerardi, C. *J. Appl. Phys.* 2004, 95(4), 2049.

[20] Spinella, C.; Lombardo, S.; Priolo, F. *J. Appl. Phys.* 1998. 84, 5383.

[21] Ratsch, C.; Venables, J. A. J. *Vac. Sci. Technol.* A 2003, 215, S96.

[22] Mulheran, P. A.; Blackman, *J. A. Phys. Rev.* B 1996, 53(15), 10261.

[23] Puglisi, R. A.; Lombardo, S.; Nicotra, G.; Spinella, C.; Ammendola, G.; Gerardi, C. *Phys. Rev.* 1995, B 71, 125322.

[24] Perniola, L.; De Salvo, B.; Ghibaudo, G.; Foglio Para, A.; Pananakakis, G.; Vidal, V.; Baron, T.; Lombardo, S. A. IEEE Trans. *Nanotechnol.* 2003 2, 277.

[25] Lombardo, S.; Puglisi, R.A.; Crupi, I.; Corso, D.; Nicotra, G.; Perniola, L.; DeSalvo, B.; Gerardi, C. IEEE Non Volatile Semiconductor Memory Workshop 2004, 69.

[26] Puglisi, R. A.; Lombardo, S.; Corso, D.; Crupi, I.; Nicotra, G.; Perniola, L.; De Salvo, B.; Gerardi, C. *Journal of Applied Physics* 2006, 100(8), 86104.

[27] Thurn-Albrecht, T.; Steiner, R.; DeRouchey, J.; Stafford, C. M.; Huang, E.; Bal, M.; Tuominen, M.; Hawker, C. J.; Russell, T. P. Adv. Mater. 2000, 12 (11), 787.

[28] Guarini, K.W.; Black, C. T.; Milkove K. R.; Sandstrom, *R. L. J. Vac. Sci. Technol.* B 2001, 19(6), 2784.

[29] Mansky, P.; Liu, Y.; Huang, E.; Russell, T. P.; Hawker, C. Science 1997, 275, 1458.

[30] Guarini, K. W.; Black C. T.; Yeung, S. H. I. *Adv. Mater.* 2002 14 (18), 1290.

[31] Thurn-Albrecht, T.; Schotter, J.; Kastle, G. A.; Emley, N.; Shibauchi, T.; Krusin-Elbaum, L.; Guarini, K.; Black, C. T.; Tuominen, M. T.; Russell, T. P. *Science* 2000, 290, 2126.

[32] Segalman, R. A.; Yokoyama, H.; Kramer, E. *J. Adv. Mater.*2001, 13, 1152.

[33] Cheng, J. Y.; Ross, C. A.; Thomas, E. L.; Smith, H. I.; Vancso, G. *J. Appl. Phys. Lett.* 2002, 81, 3657.

[34] Kim, S. O.; Solak, H. H.; Stoykovich, M. P.; Ferrier, N. J.; de Pablo, J. J.; Nealey, P. F. *Nature* 2003, 424, 411 .

[35] Black, C. T. IEEE Trans. *Nanotechnol.* 2004 3, 412.

[36] Sundrani, D.; Darling, S. B.; Sibener, *S. J. Nano Lett.* 2004, 4, 273.

[37] Guarini, K. W.; Black, C. T.; Yeung, S. H. I. *Adv. Mater.* 2002, 14, 1290.

[38] Puglisi, R.A.; La Fata, P.; Lombardo S. *Appl. Phys. Lett.* 2007, 91, 053104.

[39] Clark, L.C.; Lyons C. *Ann. N.Y. Acad. Sci.* 1962, 102, 29–45.

[40] Jain, K.K. *Trends Biotechnol.* 2000, 18, 278–280.

[41] Ramsay, G. *Commercial Biosensors: Applications to Clinical, Bioprocess, and Environmental Samples*; Wiley: New York, 1998.

[42] North, J.R. *Trends Biotechnol.* 1985, 3, 180-187

[43] Bashir, R.; Gomez, R.; Sarikaya, A.; Ladisch, M.R.; Sturgis, J.; Robinson, J.P. *Biotechn. Bioengineer.* 2001, 73, 324-328

[44] Vo-Dinh, T.; Cullum, B. Fresenius *J. Anal. Chem.* 2000, 366, 540-551.

[45] Hall, E.A.H. *Biosensors*; Open University Press, Milton Keynes, 1990.

[46] Schöning, M. J.; Schroth P.; Schütz, S. Electroanalysis 2000, 12, 645–652.

[47] Heller M.J. *Annu. Rev. Biomed. Eng.* 2002, 4, 129-153.

[48] Wang, J. Nucleic Acids Res. 2000, 28, 3011-3016.

[49] Schena, M. *Microarray Analysis*, John Wiley & Sons: New Jersey, 2003.

[50] Kim, D.-S.; Jeong, Y.-T.; Park, H.-J.; Shin, J.-K.; Choi, P.; Lee, J.-H.; Lim, G. Biosens. *Bioelectr.* 2004, 20, 69-74

[51] Bergveld, P. *IEEE Trans. Biomed. Eng.* 1970, 17, 70–71.

[52] D'Amico, A.; Di Natale, C.; Martinelli, E.; Sandro, L.; Baccarani, G. *Sens. Act.* B 2005, 106, 144-152.

[53] Sohn, B.K., Kim, C.S., *Sens. Act.* B 1996, 34, 435–440.

[54] Sakata, T.; Kamahori, M.; Miyahara, Y. Mat. *Sci. Engineer.* 2004, 24, 827-832.

[55] Estrela, P.; Stewart, A.G.; Yan, F.; Migliorato, P. Electroch. *Acta* 2005, 50, 4995-5000.

[56] Uslu, F.; Ingebrandt, S.; Mayer, D.; Böcker-Meffert, S; Odenthal, M; Offenhäusser, A. *Bios. Bioelectr.* 2004, 19, 1723-1731.

[57] Im, H.; Huang, X.-J.; Gu, B.; Choi, Y.-K. *Nature Nanotechn.* 2007, 2, 430-434.

[58] Wuag, G.; Yau, S.-T. *Appl. Phys. Lett.* 2005, 87, 25391-1 – 3.

[59] Souteyrand, E.; Cloarec, J.P.; Martin, J.R.; Wilson, C.; Lawrence, I.; Mikkelsen, S.; Lawrence, M.F. *J. Phys. Chem.* B 1997, 101, 2980–2985.

[60] Bataillard, P.; Gardies, F.; Jaffrezic-Renault, N.; Martelet, C. *Anal. Chem.* 1988, 60, 2374-2379.

[61] Shin, J.-K.; Kim D.-S.; Park, H.-J.; Lim, G. Electroanalysis 2004, 16, 1912-1918.

[62] Sakata, T.; Miyahara, Y. *ChemBioChem* 2005, 6, 703-710.

[63] Thust, M.; Schoning, M.J.; Schroth, P.; Malkoc, U.; Dicker, C.I.; Steffen, A.; Kordos, P.; Luth H. J. Molec. Cat. B: *Enzymatic* 1999, 7, 77-83.

[64] Wang, J. *Chem. Eur. J.* 1999, 5, 1681-1685.

[65] Wang, J; Rivas, G.; Cai, X.; Palecek, E.; Nielsen, P.; Shiraishi, H.; Dontha, N.; Luo, D.; Parrado, C.; Chicharro, M.; Farias, P.A.M.; Valera, F.S.; Grant, D.H.; Ozoz, M.; Flair, M.N. Analyt. *Chim. Acta* 1997, 347, 1-8.

[66] Schöning, M.J. ; Poghossian, A. *Analyst* 2002, 127, 1137-1151.

[67] Berggren, C.; Bjarnason, B.; Johansson, G. *Electroanalysis* 2001, 13, 173-180.

[68] Schöning, M.J. ; Poghossian, A. Electroanalysis 2006, 18, 1893-1900.

[69] Auroux, P.-A.; Iossifidis, D.; Reyes, D. R.; Manz, A. *Anal. Chem.* 2002, *74*, 2637-2652.

[70] Davidsson, R.; Genin, F.; Bengtsson, M.; Laurell, T.; Emnéus, *J. Lab Chip* 2004, 4, 481-487.

[71] Thevenot, D.R.; Toth, K.; Durst, R.A.; Wilson, G.S. Biosens. *Bioelectron.* 2001, 16, 121–131.
[72] Watson, J.D.; Crick, F.H.C. *Nature* 1953, 171, 737-740.
[73] Mikkelsen, S.R. *Electroanalysis* 1996, 8, 15–19.
[74] Pivadori, M.I.; Merkoci, A.; Alegret, S. Biosens. *Bioelectron.* 2001, 15, 291–303.
[75] Bontidean, I.; Berggren, C.; Johansson, G.; Csöregi, E.; Mattiasson, B.; Brown, N.; Jakeman, K.J.; Lloyd, *J. Anal. Chem.* 1998, 70, 4162-4169.
[76] Taylor, R.F.; Marenchic, I.G.; Spencer, R.H. Anal. *Chim. Acta* 1991, 249, 67-70.
[77] Janek, R.P.; Fawcett, R., Ulman, A. *J. Phys. Chem.* 1997, 101, 8550-8558.
[78] Bard, A.J.; Faulker, L.R. *Electrochemical Methods-Fundamentals and Applications*, Wiley: New York, 1980.
[79] Bhushan, B.; Tokachichu, D.R.; Keener, M.T.; Lee, S.C. *Acta Biomaterialia* 2005, 1, 327-341.
[80] Saby, C.; Jaffrezic-Renault, N.; Martelet, C.; Colin, B.; Charles, M.H.; Dehair, T.; Mandrand, B. Sens. *Actuators* B 1993, 15-16, 458-463
[81] Bard, A.J.; Faulker, L.R. *Electrochemical Methods, Fundamentals and Applications,* Wiley: New York, ed. 2, 2001.
[82] Sze, S.M. *Physics of semiconductor devices; second ed.*, Wiley: New York, 1981.
[83] Berveld, P. *Sens. Act.* B 2003, 88, 1-20.
[84] G. F. Blackburn, *Biosensors: Fundamentals and Applications*, Turner, A.P.F.; Karube, I.; Wilson, G. S., Oxford University Press: Oxford 1987, 481 – 530.
[85] Bergveld, P. Biosens. *Bioelectron.* 1991, 6, 55-72.
[86] Poghossian, A.; Cherstvy, A.; Ingebrandt, S.; Offenhäusser, A. Schöning, M. J. Sens. *Act. B* 2005, 111–112, 470-480.
[87] Herne, T. M.; Tarlov, M. J. *J. Am. Chem. Soc.* 1997, 119, 8916-8920.
[88] Piunno, P.A.E.; Krull, U.J. *Anal. Chem.* 1995, 67, 2635–2643.
[89] Subramanian, A.; Kennel, S.J.; Oden, P.I.; Jacobson, K.B.; Woodward, J.; Doktycz, M.J. *Enzyme Microb. Technol.* 1999, 24, 26-34.
[90] Longo, L.; Vasapollo, G.; Gualcito, M.R.; Malitesta, C. *Anal. Bioanal. Chem.* 2006, 385, 146-152.
[91] Beale, M.I.J.; Benjamin, J.D.; Uren, M.J.; Chew, N.G.; Cullis, A.G. *J. Crystal Growth* 1985, 73, 622-636.
[92] Libertino, S.; Fichera, M.; La Mantia, A.; Ricceri, D. *Synthetic Metals* 2003, 138, 141-144.
[93] Libertino, S.; Fichera, M.; D'Arrigo, G.; La Mantia, A.; Ricceri, D. *Synth. Met.* 2003, 138, 71-74.
[94] Lobert, P.E., Bourgeois, D.; Pampin, R.; Akheyar, A.; Hagelsieb, L.M.; Flandre, D.; Remacle, *J. Sens. Act.* B 2003, 92, 90-97.
[95] Cai, W.; Peck, J.R.; van der Weide, D.W.; Hamers, R.J. Biosens. *Bioelectron.* 2004 ,19, 1013–1019.
[96] Billard, V.; Martelet, C.; Binder, P.; Therasse, *J.Analyt. Chim. Acta* 1991, 249, 367–372.
[97] Gardies, F.; Martelet, C.; Colin, B.; Mandrand, B. *Sens. Act.* 1989, 17, 461–464.
[98] Taira, H.; Nakano, K.; Maeda, M.; Takagi, M. Analyt. *Science* 1993, 9, 199–206.

[99] Laurell, T.; Drott, J.; Rosengren, L.; Lindstroom, K. Sens. Act. B 1996, 31, 161-166.

[100] Libertino, S.; Fichera, M.; Aiello, V.; Statello, G.; Fiorenza, P.; Sinatra, F. Microelect. Eng. 2007, 84, 468-473.

[101] Libertino, S.; Scandurra, A.; Aiello, V.; Giannazzo, F.; Sinatra, F.; Renis, M.; Fichera, *M. Appl. Surf. Sci.* 2007, 253, 9116–9123.

[102] Hecht, H.J.; Kalisz, H.M.; Hendle, J.; Schnid, R.D.; Schomburg, D. *J. Mol. Biol.* 1993, 229, 153-172.

[103] Wilson, R.; Turner, A.P.F. Bios. *Bioelectron.* 1992, 7, 165-185.

[104] Malhotra, B.D.; Singhal, R.; Chaubey, A.; Sharma, S.K.; Kumar, A. *Curr. Appl. Phys.* 2005, 5, 92-97.

[105] Libertino, S.; Giannazzo, F.; Aiello, V.; Scandurra, A.; Sinatra, F.; Renis, M.; Fichera, M. Langmuir 2008, 24, 1965-1972.

[106] Libertino, S.; Fichera, M.; Aiello, V.; Lombardo, S.; Scandurra, A.; Renis, M.; Sinatra, F. accepted for publication on Sensor Letters.

[107] Ouyang, H.; Striemer, C.C.; Fauchet, P.M. *Appl. Phys. Lett.* 2006, 88, 163108- 1-3.

[108] Briggs, D., Seah, M.P., Eds. Practical Surface Analysis, Vol. 1, Second Edition John Wiley & Sons: Chichester, 1990.

[109] Beamson, G., Briggs, D., Eds. The XPS of Polymers Database, Surface Spectra Publisher, 2000.

[110] Karadas, F.; Ertas, G.; Suzer, S. *J. Phys. Chem.* B 2004, 108, 1515-1518.

[111] Sakata, T.; Miyahara, Y. Biosens. *Bioelectr*. 2005, 21, 827-832.

In: Biomimetic and Supramolecular Systems Research ISBN: 978-1-60456-405-1
Editor: Arturo H. Lima

Chapter IV

Optical Properties of Self-Assembled Colloidal Photonic Crystals Doped with Citrate-Capped Gold Nanoparticles

***Marco Cucini, Marina Alloisio, Anna Demartini, and Davide Comoretto*[1]**

Dipartimento di Chimica e Chimica Industriale, Università degli Studi di Genova, via Dodecaneso 31, 16146 – Genova,Italy

Abstract

Photonic crystals are composite materials possessing a periodical modulation of their dielectric constant on the scale length of the visible wavelength. This dielectric periodicity affects light propagation creating allowed and forbidden photon energy regions. Among different types of photonic crystals, we addressed our attention to artificial opals i.e. three dimensional self-assembled arrays of spheres packed in a face centred cube crystal lattice. In order to tune the optical properties of such photonic crystals, we inserted gold nanoparticles in the interstices between the spheres. We call this process doping. In this work we report on the optical properties of artificial opals doped at various levels with gold nanoparticles (NpAu) stabilized with citrate molecules. By increasing the NpAu doping level, the opal photonic stop band was bathochromically shifted as well as its full width half maximum and intensity were reduced. No effects of nanoparticle coalescence was observed as previously detected for NpAu prepared by laser ablation. In order to improve the mechanical stability of NpAu doped opals, we started a preliminary thermal annealing study with in situ monitoring of optical properties. We found that thermal degradation of doped samples was reduced with respect to bare opals. In addition, the annealing temperature of NpAu doped opals was much higher than that observed for bare ones.

[1] Corresponding author, e-mail comorett@chimica.unige.it.

Keywords: *Photonic crystals, artificial opals, gold nanoparticles, optical properties*

1. Introduction

Nanostructured materials are attracting much attention due to their application in photonics, electronics, biology and catalysis. Among these variegated systems, materials possessing a periodical modulation of the dielectric constant on the wavelength scale of the visible light (photonic crystals, PC) are currently extensively investigated since they behave as light semiconductors and then find application in photonics for low threshold laser action, high bending angle waveguides, superprism effect, sensors and optical switches [1-4]. Photonic crystals are grown both with top-down and bottom-up approaches. The first method is very powerful and allows detailed and variegated structures to be prepared [4]. However, it is very expensive since requires an extensive application of UV or electron beam lithographic techniques. On the other hand, bottom-up approaches are very interesting, cheap and easy methods since exploit the self-assembling properties of molecules, supramolecules and meso-objects like colloids which move spontaneously towards quasi-equilibrium steps, building an ordered structure [5]. Bottom-up methods are extensively adopted in nature to create nanostructures possessing unusual optical properties suitable for biomimetic purposes [6,7] or mating [8,9]. For instance, it has been noticed that the presence of ripples on arthropoda ommatidia or on large wings of diurnal moths provides a sort of anti-reflection layer, which increases transparency thus improving vision or develops mimetism against predators attacks [10]. Very well known is the PC nature of nanostructures inside butterflies wings [6,8-14] or peacock feathers [15], which are responsible for their wonderful colours. In sea mouse, the iridescent needles observed at scanning electron microscope (SEM) display a photonic fiber-like structure [16,17] also observed in edelweiss bracts [18]. The chitin exocuticol of Chrysochroa vittata is an example of one dimensional (1D) PC [19], but very surprising is the case of an Australian beetle, which artificially produces opals over its exoskeleton in order to improve its mimetism [20]. As reported in the Sanders' pioneer work, the structure of opals is made of silica nanospheres packed in a face centred cubic (FCC) structure. Diffraction of light from crystallographic planes provides the amazing colour of these gemstones [21,22]. Different preparation methods of opals starting from monodisperse silica or polymeric microspheres ad hoc synthesized are currently extensively pursued in the scientific community [23,24]. As a matter of fact, opals and inverse opals are very interesting since provide a useful playground for photonic band gap engineering [23,25-28]. This can be done both by introducing controlled structural defects [29,30] into the opal PC structure or by infiltration of sphere interstices with photoactive materials [23,31-40].

Recently, the use of metallic nanoparticles as a medium for opal infiltration or inverse opal preparation has attracted much attention [29,41-47]. As a matter of fact, nanoparticle optical properties due to surface plasmon resonance are strongly dependent on the environment, thus being used for sensing and sensitizing applications [48]. Moreover, NpAu also show a high and fast nonlinear optical (NLO) response [49]. In our previous works [28,50], NpAu used for opal infiltration were fabricated by laser ablation [31,51], a technique which requires the use of a surfactant dissolved in the liquid where NpAu are prepared in order to prevent their coalescence. Those opals showed interesting nanosecond optical switching properties when resonantly photoexcited [28]. In this work, we would like to extend our experience to NpAu obtained by chemical reduction of a gold salt and stabilized with a citrate outer shell in order to improve the anti-aggregation effects even in the solid state.

The paper is organized with an experimental section where NpAu synthesis and opal preparation are described together with the experimental techniques used for their morphological and optical characterization. In the results and discussion section, spectroscopic properties of bare and NpAu doped opals are discussed in the framework of the dielectric response. Suggestions to improve the mechanical stability of these systems by thermal annealing are also reported.

2. Experimental

NpAu were prepared following a modified version of the method developed by Turkevich and co-workers [52]. Briefly, a weighed amount of $HAuCl_4$ was dissolved in MilliQ H_2O, heated at boiling temperature under vigorous stirring and added dropwise of a fixed amount of citrate reducing/stabilizing agent. The final reagent concentrations were 0.28 and 2.8 mM for the metal and citrate ions, respectively. The aqueous mixture was let to boil until the solution colour turned from yellow to red (about 1 hour) and then maintained under stirring at room temperature for other 3 hours. At reaction completed, a red-purple solution composed by gold nanoparticles stabilized with citrate (citrate-NpAu, Figure 1a) was collected and stored in the dark at room temperature. Insignificant modifications were observed for over 6 months, indicating that the as-prepared colloids had good stability in aqueous solutions. All chemical reagents and solvents were spectroscopic grade commercial products used as received. $AuCl_3$, HCl and trisodium citrate were purchased from Lancaster.

Electronic absorption spectra were recorded at room temperature on a Perkin-Elmer Lambda 9 spectrophotometer in the UV-Vis range with spectral resolution ± 0.2 nm).

Bright-field transmission electron microscopy (TEM) images were obtained with a Jeol electron microscope, model IEM-2010, operating at 200 kV. Specimens were prepared by evaporating a drop on a 300 mesh carbon-coated copper grid (Lacey). Qwin software V3 for digital image processing and analysis was used for measurement of particle size employing multiple pictures from different areas.

Opals were grown from commercial polystyrene (PS) monodisperse microspheres (260, 300 and 340 nm diameters, standard deviation $< 5\%$) water suspension 10% by weight (Duke Scientific; refractive index, $n_{PS}=1.59$). For each microsphere diameter, we prepared six different suspensions having different concentration of both PS microspheres and NpAu. We used two different concentrations of microspheres and three different concentrations of NpAu. We mixed 0.06 (0.3) ml microsphere suspension with 0.5 ml (Low Load, LL), 1.0 ml (Medium Load, ML) and 2.0 ml (High Load, HL) NpAu suspension and then we further diluted to 5.0 ml with distilled water. The amount of PS microspheres determined the average thickness of samples, namely thin (thick) samples. Samples were named according to sphere diameter, thickness and NpAu load (ex. 260 thin LL). With these six suspensions we grew opal films having different degrees of infiltration (NpAu doping) and different thickness by using the meniscus technique [43].

Transmittance (T) and reflectance (R) spectra were recorded with an Avantes Avaspec-2048 compact spectrometer (300-1100 nm spectral range, 1.4 nm spectral resolution). Light from a tungsten-halogen lamp was guided by an optical fiber to a proper focusing optics (spot diameter 100 μm or variable in the range 0.5-5 mm) on the sample, eventually mounted on a goniometer; transmitted light was then collected and driven by optical fibers to the spectrometer. For the 100 μm diameter spot, the sample was mounted on a XY micrometric translator and probed area was selected and observed by a Sony XCD-X710CR video-camera. Normal incidence reflectance was measured on a 2 mm diameter spot by an Y reflection probe bundle fiber. Sample heating was provided by a hot plate; the temperature was monitored with a Testo 110 thermometer equipped with a flat head probe Testo AG 630.

3. Results and Discussion

The citrate-NpAu morphological and optical properties were investigated by TEM and UV-Vis techniques. Figure 1b presents a TEM micrograph of a drop-cast film of the colloids with the corresponding histogram (Figure 1c) of the particle size distribution alongside. The nanoparticles average diameter is 16.6±3.0 nm and the axial ratio is 1.14 as measured from the TEM image. It is evident that quite homogeneous, well-separated, nearly spherical gold nanoparticles are obtained. By assuming all particles as spherical, a final colloid concentration 1.94 nM was evaluated from the mean diameter value. As far as concerned the particles arrangement, the nanohybrids tend to hold together, despite the negatively charged citrate chains, in linear or two-dimensional structures quite ordered in shape.

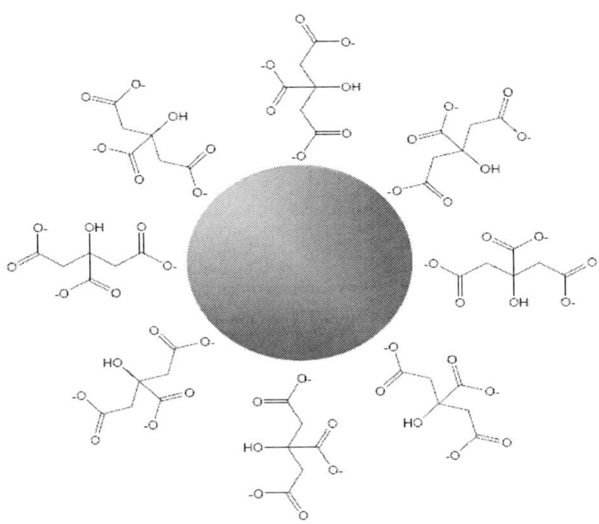

(a)

Figure 1. (Continued on next page.)

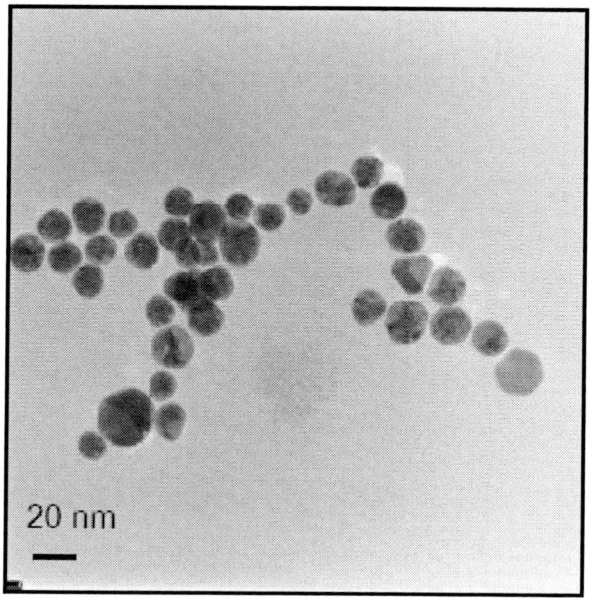

(b)

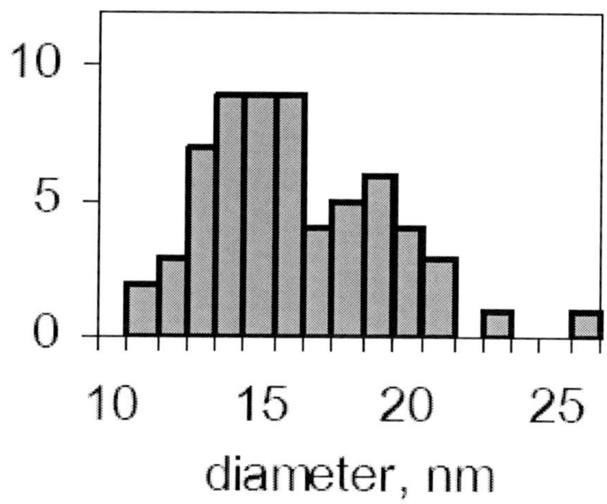

(c)

Figure 1. Schematic structure of gold nanoparticles (NpAu) stabilized with citrate (a); bright field TEM image (b) and its corresponding size distribution histogram (c) for citrate-capped NpAu.

The state of citrate-coated NpAu was confirmed by means of electronic absorption spectroscopy. In Figure 2a, the electronic absorption spectrum of the aqueous suspension is reported. A well-resolved, sharp surface plasmon band (SPB) at 524.0 nm, typical of gold nanosized spherical clusters and responsible for the red-purple colour of the colloidal solution, is observed. The band is attributable to the SPB transverse mode, perpendicular to the particle long axis [53]. The absence of distinct absorptions in the 650-700 nm range, corresponding to the longitudinal plasmon resonance band, confirms that in these conditions the colloids do not aggregate, in agreement with the TEM data.

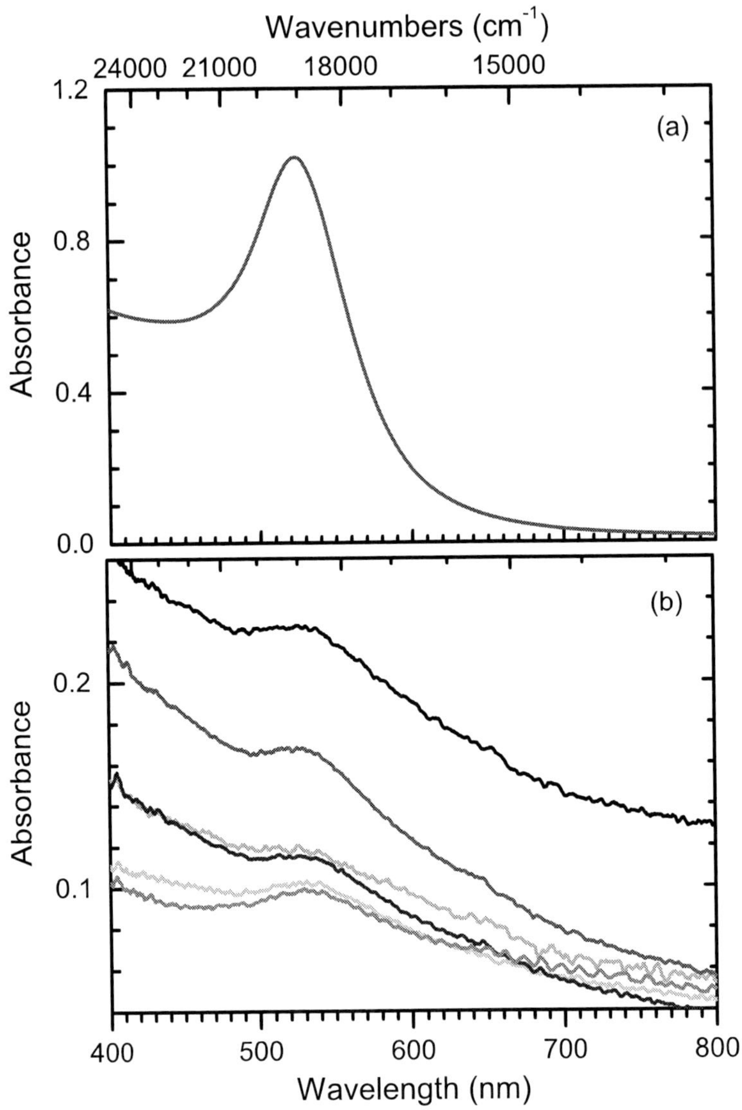

Figure 2. Electronic absorption spectra of citrate-NpAu aqueous suspension used in this work (a) and of a 300 nm dissolved doped opal film in different positions (b).

We turn now to the discussion of the optical properties of opals infiltrated with citrate-capped NpAu. Figure 3 shows the T (a) and R (b) spectra of opals infiltrated with NpAu at three different doping levels compared with those of a bare one.

The of transmission/reflectance spectrum of bare opals presents a minimum/maximum at 595/590 nm. Its origin is due to the photonic stop band of the photonic crystal which does not allow the light to propagate inside the opal. As a matter of fact, light of such wavelength is backward diffracted by opals (111) crystallographic planes giving rise to a prominent transmission dip/reflectance peak. It is evident that the spectral position of these structures bathochromically shifts upon increasing the NpAu doping level. In particular, for LL, ML, and HL samples, the transmission spectra stop band shifts to 610, 615 and 630 nm respectively (Figure 3a), while in reflectance peaks are observed at 615, 625 and 630 nm.

Notice that in these spectra, no evidence of NpAu plasmon absorption is detected. The minor wavelength difference between the stop bands recorded in T spectra with respect to the R ones is probably due to light scattering which particularly affects the background of transmittance. However, we cannot exclude that the sample area probed by the two set-ups could be slightly different.

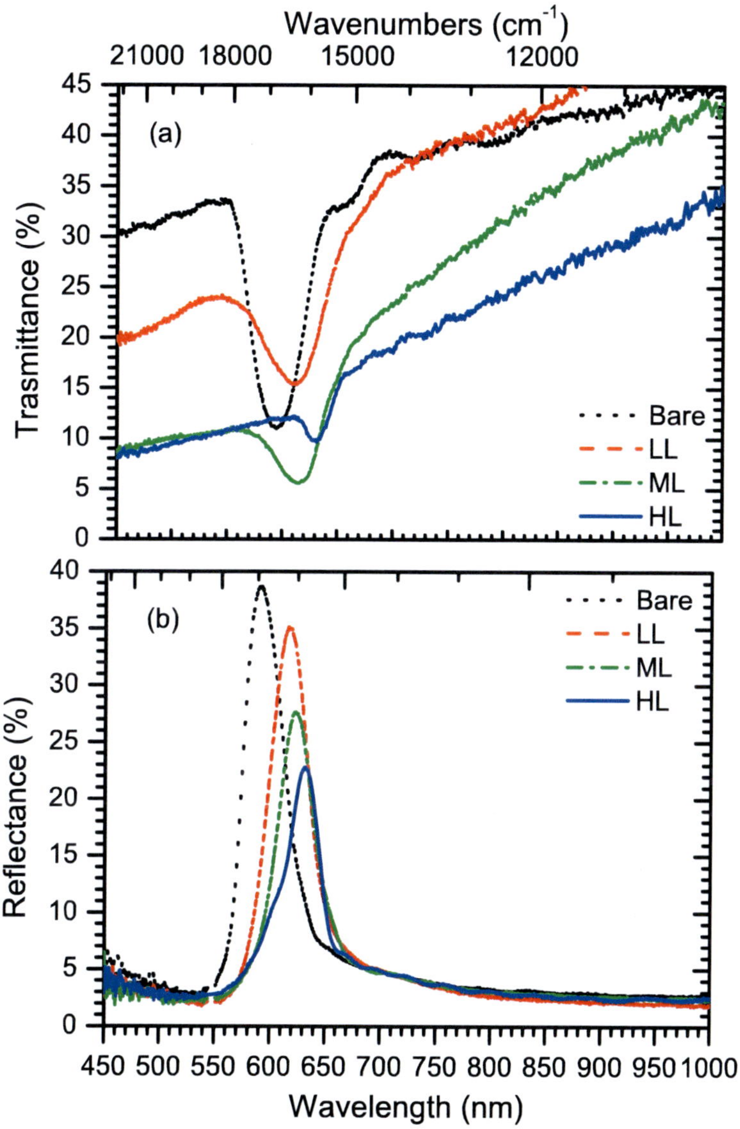

Figure 3. Transmittance (a) and reflectance (b) spectra of thin 260 nm opals bare (black dot line) and NpAu doped at LL (red dash line), ML (green dot dash line) and HL (blue solid line) levels.

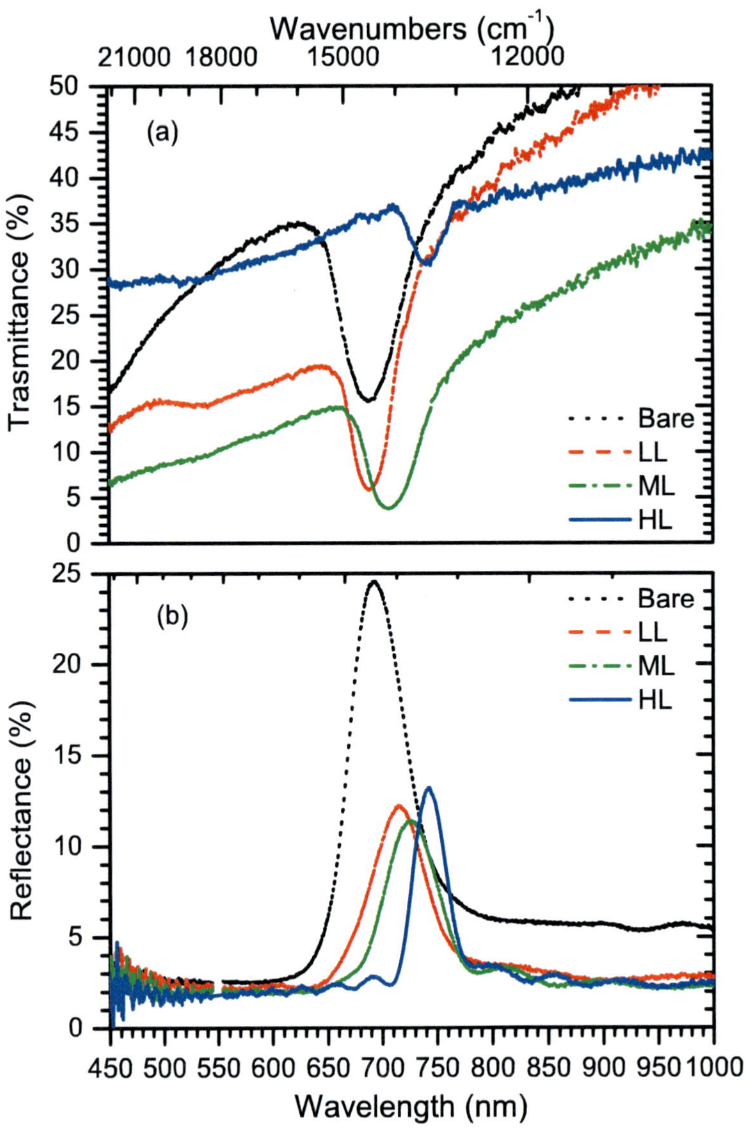

Figure 4. Transmittance (a) and reflectance (b) spectra of thin 300 nm opals bare (black dot line) and NpAu doped at LL (red dash line), ML (green dot dash line) and HL (blue solid line) levels.

Reflectance spectra, in particular for thin samples, show a wavy background due to interference fringes pattern caused by multiple reflection inside the sample. The presence of interference fringes in opal spectra is a signature for their good optical quality.

Additional effects of NpAu doping level are related to the peak width and intensity. From spectra reported in Figure 3 we observe a reduction of both the full width half maximum (FWHM) and signal intensity upon increasing the doping level. This is particularly evident in the R spectra since they show a flat background. In T spectra, even if the effect is still detectable, the background is less regular since strongly dependent on the surface roughness and scattering processes. Deviation from these trends are due to sample inhomogeneities.

Figure 4 shows the transmittance (a) and reflectance (b) spectra for thin 300 nm opals doped with NpAu at the same levels previously used for the 260 ones. Bare opals show the transmittance (reflectance) stop band at 690 nm (695 nm), while LL, ML and HL NpAu infiltrated opals at 690 (720), 710 (730) and 740 (745) nm, respectively.

If we further increase the sphere diameter at 340 nm (Figure 5) a more relevant bathochromic shift is observed in T/R spectra for bare (780/785 nm), LL (820/820 nm), ML (830/835 nm), and HL (850/850 nm) opals. Moreover, interference fringes in the transparent region are found for all samples. Notice also that for 300 and 340 nm opals a progressive intensity and FWHM reduction of the stop band upon the increasing doping level is observed, as previously detected for the 260 nm ones. Table I summarized the main optical and morphological parameters evaluated for all prepared samples.

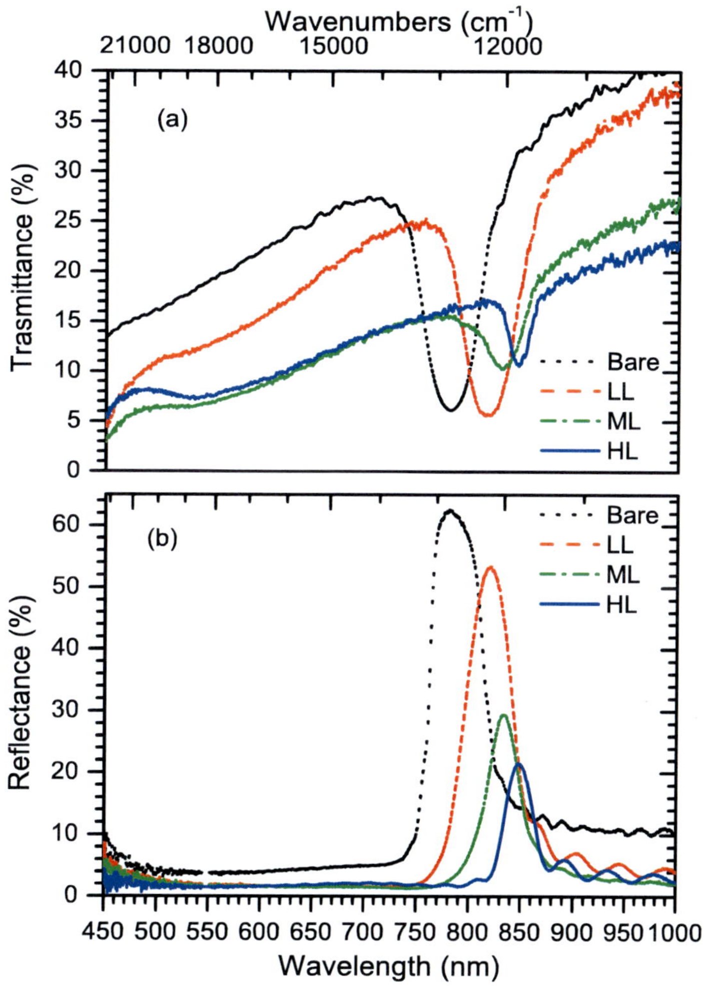

Figure 5. Transmittance (a) and reflectance (b) spectra of thin 340 nm opals bare (black dot line) and NpAu doped at LL (red dash line), ML (green dot dash line) and HL (blue solid line) levels.

Table I. λ_{Bragg}, FWHM, n_{eff} (both from shift and dispersion when available), n_i, ε_i, T, R, thickness (d), and number of sphere layers (p)deduced from the optical spectra

Sample	λ_{Bragg} (nm)	FWHM (cm^{-1})	n_{eff} shift	n_{eff} disp	n_i	ε_i	T (%)	R (%)	d (μm)	p
260 bare	594	1537	1.40		1.0	0.92	24	36	-	
260 thin LL	613	1713	1.45	1.45	1.1	1.21	12	33	-	
260 thin ML	616	1084	1.45		1.1	1.21	7	26	5.5	11
260 thin HL	630	509	1.49		1.2	1.49	3	20	5.5	11
260 thick LL	588	1284	1.39		0.9	0.86	17	33	-	
260 thick ML	590	1265	1.39		0.9	0.86	16	31	-	
260 thick HL	637.5	411	1.50		1.3	1.56	2	21	7.5	15
300 bare	688	1312	1.41		1.0	0.96	23	22	-	
300 thin LL	689	779	1.41		1.0	0.96	16	10	-	
300 thin ML	706	1015	1.44		1.1	1.14	13	10	-	
300 thin HL	741	575	1.52		1.3	1.74	7	11	-	
300 thick LL	705	1010	1.44		1.1	1.14	5	27	-	
300 thick ML	726	1060	1.48		1.2	1.42	3	25	8.5	15
300 thick HL	740	288	1.51		1.3	1.66	3	17	-	
340 bare	780	976	1.39		0.9	0.86	24	58	-	-
340 thin LL	818	839	1.46		1.1	1.28	23	52	7	11
340 thin ML	832	675	1.48		1.2	1.42	8	28	-	
340 thin HL	848	338	1.51		1.3	1.66	8	20	7	11

The bathochromic shifts observed in spectra reported in Figs. 3-5 can be generated by two factors, one of which is the increase of the microsphere diameter and the other is the infiltration process. We now discuss these important features in details.

The shift of the stop band upon changing the sphere diameter can be interpreted according to the scaling laws of photonic crystals [54]. The master equation governing the properties of photonic crystals is

$$\nabla \times \left(\frac{1}{\varepsilon(\mathbf{r})} \nabla \times \mathbf{H}(\mathbf{r}) \right) = \left(\frac{\omega}{c} \right)^2 \mathbf{H}(\mathbf{r}) \tag{1}$$

where ε(**r**) is the periodical real part of the dielectric constant, H(r) is the macroscopic magnetic field, ω the frequency, and c the speed of light. When the spatial distribution of the dielectric constant of the system is scaled through a generic factor s, the dielectric constant assumes the form

$$\varepsilon'(\mathrm{r}) = \varepsilon(\mathrm{r}/\mathrm{s}) \tag{2}$$

By simply changing variables (**r'=sr**) the new master equation can be solved with mode profile H'(r')=H(r'/s) and frequency becomes

$$\omega' = \omega/s \tag{3}$$

Since in our case s>1, we expect a reduction (increase) of energy (wavelength) of the photonic band structure upon increasing the PS microsphere diameter as experimentally observed.

Let's now discuss the effects of infiltration. The spectral position of the stop band or Bragg peak (λ_{Bragg}) as well as its dependence on the incidence angle off from the normal (θ) and effective refractive index ($n^2_{eff} = \varepsilon_{eff}$, effective dielectric constant) of the system depends on the details of the photonic band structure, i.e. on the eigenvalues of Eq. 1 [25]. If we are interested only in the low energy properties of this band structure (close to the stop band around the Γ-L direction) we can use the much simpler Bragg-Snell formula [55] which can be derived by the Bragg law joined to the Snell one

$$m\lambda_{\mathrm{Bragg}} = 2D\sqrt{n_{eff}^2 - \sin^2 \vartheta} \tag{4}$$

where m is the diffraction order, D is the interplanar spacing in the [111] growth direction which is perpendicular to the opal surface. We can notice that Eq. 4 predicts a bathochromic shift of λ_{Bragg} upon increasing n_{eff}. This explains the spectral position of the Bragg peak upon increasing the NpAu doping level of opals. As a matter of fact, upon adding NpAu into the interstices, their dielectric constant is increased respect to the value for void (ε=1) and then λ_{Bragg} increases. However, Bragg-Snell formula (Eq. 4) cannot be used to understand the nature of infiltrated material. This is a very important item since NpAu might give rise to aggregates or to bulk gold, if they loose their nanostructure. Since properties of variously (nano)structured gold are significantly different each other, it is fundamental to address this point. In order to investigate on the real state of the infiltrated NpAu, we dissolved our doped opals with few drops of toluene. The absorption spectra for a 300 nm doped dissolved opal film were recorded in different sample positions and are reported in Figure 2b. We notice that the stop band at 690 nm disappears since the PC has been destroyed, while a new not previously seen absorption peak is now evident at 530 nm. The peak position corresponds to that of surface plasmon for spherical NpAu in solution (Figure 2a). First of all, this result demonstrates that infiltration of opals was actually achieved with spherical nanoparticles. Moreover, since no evidence of absorption band at 650-700 nm is detected, no NpAu aggregates or bulk gold are formed during opal preparation because the citrate-capping shell preserves the NpAu individuality. We also remark that when similar studies were carried out for opals doped with NpAu obtained by laser ablation, where the capping layer is provided by a surfactant dissolved in the solution, strong aggregation effects have been observed instead [28].

We now turn back our attention to n_{eff}. The value of n_{eff} determined form the optical spectra is important to obtain a quantitative evaluation of the dielectric constant (ε_i) or refractive index (n_i) of interstices. As a matter of fact, it has been recently shown [28] that n_{eff} is connected to n_i by the Lorentz-Lorenz formula for the effective medium [44]

$$\frac{\varepsilon_{eff}-1}{\varepsilon_{eff}+2}=f_{PS}\frac{\varepsilon_{PS}-1}{\varepsilon_{PS}+2}+(1-f_{PS})\frac{\varepsilon_i-1}{\varepsilon_i+2} \tag{5}$$

where $\varepsilon_{eff}=(n_{eff})^2$, $\varepsilon_{PS}=2.53$ is the dielectric constant of polystyrene sphere, ε_i is the dielectric constant of the interstices ($\varepsilon_i=1$ for bare opals), and f_{PS} is the volume fraction of the unitary cell filled with nanospheres (0.74 for a close packed FCC crystal). By inverting Eq. 5, we can obtain ε_i (n_i) values. As an input for Eq. 5 we used n_{eff}, which can be obtained in two ways from Eq. 4. We can first fit n_{eff} in order to match the observed position of the stop band(λ_{Bragg}). On the other hand, we can modify the incidence angle, record the spectra and then fit the dispersion data with Eq. 4. This last procedure is well established [28], but in this work was used only for 260 thin LL sample. The calculated data for n_{eff}, n_i and ε_i are reported in Table I. We notice that both procedures provide results in good agreement thus confirming its internal consistency as well as the quality of our samples.

It is also interesting to give a look to the absolute values of n_{eff} (ε_{eff}) or n_i (ε_i). For bare opals n_{eff} ranging from 1.39 to 1.41 is obtained in good agreement with the value of 1.41 deduced from the effective medium theory. Upon increasing the doping level ε_i increases and its average values, calculated among the available samples, become 1.09, 1.21 and 1.62 for LL, ML, and HL samples, respectively. This means that our doping procedure provides a very strong variation of ε_i thus deeply modifying the dielectric properties of the system.

As far as n_{eff} is known, we can obtain additional information from their spectroscopic characterization. First of all, the interference fringes pattern observed in the R spectra allows the determination of the sample thickness d from the formula [56]

$$d=\frac{N_f}{\frac{1}{\lambda_k}-\frac{1}{\lambda_{k+N_f}}} \tag{6}$$

where N_f is the number of fringes observed in the spectral range λ_k-λ_{k+Nf} , k labelling the generic fringe. On this basis, starting from the spectra in Figure 3 to 5 and from Eq. 6, we derived the thickness of our samples that is also reported in Table I. Thickness d can be related to the number of sphere planes (p) in the samples. As a matter of fact, for FCC packing the interplanar spacing D along the [111] direction is given by $D=a(2/3)^{0.5}$ where a is the PS microsphere radius. As a consequence p=d/D corresponds to 11 or 15 microsphere planes for thin and thick samples, respectively.

Additional interesting properties of the NpAu doping process are the clear reduction of the Bragg peak FWHM and intensity. They can be interpreted as the effect of the reduction of the dielectric contrast between nanospheres and interstices, where the presence of few NpAu strongly modifies ε_i (9-62 %) with respect to the case of bare opals (ε_i =1). The high sensitivity of the photonic band structure upon partial infiltration of interstices with highly polarizable materials envisage their use not only for photonics but also for sensing [49]. The dependence of the FWHM on the NpAu doping level can be understood recalling the expression for the stop band width (W) for a one dimensional (1D) photonic crystal [57]

$$W = \frac{4hc}{\pi} \frac{1}{\lambda_{Bragg}} \frac{\left| n_i - n_{PS} \right|}{n_i + n_{PS}} \tag{7}$$

Even though Eq. 6 cannot be directly used to determine the FWHM of infiltrated opals, since they posses a much more complicated photonic structure with respect to a one-dimensional PC, it is interesting to analyze the behaviour of reduced FWHM (W/W_0), i.e. the FWHM for infiltrated opals divided by that for a bare one (W_0). The reduced FWHM as a function of n_{eff} or n_i for all our samples is reported in Figure 6. Both n_{eff} and n_i are representative of the NpAu doping level, through Eq. 5. It is evident that a very good agreement between experimental data for all samples and theory is achieved.

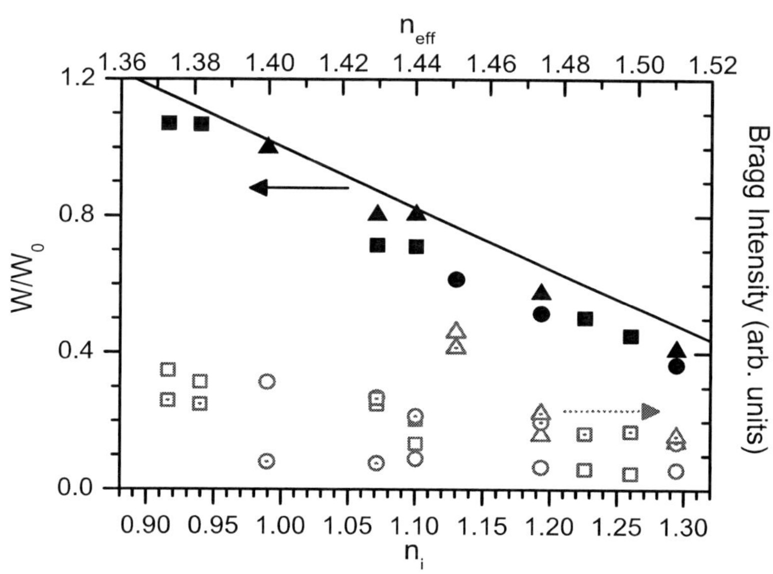

Figure 6. Reduced FWHM (left axis, full symbols) and intensity of the Bragg peak (right axis) as measured both in transmittance (open symbols) and reflectance (marked open symbols) spectra for bare and infiltrated opals of different diameters (squares, 260 nm; circles, 300 nm; diamonds, 340 nm) as a function of n_i (bottom axis) or n_{eff} (top axis). Full line shows theoretical values for reduced FWHM derived from Eq. 7.

An additional spectral effect observed in Figs. 3-5 and again related to the infiltration process concerns the intensity of the reflectance peak. The increase of n_i upon increasing the doping level reduces the dielectric contrast with respect to polystyrene spheres. This implies a reduction of light diffraction efficiency by the sphere planes along the [111] direction thus decreasing the intensity of the Bragg peak. Figure 6 also shows the reflectance and transmittance intensities measured at the Bragg peak for our samples as a function of n_i. Again a clear trend of intensity reduction upon increasing doping is observed. We cannot however exclude that an additional effect might contribute to the experimental intensity reduction. In fact, doping with absorbing media induces a strong light scattering due to NpAu, which increases upon increasing their loading. Photons scattered in all directions reduces the signal detected in specular reflectance and transmittance configurations. However, at better of our knowledge however, no analytical theoretical models accounting for these effects have been reported so far.

As previously observed for bare opals, these photonic crystals are brittle (in particular when infiltrated from solutions) and their mechanical robustness sometimes limited [40]. In order to improve the mechanical properties of these opals, a specific annealing procedure has been developed. Briefly, a differential scanning calorimetric characterization of bare opals showed a thermal behaviour similar to that of standard polystyrene i.e. a glass transition temperature (T_g) at about 80 °C and a melting temperature (T_m) of 110 °C, independently on the sphere diameter. On the basis of these results, annealing was performed in three temperature ranges, namely $T<T_g$, $T \leq T_g$ and $T_g<T<T_m$. Bare opals annealed at $T_g<T \leq T_m$ suddenly loose their photonic properties even though they preserve a pale reminiscence of their original nanostructure [40]. Opals annealed at $T<T_g$ (about 50 °C) increases their robustness and sometimes an improvement in the optical properties (evaluated both as a reduction of the Bragg peak FWHM and of the scattering background) is observed [40]. In Figure 7a we report the effect of annealing at $T \leq T_g$ (nominal temperature 75°C) for bare opals on the Bragg peak spectral position and intensity. Notice that these measurements are in situ performed during annealing on a hot plate. It is well known that hot plates does not provide a stable temperature; as a matter of fact we measured a periodical (about 10 minutes) variation of temperature in the 65-80 °C range. Data in Figure 7a show a dramatic change of the bare opals optical properties following the periodical heating of the hot plate. In particular, even though λ_{Bragg} is almost insensitive to temperature, within the spectral resolution of our set-up, the reflectance intensity during the heating phase of the hot plate suddenly decreases thus demonstrating a progressive and irreversible damaging of the opal. Damaging was also confirmed by scanning electron microscopy analysis showing an anisotropical deformation of PS microspheres that destroys the periodical nanostructure of the photonic crystal. Notice that annealing effects are similar for any opals at a given temperature and do not depend on the polystyrene microsphere diameter.

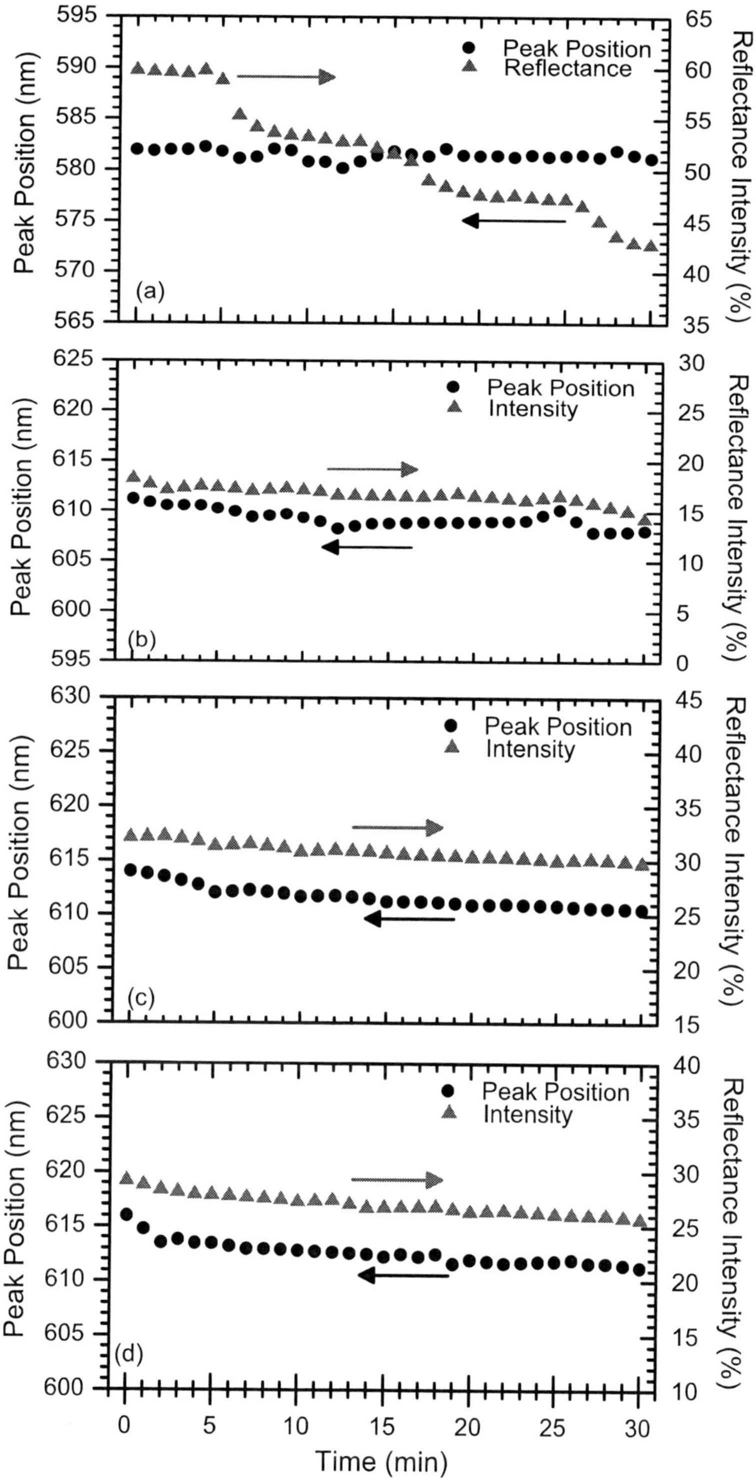

Figure 7.Time evolution of the reflectance Bragg peak wavelength (circles) and intensity (triangles) during thermal annealing at nominal temperature 75 °C for 260 nm bare (a), LL (b), ML (c), and HL (d) opals.

Since even NpAu doped opals show the poor mechanical stability observed in the bare ones, we repeated a thermal annealing study on these samples in order to address the problem. Again for $T_g<T \leq T_m$ the photonic crystal get lost even though not as suddenly as observed for bare ones. For $T< T_g$ no change in the optical properties and no significant variation of mechanical robustness are observed. In Figure 7b, c, and d the effect of annealing at $T\leq T_g$ (T=75°C) on 300 nm LL, ML, and HL opals, respectively, is reported. While in bare opals the heating cycle generated irreversible degradation of the photonic crystal, no such an effect is observed for NpAu doped opals independently on the doping level even after several cycles. Moreover, the mechanical stability of these opals seems to be slightly increased. From these results, we deduce that the in doped opals heating flux does not affect polystyrene microspheres but it is mainly directed through NpAu which become an heating sink preventing polystyrene melting. As a consequence, a proper annealing procedure for metallic nanoparticles doped polymeric opals different with respect to that used for bare ones have to be developed in order to optimize their mechanical properties.

4. Conclusion

We demonstrated that artificial opals can be successfully infiltrated in a controlled manner (doped) by citrate-capped NpAu. The citrate shell around NpAu prevents nanoparticles coalescence previously observed in opals doped with NpAu obtained by laser ablation. A detailed optical characterization of opals doped with NpAu shows a dramatic change of their dielectric properties which induces a bathochromic shift of the stop band as well as a reduction of its FWHM and intensity with respect to bare ones. These effects have been accounted for by a simple dielectric model. NpAu also affect thermal behaviour of doped opals. Preliminary results on the temperature dependence of the optical properties of NpAu doped opals indicates that their annealing temperature is significantly higher than that of bare ones.

Acknowledgments

This work is supported by the Italian Ministry of Instruction, University and Research through projects PRIN 2006031511 and by Progetti di Ricerca di Ateneo 2006 (Università di Genova).

References

[1] Painter, O.; Lee, R.K.; Scherer, A.; Yariv, A.; O'Brien, J.D.; Dapkus, P.D.; Kim, I. *Science* 1999, *284*, 1819.

[2] Mekis, A.; Chen, J.C.; Kurland, I.; Fan, S.; Villeneuve, P.R.; Joannopulos, J.D. *Phys. Rev. Lett* 1996, *77*, 3787.

[3] Serbin, J.; Gu, M. *Adv. Mater* 2006, *18*, 221.

[4] *Photonic Crystals: Advances in Design, Fabrication, and Characterization*; Busch, K.; Lölkes, S.; Wehrspohn, R.B.; Föll, H., Eds.; Wiley: Weinheim, 2004.

[5] Jin, F.; Li, C.-F.; Dong, X.-Z.; Chen, W.-Q.; Duan, X.-M. *Appl. Phys. Lett.* 2006, *89*, 241101.

[6] Huang, J.; Wang, X.; Wang, Z.L. *Nan Lett.* 2006, *6*, 2325.

[7] Vukusic, P.; Sambles, R. *Nature* 2003, *424*, 852.

[8] Sweeney, A.; Jiggins, C.; Johnsen, S. *Nature* 2003, *423*, 31.

[9] Biro´, L.P.; Ba´lint, Z.; Kerte´sz, K.; Ve´rtesy, Z.; Ma´rk, G.I.; Horva´th, Z.E.; Bala´zs, J.; Me´hn, D.; Kiricsi, I.; Lousse, V.; Vigneron, J.-P. *Phys. Rev. E* 2003, *67*, 021907.

[10] Vukusic, P.; Sambles, J.R.; Lawrence, C.R.; Wootton, R.J. *Nature* 2001, *410*, 36.

[11] Vukusic, P.; Hooper, I. *Science* 2005, *310*, 1151.

[12] Ghirardella, H. *Appl. Opt.* 1991, *30*, 3492.

[13] Kertész, K.; Bálint, Z.; Vértesy, Z.; Márk, G.I.; Lousse, V.; Vigneron, J.P.; Rassart, M.; Biró, L.P. *Phys. Rev. E* 2003, *74*, 021922.

[14] Vukusic, P.; Sambles, J.R.; Lawrence, C.R. *Nature* 2000, *404*, 457.

[15] Zi, J.; Yu, X.; Li, Y.; Hu, X.; Xu, C.; Wang, X.; Liu, X.; Fu, R. *Proceedings of the National Academy of Science* 2003, *100*, 12576.

[16] Tayeb, G.; Gralak, B.; Enoch, S. *Opt. Phot. News* 2003, *February*, 38.

[17] McPhedran, R.C.; Nicorovici, N.A.; McKenzie, D.R.; Rouse, G.W.; Botten, L.C.; Welch, V.; Parker, A.R.; Wohlgennant, M.; Vardeny, V. *Physica B* 2003, *338*.

[18] Vigneron, J.P.; Rassart, M.; Vértesy, Z.; Kertész, K.; Sarrazin, M.; Biró, L.P.; Ertz, D.; Lousse, V. *Phys. Rev. E* 2005, *71*, 011906.

[19] Vigneron, J.P.; Rassart, M.; Vandenbem, C.; Lousse, V.; Deparis, O.; Biró, L.P.; Dedouaire, D.; Cornet, A.; Defrance, P. *Phys. Rev. E* 2006, *73*, 041905.

[20] Parker, A.R.; L.Welch, V.; Driver, D.; Martini, N. *Nature* 2003, *426*, 786.

[21] Sanders, J.V. *Nature* 1064, *204*, 1151.

[22] Sanders, J.V.; Darragh, P. *J. Min. Rec. (1971)* 1971, *2*, 261.

[23] López, C. *Adv. Mater.* 2003, *15*, 1679.

[24] Jiang, P.; Bertone, J.F.; Hwang, K.S.; Colvin, V.L. *Chem. Mat.* 1999, *11*, 2132.

[25] Pavarini, E.; Andreani, L.C.; Soci, C.; Galli, M.; Marabelli, F.; Comoretto, D. *Phys. Rev. B* 2005, *72*, 045102.

[26] Palacios-Lidon, E.; Galisteo-Lopez, J.F.; Juarez, B.H.; Lopez, C. *Adv. Mater.* 2004, *16*, 341.

[27] Tetreault, N.; Mihi, A.; Miguez, H.; Rodriguez, I.; Ozin, G.A.; Meseguer, F.; Kitaev, V. *Adv. Mater* 2005, *17*, 1912.

[28] Morandi, V.; Marabelli, F.; Amendola, V.; Meneghetti, M.; Comoretto, D. *Adv. Funct. Mater.* 2007, *17*, 2770.

[29] Tan, Y.; Qian, W.; Ding, S.; Wang, Y. *Chem. Mater.* 2006, *18*, 3385.

[30] Antoine, R.; Brevet, P.F.; Girualt, H.H.; Bethell, D.; Schiffrin, D. *Chem. Commun.* 1997, 1901.

[31] Amendola, V.; Polizzi, S.; Meneghetti, M. *J. Phys. Chem. B* 2006, *110*, 7232.

[32] Markowicz, P.P.; Tiryaki, H.; Pudavar, H.; Prasad, P.N.; Lepeshkin, N.N.; Boyd, R.W. *Phys. Rev. Lett* 2004, *92*, 083903.

[33] Markowicz, P.; Friend, C.; Shen, Y.; Swiatkiewicz, J.; Prasad, P.N.; Toader, O.; John, S.; Boyd, R.W. *Opt. Lett.* 2002, *27*, 351.

[34] Polson, R.C.; Chipouline, A.; Vardeny, Z.V. *Adv. Mater.* 2001, *13*, 760.

[35] Shkunov, M.N.; Vardeny, Z.V.; DeLong, M.C.; Polson, R.C.; Zakhidov, A.A.; Baughman, R.H. *Adv. Funct. Mater.* 2002, *12*, 21.

[36] Eradat, N.; Sivachenko, A.Y.; Raikh, M.E.; Vardeny, Z.V.; Zakhidov, A.A.; Baughman, R. *Appl. Phys. Lett.* 2002, *80*, 3491.

[37] Sumioka, K.; Nagahama, H.; Tsutsui, T. *Appl. Phys. Lett.* 2001, *78*, 1328.

[38] Marabelli, F.; Comoretto, D.; Bajoni, D.; Galli, M.; Fornasari, L. *Mat. Res. Soc. Symp. Proc.* 2002, *708*, BB10.19.1.

[39] Comoretto, D.; Marabelli, F.; Soci, C.; Galli, M.; Pavarini, E.; Patrini, M.; Andreani, L.C. *Synth. Met.* 2003, *139*, 633.

[40] Cucini, M.; Narizzano, R.; Comoretto, D.; Morandi, V.; Marabelli, F. 2007, *Paper presented at 7th International Conference on Optical Probes of p-conjugated polymers and functional self-assemblies, Turku (FI) 11-15 June 2007. Manuscript in preparation.*

[41] Kulinowski, K.M.; Jiang, P.; Vaswani, H.; Colvin, V.L. *Adv. Mater.* 2000, *12*, 833.

[42] Rodriguez-Gonzalez, B.; Salgueiriño-Maceira, V.; Garcia-Santamaria, F.; Liz-Marzán, L.M. *Nano Lett.* 2002, *2*, 471.

[43] Miclea, P.T.; Susha, A.S.; Liang, Z.; Caruso, F.; Sotomayor-Torres, C.M.; Romanov, S.G. *Appl. Phys. Lett.* 2004, *84*, 3960.

[44] Romanov, S.G.; Susha, A.S.; Sotomayor-Torres, C.M.; Liang, Z.; Caruso, F. *J. Appl. Phys.* 2005, *97*, 086103.

[45] Wang, D.; Salgueirño-Maceira, V.; Liz-Marzán, L.M.; Caruso, F. *Adv. Mater.* 2002, *14*, 908.

[46] Wang, D.; Li, J.; Chan, C.T.; Salgueirño-Maceira, V.; Liz-Marzán, L.M.; Romanov, S.; Caruso, F. *Small* 2005, *1*, 122.

[47] Tessier, P.M.; Velev, O.D.; Kalambur, A.T.; Lenhoff, A.M.; Rabolt, J.F.; Kaler, E.W. *Adv. Mater* 2001, *13*, 396.

[48] Garcia-Santamaria, F.; Galisteo-Lopez, J.F.; Braun, P.V.; Lopez, C. *Physical Review B (Condensed Matter and Materials Physics)* 2005, *71*, 195112.

[49] Xia, Y.; Gates, B.; Li, Z.-Y. *Adv. Mater* 2001, *13*, 409.

[50] Comoretto, D.; Morandi, V.; Marabelli, F.; Amendola, V.; Meneghetti, M.; SPIE: 2006; Vol. 6182, p 6182D.

[51] Amendola, V.; Rizzi, G.A.; Polizzi, S.; Meneghetti, M. *J. Phys. Chem. B* 2005, *109*, 23125.

[52] Turkevich, J.; Stevenson, P.L.; Hillier, J. *Discuss. Faraday Soc.* 1951, *11*, 55.

[53] Yu, Y.Y.; Chang, S.S.; Lee, C.L.; Wang, C.R.C. *J. Phys. Chem. B* 1997, *101*, 6661.

[54] Joannopulos, J.D.; Meade, R.D.; Win, J.N. *Photonic Crystals: Molding the Flow of the Light*; Princeton University Press: Princeton, 1995.

[55] Lee, W.; Pruzinsky, S.A.; Braun, P.L. *Adv. Mater.* 2002, *14*, 271.

[56] Comoretto, D.; Dellepiane, G.; Marabelli, F.; Cornil, J.; dos Santos, D.A.; Bredas, J.L.; Moses, D. *Phys. Rev. B* 2000, *62*, 10173.

[57] Haroche, S. In *Fundamental Systems in Quantum Optics*; Zinn-Justin, J.D.J.M.R.J., Ed.; Elsevier: Amsterdam, 1992.

In: Biomimetic and Supramolecular Systems Research
Editor: Arturo H. Lima
ISBN: 978-1-60456-405-1

Chapter V

Tuneable Micro- and Nano-Periodic Structures in Urethane/Urea Networks

***M. H. Godinho*[1]*, J. L. Figueirinhas*[2,3]*, P. Brogueira*[3,4] *and P. I. C. Teixeira*[5,6]**

[1]Departamento de Ciência dos Materiais and I3N/CENIMAT Faculdade de Ciências e Tecnologia, Universidade Nova de Lisboa, P-2829-516 Caparica, Portugal
[2]Centro de Física da Matéria Condensada da Universidade de Lisboa Avenida Professor Gama Pinto 2, P-1649-003 Lisboa, Portugal
[3]Departamento de Física, Instituto Superior Técnico Avenida Rovisco Pais, P-1049-001 Lisboa, Portugal
[4]Instituto de Ciência e Engenharia de Materiais e Superfícies, Instituto Superior Técnico Avenida Rovisco Pais, P-1049-001 Lisboa, Portugal
[5]Instituto Superior de Engenharia de Lisboa Rua Conselheiro Emídio Navarro 1, P-1950-062 Lisboa, Portugal
[6]Centro de Fisica Teórica e Computacional da Universidade de Lisboa Avenida Professor Gama Pinto 2, P-1649-003 Lisboa, Portugal

Abstract

Micro- and nano-patterned materials are of great importance for the design of new nanoscale electronic, optical and mechanical devices, ranging from sensors to displays. A prospective system that can support a designed functionality is elastomeric polyurethane thin films with nano- or micro-modulated surface structures ("wrinkles"). These wrinkles can be induced on different lengthscales by mechanically stretching the films, without the need for any sophisticated lithographic techniques. In the present article we focus on the experimental control of the wrinkling process. A simple model for wrinkle formation is also discussed, and some preliminary results reported. Hierarchical assembly of these tunable structures paves the way for the development of a new class of materials with a wide range of applications, from electronics to biomedicine.

1. Introduction

Many technologies, both existing and planned, rely on exquisitely prepared solid surfaces. Refined, next-generation micromechanical systems require "smart" surfaces that can not only sense or respond to environmental stimuli, but also be tough and possess tailored, on-demand topographic properties [1-4]. Techniques for the micropatterning of soft materials are therefore of great importance for the design of new nanoscale electronic, optical and mechanical devices, ranging from displays [5,6] and actuators [7] to microfluidics [8] to biomedical materials [9]. Such tecniques should be simple and low-cost, and allow a very precise (and ideally reversible) tuning of system properties. Earlier attempts have included the deposition of metal films on elastomeric polymer substrates [5,7,10-13], swelling of a supported polymer layer [14], heating of a polymer bilayer [15], and layer-by-layer assembly of gold nanoparticle sandwiches [16,17]. These, however, perforce combine different types of material in sophisticated multi-step processing routines. Ideally one would like to use a single compound – a polymer if possible, in view of its mechanical compliance – and standard laboratory procedures.

Block copolymers (BCPs) are composed of long sequences ("blocks") of the same monomer unit, which are covalently bonded to other sequences of identical monomer units, of one or more different types. BCPs have been extensively investigated [18] and have proven to be eminenty suitable as multipurpose lively materials in nanoscience and nanotechnology [19-22]. They are known to self-assemble at the nanometer length scale with the formation of a variety of highly ordered phases [23], not only in bulk but also upon dilution by a solvent that interacts selectively with one of the constituent blocks [24-26]. This is a consequence of the fact that the different blocks may be thermodynamically incompatible, but may not bulk phase-separate because they are chemically linked.

In particular, segmented polyurethanes that consist of soft and hard segments [27] may exhibit segregation into micro-domains [28]. The development by Zhao and Pinho [29] of cross-linked urethane/urea polymers with two soft segments, polypropylene oxide and polybutadiene, opened wide horizons for the tailoring of polymer films that display different degrees of phase separation, and consequently very different optical and mechanical properties. Very clear films can be produced of polypropylene oxide-based urethane prepolymers with three isocyanate terminal groups (PU); upon incorporation of the more hydrophobic soft segment, polybutadiene (PBDO), the films become slightly translucent, which suggests some degree of microphase separation [29]. Characterisation of these films by infrared spectroscopy (IR) and dynamic mechanical thermal analysis (DMTA) confirms that varying the type and proportion of the two segments indeed drives different hydrogen bonding environments, hence different degrees of segregation. Zhao and Pinho proposed that increasing the PBDO content weakens the hydrogen bonding between urethane/urea groups, thus promoting their freedom to mix with PU [29].

In addition, the dynamical mechanical properties of the bi-soft segment urethane/urea polymers show a strong stoichiometry dependence. When the PBDO content is less than 50%, two peaks appear in the DMTA spectra, corresponding to the glass transitions of PBDO and PU segments. The two transitions get closer to each other with increasing PBDO content, and only one glass transition was observed when this content was equal to or higher than 60%, for which a molecularly mixed phase is obtained [29].

In this chapter we review our work on a mechanico-optical effect exhibited by free-standing thin films of urethane/urea elastomers prepared with PU and PBDO in varying proportions. The effect is the appearance of two very distinct optical states, one transparent (on) and one translucent (off), upon uniaxially stretching films that have previously been subjected to daylight for several months, or to ultra-violet (UV) radiation in a controlled environment for several days. These two optical states are associated with two different surface textures of the films. In the stretched (on) state, the periodic modulation ("stripes") has wave vector perpendicular to the direction of deformation, and a small amplitude; if the imposed stress is then removed (off), the stripes vanish and a new periodic modulation develops, with wave vector parallel to the direction along which the film was previously deformed, and an amplitude larger than the stripes ("bands") [30,31]. In both cases, the wavelength of the modulation is of the order of a few μm, and their amplitudes of the order of tens to hundreds of nm, hence much smaller than the film thickness, which is typically 60-100 μm. Such textures are highly reproducible and contain remarkably few defects. The process of band formation is completely reversible in cycles of increasing and decreasing, strains, with recovery times of less than 2 s [30].

The translucent/transparent switching driven by the mechanical field resembles the polydomain/monodomain transition of liquid crystal elastomers [32] (although our films are not liquid crystalline, as evidenced by their very small birefringence). Moreover, the surface topography of the translucent state after stress removal can be micro-tuned by applying/removing an external uniaxial mechanical field along different directions [33], by varying the relative proportions of PU and PBDO segments present in the elastomer and developed with the solvent [34], or the UV irradiation times [35].

This chapter is organised as follows: in section 2 we describe the chemical structure, and the synthesis and preparation, of our elastomer films. Section 3 summarises the different experimental methods used to study these films, and what information can be obtained from each of them.

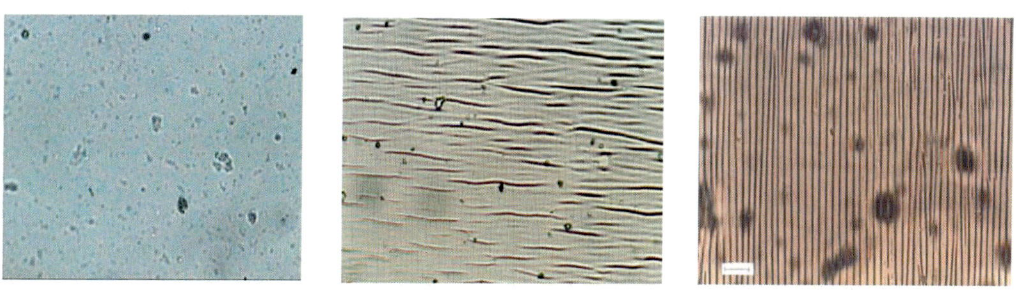

(a) (b) (c)

Figure 1. Optical micrographs of a 40% PU / 60% PBDO film (a) before stretching; (b) during stretching (showing the stripes); and (c) after stretching (showing the bands). The arrow is the direction along which stress was applied; the white bar denotes a length of 20 μm.

As mentioned above, elastomer properties are very strongly dependent on their composition; thus in section 4 we start by presenting detailed results for the two most thoroughly studied formulations, 40% PU / 60% PBDO [30,31.36] and 60% PU / 40% PBDO [35,37,38], and then briefly discuss the properties of a few others [39,40]. In section 5 we propose a simple (zeroth-version) theoretical model to account for the periodic surface modulations (especially bands) and recapitulate some preliminary results [38]; work on expanding these is still in progress. Then in section 6 we evaluate some prototypical applications of textured films, namely as masks for soft lithography [41] and substrates for cell adhesion [42]. We conclude in section 7 by summarising our work and outlining directions for further research.

2. Synthesis

The elastomer was synthesized via two chemical steps between the prepolymers PU and PBDO, shown in figure 2, according to the procedure described in [29,31]

The PBDO is a mixture of *cis* and *trans* isomers, with glass transition temperatures –108°C and –78°C respectively, and a number-average molecular weight of 2800 g/mol; it was supplied by Aldrich. The PU has a molecular weight of approximately 3500 g/mol, its glass transition temperature is approximately –75°C in bulk, and –25°C in a crosslinked film; it was purchased from the Portuguese petrochemical industry (CPB).

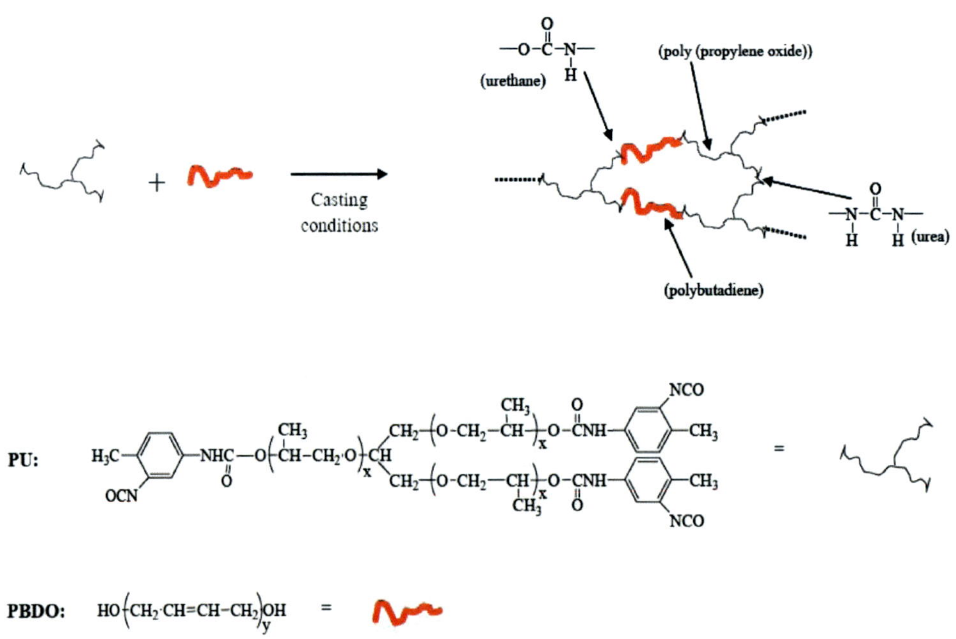

Figure 2. First chemical synthesis step to form urethane linkages from reaction of the end groups of the tri-functional polypropylene oxide-based isocyanate prepolymer (PU), x=20, with the end groups of polybutadiene diol (PBDO), y=50, catalysed by dibutyltin dilauryate (DBTDL).

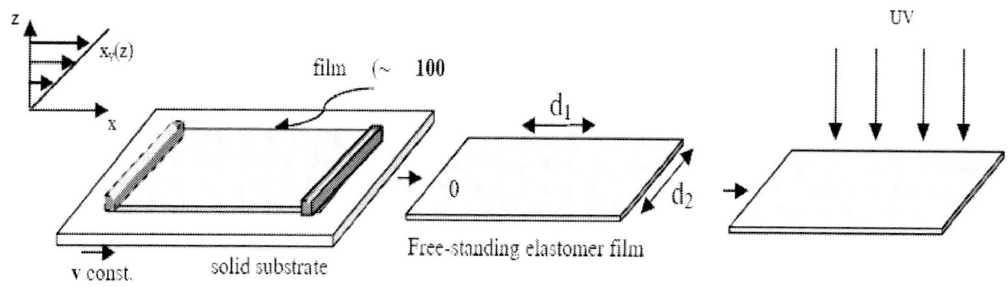

Figure 3. Schematics of the preparation procedure of poly(propylene oxide)/polybutadiene bi-soft segment urethane/urea thin solid film (~60-100 μm) samples. From left to right: Initial solid film is prepared, after 30 min of chemical reaction (see figure 2, from a mixture with a solid content of 40%wt in toluene. The mixture is cast onto a treated glass (solution of 20% wt of silane in ethanol) and sheared by moving a casting knife at a controlled shear rate (v = 5 mm/s), the direction of shear is d1. The free-standing solid elastomer film is then removed from the treated glass substrate after curing (72 h) in atmospheric moisture. d_2 is the direction perpendicular to shear (d_1).

Prepolymers PU and PBDO, in different weight ratios (see below) were dissolved in toluene (solid content 40 wt%) and one drop of dibutyltin dilauryate (DBTDL) catalyst was added. Reaction of the cyanate end groups of PU with the hydroxyl end groups of PBDO was allowed to proceed for 30 minutes, under a nitrogen atmosphere; a weakly cross-linked urethane network was obtained. The elastomer was then cast and sheared simultaneously onto a glass plate at room temperature, by moving a casting knife at a controlled shear rate (v = 5 mm/s) along a direction henceforth denoted d_1, as shown in figure 3; we also define d_2 as the direction at 90° to d_1 in the plane of the film.

The resulting elastomer film was then cured in an oven at 70-80°C for 3.5 hours, and subsequently allowed to continue curing for at least 72 hours in air. The cross-linking density was thus modified by reaction of the remaining cyanate groups with atmospheric moisture to give urea bonds between the PU units. This was further promoted by the casting-shearing technique. The film was then exposed to UV radiation (λ = 254 nm) for, in most cases, 4 days (but see below). After solvent evaporation and cross-linking the film was carefully removed from the glass-treated substrate. Its final thickness was measured using a Mitutoyo digital micrometer; for all films this was in the range 60-100 μm. Finally, films were cut into 5 cm x 2 cm samples.

3. Summary of Experimental Methods

Measurements of mechanical properties (Young's moduli and stress-strain curves) were carried out on a Rheometric Scientific (Minimat-Firm ware Version 3.1) testing machine. A mechanical property of a given sample was taken to be the average of the results of (depending on sample composition) between four and six successful measurements.

Atomic force microscopy (AFM) was used to characterise quantitatively the topographical features of the films, before and after application of the mechanical field, and before and after extraction with toluene. A Dimension 3100 SPM with a Nanoscope IIIa controller from Digital Instruments (DI) was employed. Measurements were performed in tapping mode TM under ambient conditions, using a commercial tapping-mode etched silicon probe from DI and a 90 x 90 μm² scanner. Besides allowing the capture of 3D images, AFM can also measure a number of parameters that characterise surface topography: if $h(\vec{r})$ is the surface height at point $\vec{r}=(x,y)$, one defines the surface *mean roughness* R_a as

$$R_a = \frac{1}{L_x L_y} \int_0^{L_x} \int_0^{L_y} |h(x,y)| \, dx dy \,,$$

and the *root mean square roughness* R_q, as

$$R_q = \sqrt{\frac{1}{L_x L_y} \int_0^{L_x} \int_0^{L_y} h^2(x,y) \, dx dy} \,,$$

where L_x, L_y are the scanning window dimensions along x and y, respectively.

A suite of optical techniques were also employed. Polarising optical microscopy (POM) images were taken with an Olympus polarising microscope equipped with a photographic camera. Small-Angle Light Scattering (SALS) data were recorded with a CCD video camera on an optical bench equipped with a green (λ = 543.5 nm) helium-neon laser; the axis of the applied stress was perpendicular to the laser beam. POM and SALS are complementary in that POM, like AFM, only reveals surface features, whereas SALS probes the whole film thickness.

Extraction in toluene was performed using a Soxhlet apparatus for up to 72 hours at 80°C, to remove any unreacted products.

4. Results

4.1. 40% PU / 60% PBDO System

4.1.1. Mechanical Properties

Figure 4 shows the stress-strain curves and Young's moduli before and after UV irradiation, along directions parallel and perpendicular to the casting shear [31].

The films are clearly anisotropic, being substantially stiffer along the $\mathbf{d}_2$, the direction perpendicular to the casting shear. This persists after UV irradiation, which has the effect of making the film stiffer, more noticeably so along $\mathbf{d}_2$ (~75%) than along $\mathbf{d}_1$ (~14%). This behaviour is compatible with a scenario suggested by infrared spectroscopy, where the secondary amines in the rigid part of the copolymer were converted to tertiary amines, thereby promoting new cross-links between the hard parts of the copolymer.

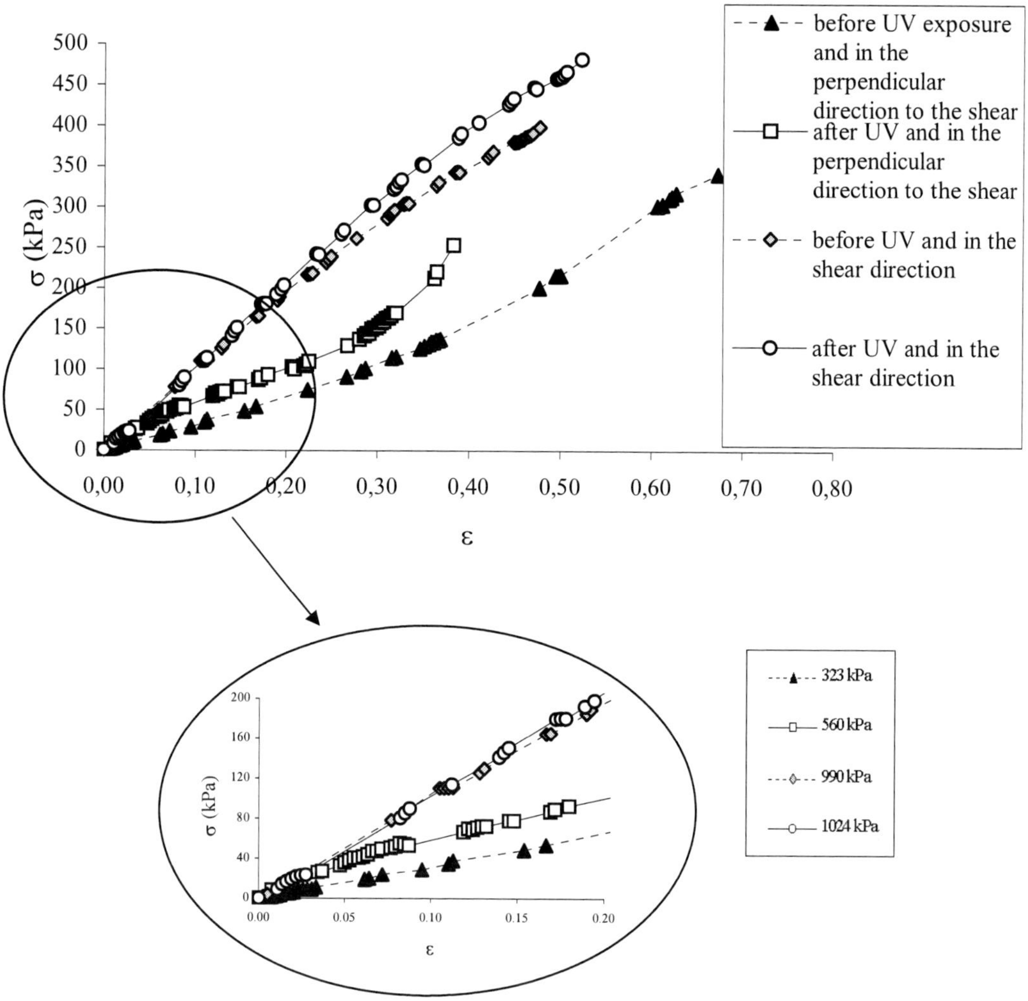

Figure 4. Stress-strain curves of the 40% PU / 60% PBDO sample before and after UV exposure. The Young's moduli are also given.

4.1.2. Morphology

Figures 5 and 6 show, respectively, the SALS patterns and the light transmission curve recorded while the sample was being subjected to a periodic rate of extension of +0.0286 s^{-1} in the first half of a cycle, and -0.0285 s^{-1} in the second half; a maximum elongation L/L_0=1.23 was reached, for an applied stress of σ = 0.45 MPa [30]. Data were collected after the sample had undergone at least 100 cycles. The unstretched elastomeric film is optically isotropic and slightly translucent; upon stretching it becomes transparent, with a maximum transmittance around 55%. The transparent stretched film was observed by tramsission optical microscopy and found to be weakly birefringent. The two-lobe pattern in figure 5a implies a periodicity with wave vector perpendicular to the direction of the applied stress ("stripes", see figure 1b) at maximum elongation, L/L_0=1.23.

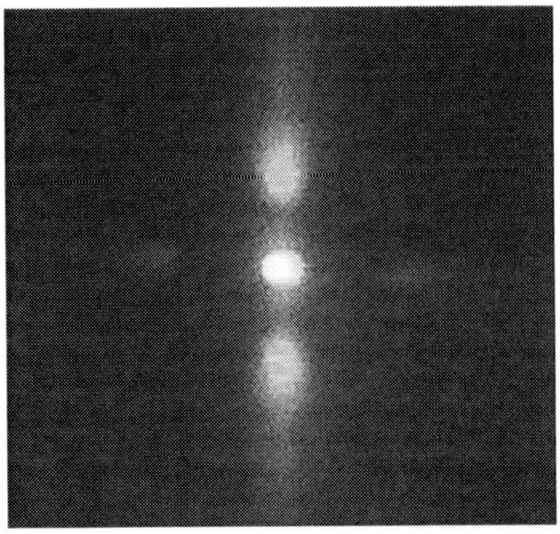

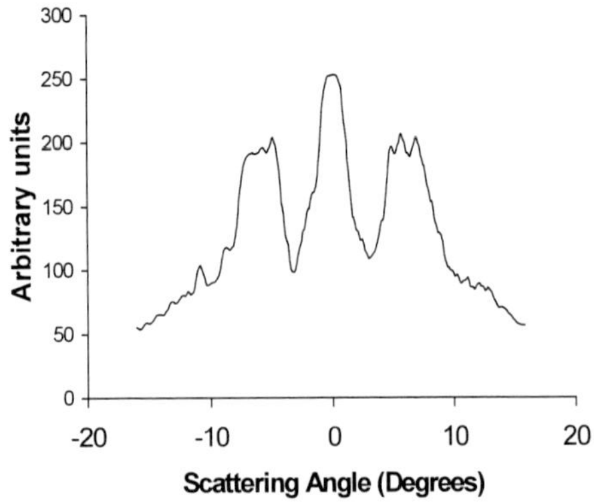

I (a) L/L_o=1.23 II(a)

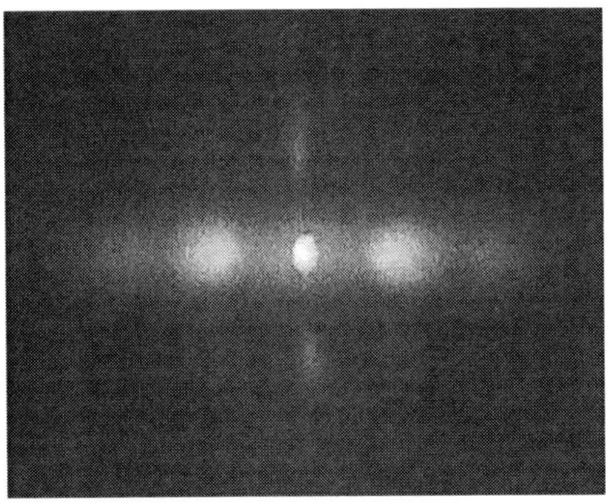

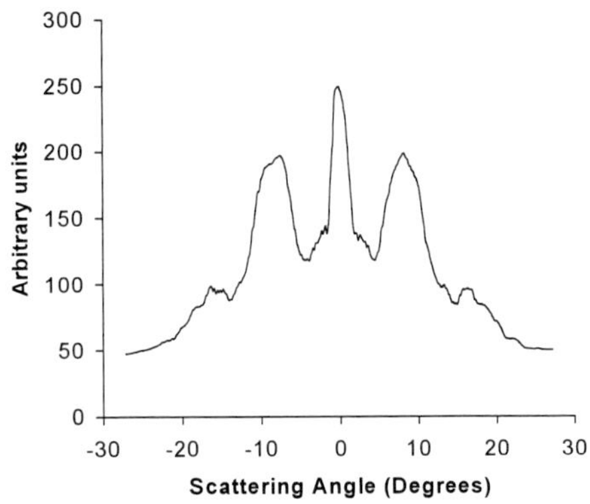

I(b) L/L_o=1.11 II(b)

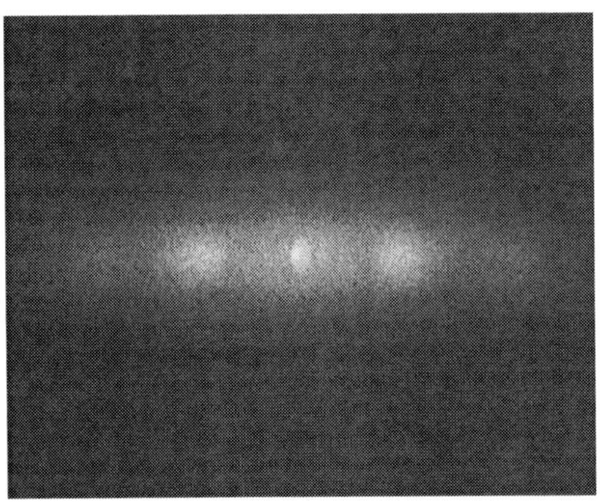

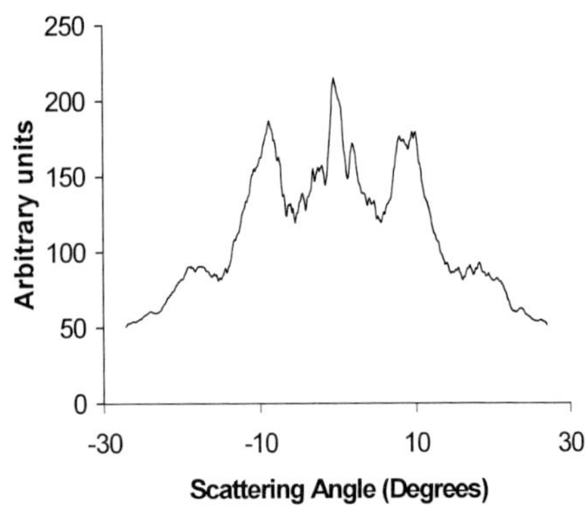

I(c) L/Lo=1 II(c)

Figure 5. SALS patterns for the 40% PU / 60% PBDO sample at theree different elongations: (a) L/L_0=1.23; (b) L/L_0=1.11; (c) L/L_0=1.0.

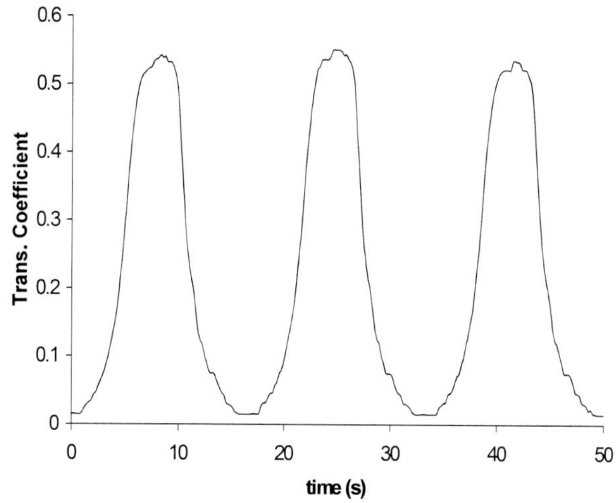

Figure 6. Transmission coefficient *vs* time, after 100 cyclesw, for the 40% PU / 60% PBDO sample. The minimum and maximum extensions are L/L_0=1.0 and L/L_0=1.23, for which the transmittance is 1.4% (translucent) and 55% (transparent), respectively.

When the stress is removed, the elastomer recovers its undeformed dimensions, L/L_0=1.0, and a different periodicity ("bands", see figure 1c) appears, with wave vector parallel to the direction of the previously applied stress, i.e., perpendicular to the stripes. The film now scatters light very strongly, with a transmittance of only 1.4%; furthermore, it is dark when viewed between cross polars (no . The corresponding SALS pattern is in figure 5c: again two lobes, now at 90° to those of the stripes in figure 1a. The band periodicity as determined from SALS and optical microscopy is about 4 μm. The process of stripe/band formation is completely reversible and the sample can be converted between the two states continuously in time, see figure 6. This holds for strain rates up to at least 0.5 s^{-1} and a maximum elongation of L/L_0=1.3 (samples break at L/L_0=1.6), whence the recovery time is less than 2 s. Finally, the four-lobe SALS pattern in figure 5b for a state of elongation L/L_0=1.11 corresponds to the transition between bands for L/L_0=1.0 and stripes for L/L_0=1.23. Under optical microscopy, stripes differ from bands in their orientation, their length and their width: they are shorter and wider, and only appear in stretched states. Macroscopically, a striped sample is transparent, whereas a banded one is translucent.

We have investigated the effects of varying the maximum elongation, the rate of extension and the temperature upon the optical properties of the films [36.39]. Results for stripes are shown in figure 7, and for bands in figures 8 and 9.

Increasing the maximum elongation increases the scattering intensity and shortens the stripe/band wavelength, although there are too few experimental points to be able reliably to resolve a functional dependence. The rate of extension does not seem to affect the wavelengths of stripes or bands very much, at least for most elongations; this suggests that the periodicity is determined by the (equilibrium) elastic properties of the material, and that dynamical properties (e.g., viscosity) play no role here. On the other hand, at large elongations the scattering intensity of stripes decreases with increasing extension rate, whereas that of bands increases. In addition (data not shown) [36], a faster extension rate causes the bands to relax faster to their final configuration. Finally, bands can be erased by heating the sample to about 60oC; their wavelength, however, remains fairly constant throughout their range of existence.

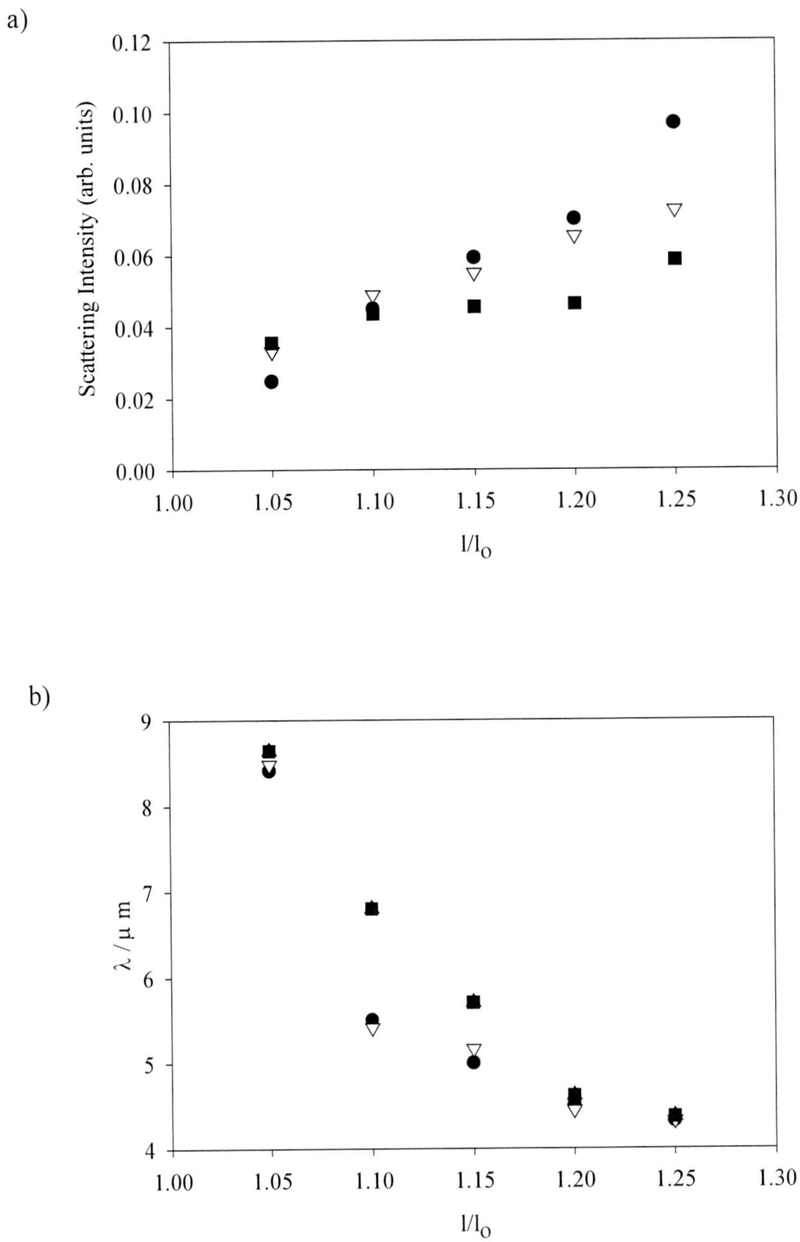

Figure 7. SALS data for the striped (transparent) state of the 40% PU / 60% PBDO sample *vs* maximum elongation for three different rates of extension (■: $5.40 \times 10^{-3}\ s^{-1}$, ▿ : $2.70 \times 10^{-3}\ s^{-1}$, and ● : $1.34 \times 10^{-3}\ s^{-1}$). (a) Maximum scattering intensity; (b) stripe wavelength.

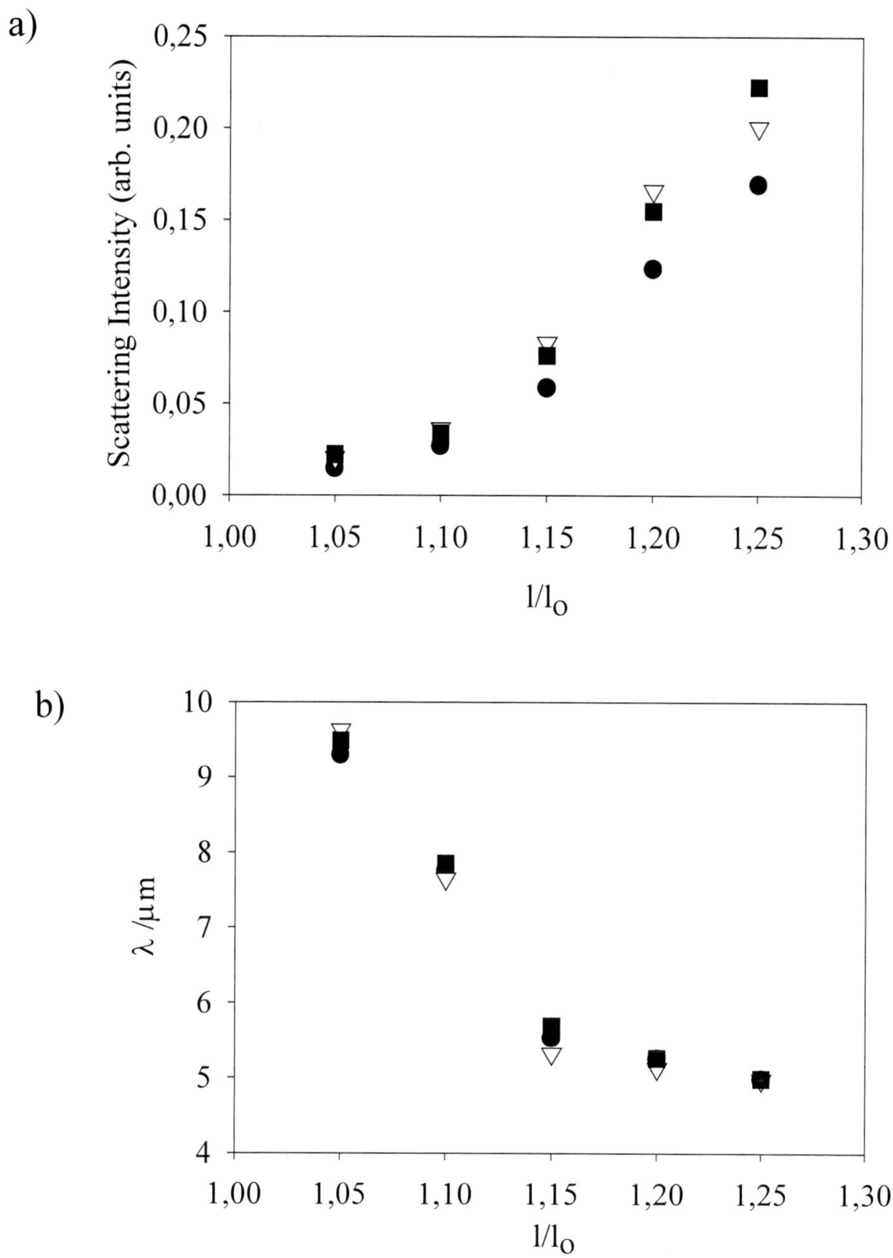

Figure 8. Same as figure 8, but for the banded (translucent) state.

4.2. 60% PU / 40% PBDO System

For this formulation we investigated in considerably more detail how elastomer morphology depends on (i) the mechanical treatment of the initial sample; and (ii) whether sol/gel extraction was performed or not [38].

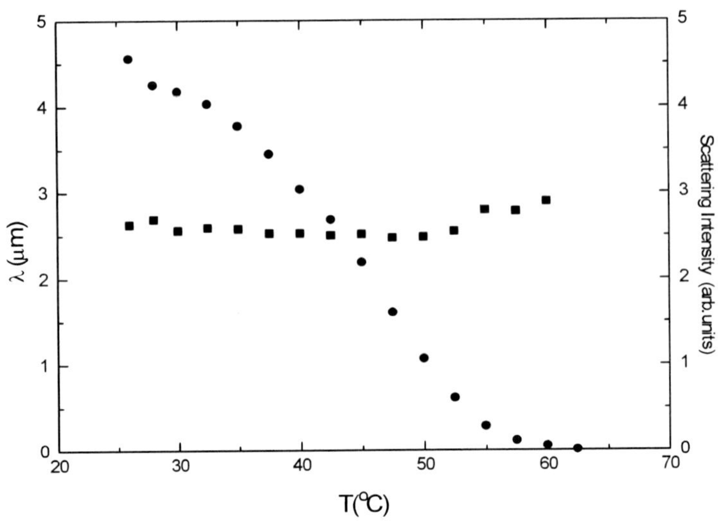

Figure 9. Maximum scattering intensity (●) and band periodicity (■) *vs* temperature, for the 40% PU / 60% PBDO sample.

Both are standard laboratory procedures. We use an excess of PU relative to PBDO in order to obtain a network with two kinds of covalent bonds, involving urethane and urea groups; the resulting elastomer is easier to manipulate.

Before going on to discussing our results it is crucial to introduce a convenient notation for describing the various treatments that films have been subjected to.

- Type I films are films as cast, i.e., without any further treatment.
- Type II films are UV-irradiated films.
- Type III films have been extracted in toluene for 72 h and then dried. They divide into two subtypes:
  - Type III-1 films are films extracted in toluene but not UV-irradiated (i.e., they were obtained by extracting type I films in toluene).
  - Type III-2 films are films extracted in toluene after UV irradiation (i.e., they were obtained by extracting type II films in toluene).

Films of any type can be in one of three different states:

- State 0: the film has not been subjected to any mechanical stress or deformation.
- State A: the film was subjected to a plane uniaxial stress along d_1, this was removed, then a plane uniaxial stress was applied along d_2, finally this was also removed.
- State B: the film in state A was subjected to a plane uniaxial stress along d_1, which was subsequently removed.

By combining the above, we are henceforth able to reference our films in a particularly compact and convenient manner: thus *sample I0* means a type I film in state 0, *sample IIA* is a type II film in state A, *sample III-2B* is a type III-2 film in state B, etc.

4.2.1. Effect of UV Irradiation

The shear-induced anisotropy of the as-cast film *(sample I0)* is reflected in the different values of its Young's modulus measured in the direction of casting shear (i.e., along d_1), and perpendicular to the direction of casting shear (i.e., along d_2). UV irradiation of this sample yields *sample II0*: the resulting higher cross-linking density leads to the moduli being larger than for (unirradiated) sample I0, see Table 1.

It was recently pointed out [22] that UV exposure, by reinforcing the links between the hard and soft parts of the segregated copolymer for these kinds of elastomer, gives rise to a different molecular organisation that is reflected in their mechanical properties.

Table 1. Elastic modulus determined by means of stress-strain measurements at room temperature (~22°C) (nominal stress *vs* strain) for some of our urethane/urea elastomer (60% PU / 40% PBDO). Sample I0 is an elastomer film before UV irradiation, sample II0 originates from sample I0 after UV irradiation, sample IIA originates from sample II0 after extension along d_1 and d_2, sample IIB originates from sample IIA after extension along d_1. Sample III-20 originates from sample II0 after extraction in toluene, samples III-2A and III-2B originate from, respectively, samples IIA and IIB after extraction in toluene (see also figures 3 and 16).

Urethane/Urea Elastomer Sample	UV irradi-ation	(A) Stresses applied along **d_1** and **d_2** consecutively and removed.	(B) Stress applied along **d_1** after cycle A	Extraction in toluene	Elastic modulus along **d_1** (MPa)	Elastic modulus along **d_2** (MPa)
Sample I0	No	No	No	No	4.22 ±0.01	2.84 ±0.01
Sample II0	Yes	No	No	No	4.39 ±0.02	3.29 ±0.05
Sample III-20	Yes	No	No	Yes	3.73 ±0.02	3.76 ±0.02
Sample III-2A	Yes	Yes	No	Yes	3.93 ±0.05	3.70 ±0.01
Sample III-2B	Yes	Yes	Yes	Yes	3.57 ±0.03	2.47 ±0.02

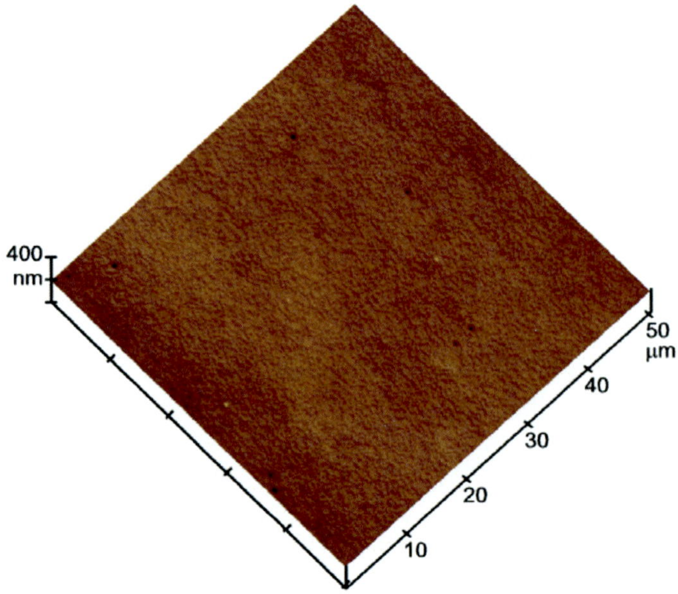

Figure 10. AFM 3D topography image (50x50 μm^2 scan) of the free surface of sample II0, 60% PU / 40% PBDO. This sample has been UV irradiated but not subjected to any mechanical stress. Here and in all topography images that follow, the scale on the right is in the plane of the sample, and that on the left is the vertical scale coming out of the plane of the sample. The surface is nanometrically flat, as can be concluded from its roughness parameters listed in Table 2.

The free surface of the films was observed before (*sample I0*) and after *(sample II0*) UV irradiation, in the absence of applied stresses; figure 10 shows the AFM 3D topography image of the nanometrically flat surface of *sample II0.* This is characterised by the roughness parameters listed in Table 2. A similarly flat topography was obtained for elastomer films removed from the substrate and subjected to plane uniaxial stresses, but not exposed to UV radiation (*samples IA or IB).*

Table 2. Roughness parameters as determined by AFM for some of the urethane/urea elastomer 60% PU / 40% PBDO films in figure 10 (see also figure 3)

Urethane/Urea Elastomer Sample	Mean roughness R_a (nm)	Root mean square roughness R_q (nm)
Sample II0	0.57	0.75
Sample III-20	217	270
Sample III-2A	170	205
Sample IIB	61	73
Sample III-2B	294	392

4.2.2. Effect of Solvent Extraction and of Mechanical Deformations

The irradiated film (sample II0) was extracted in toluene for 36 hours, yielding sample III-20. Its AFM topography image can be seen in figure 11a: the surface is corrugated, with μm-sized features along all axes, resulting in a dramatic increase in all roughness parameters, see Table 2. It is also much less elastically anisotropic than samples I0 or II0, see Table 1.

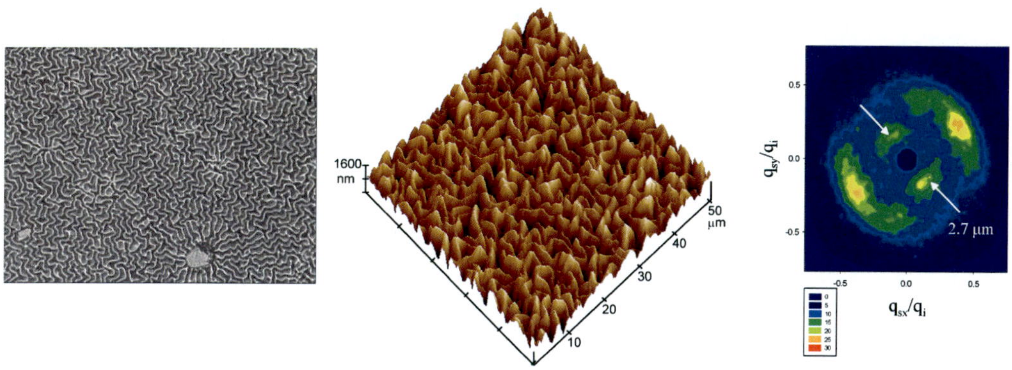

Figure 11. Sample III-20, 60% PU / 40% PBDO, which was obtained from sample II0 after extraction with toluene and drying. (a) AFM 3D topography image (50x50 μm^2 scan) of the free surface; (b) POM image; (c) SALS pattern.

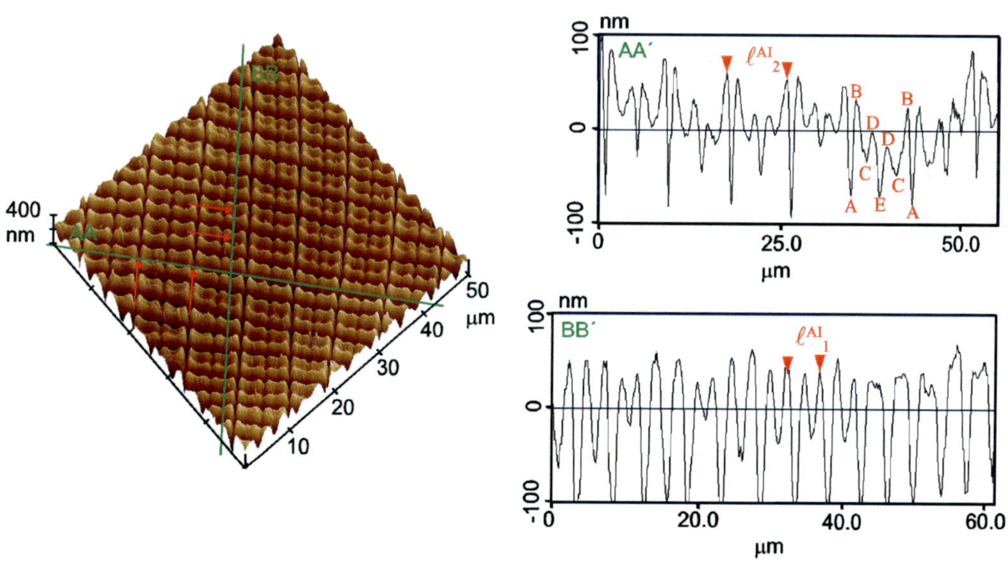

Figure 12. AFM 3D topography image (50x50 μm^2 scan) of the free surface of sample IIA, 60% PU / 40% PBDO, which was obtained by subjecting sample II0 to uniaxial plane stresses along d_1 and d_2 consecutively. The AA' and BB' cross-sections were taken along the lines marked on the 3D image, which are parallel to d_2 and d_1, respectively. See the text for details.

This sample was also observed under parallel polars (see figure 11b) and the POM image, obtained with the microscope focused near the surface, reveals a landscape of long, winding, randomly-distributed line-like textures. The spacing between these structures has a periodicity of about 2.7 μm, in agreement with the AFM topography. The rich pattern revealed by SALS (see figure 11c) exhibits two well-defined periodicities, one corresponding to that observed by POM and another (about 1.3 μm) that might be attributed to ordering of the material in the bulk, also induced by the strain generated during swelling. It is not detected by AFM, which probes the surface only, but exclusively by SALS, which probes the whole thickness of the film. SALS measurements performed on different regions of the sample show that while the characteristic periodicity length scales remain in the same range, the preferential orientation changes from region to region. This behaviour suggests that although orientational order exists over one typical diameter of the SALS laser spot (around 1.5 mm), orientational correlations do not extend across the whole sample. The orientational order suggested by SALS at the sub-mm scale is visible in the POM image.

To single out the effect of UV irradiation, the film with no UV exposure (sample I0) was also extracted in toluene, giving a different sample III-10, with a flat surface. We are thus led to the conclusion that UV irradiation changes the properties of a thin layer on the upper surface of the elastomer film that is responsible for the periodic modulations described above (recall that the film thickness is much greater than the typical amplitude of the modulations: there is no question of the whole film wrinkling, just a thin surface layer). The bottom surface of the film, which was not exposed to UV radiation, is indeed flat.

In order to investigate the effect of mechanical fields upon the surface topography of the irradiated samples, sample II0 was subjected to a (elastic) uniaxial deformation $\varepsilon_1 = 0.2$ along d_1 by applying stress $\sigma_1 = 0.88$ MPa (see figure 4 for typical stress-strain curves). After removal of this stress a pattern of periodic bands developed on the elastomer film surface with wave vector parallel to d_1. Then an (elastic) extensional deformation $\varepsilon_2 = 0.2$ was imposed along d_2 by means of stress $\sigma_2 = 0.66$ MPa, which was subsequently removed, yielding sample IIA. The surface topography of sample IIA is shown in figure 12. It is characterised by two periodic modulations with wave vectors parallel to d_1 and d_2. The AFM height profile along d_1 (figure 12, cross section BB') has spatial periodicities ℓ_1 ranging from 4 to 5 μm; the peak-to-valley heights are alternately $h_{11} = 65\pm16$ nm and $h_{12} = 160\pm30$ nm. In contrast, the AFM height profile along d_2 (figure 12, cross section AA') reveals a complex periodic structure that repeats with a spatial periodicity ℓ_2 in the range 7.9 to 8.6 μm. Within this period a symmetric sequence of nine peaks is always observed with the order ABCDEDCBA (shown in the AA' cross-section), where the peak-to-valley height varies between 30 and 60 nm.

Next we looked at how extraction in toluene changes the topography of a film that was previously deformed. Sample IIA was extracted in toluene yielding sample III-2A. The AFM 3D topography image is shown in figure 13a, and corresponding roughness parameters are collected in Table 2. This was also observed under parallel polars (figure 13b) and its POM image reveals a randomly distributed line-like corrugated texture with a similar periodicity (2 μm) as found in sample III-20. Besides this, another periodicity is present (11 μm) with wave vector along d_1.

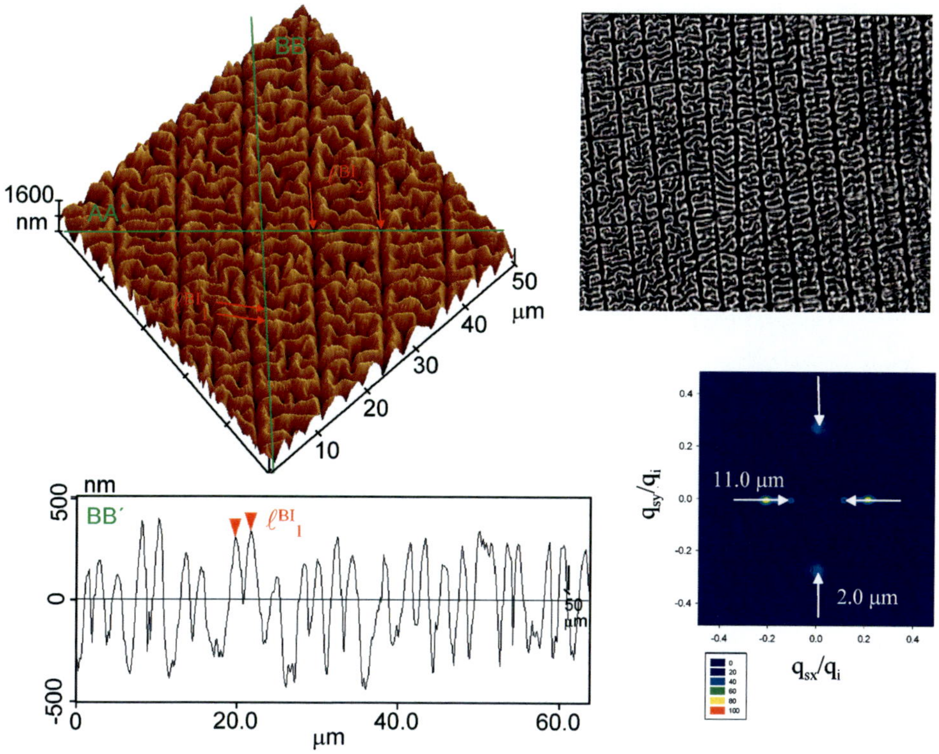

Figure 13. Sample III-2A, 60% PU / 40% PBDO, which was obtained by extracting sample IIA in toluene. (a) AFM 3D topography image (50x50 μm^2 scan) of the free surface: the BB' cross-section was taken along the lines marked on the 3D image, which is parallel to d_1; (b) POM image; (c) SALS pattern. See the text for details.

In the SALS pattern (figure 13c) the arrows indicate the position of the first-order scattering peaks corresponding to the two main periodicities previously observed by POM. The second periodicity (11 μm) may be attributed to the mechanical treatment the sample was subjected to before extraction in toluene: it is not present in the unstrained samples, and is detected by both POM and SALS. The two main periodicities found by POM and SALS can also be found by AFM, $\ell_1 = 2\pm15$ μm and $\ell_2 = 10\pm0.5$ μm (figure 13a, cross sections BB' and AA', respectively). As can be seen from Table 1, sample III-2A is also elastically less anisotropic than sample II0, but more so than sample III-20. This is compatible with the fact that the precursor of sample III-2A, sample IIA, has residual order induced by the mechanical fields, whereas the precursor of sample III-20, sample II0, does not.

It is also interesting to research whether such mechanically imprinted textures are permanent, or can be further modified. To this end we obtained samples in state B by applying an extension along d_1 to samples in state A. Figure 14 shows the AFM 3D topography image of the free surface of sample IIB, which is characterized by an increase in the roughness parameters relative to sample II0, see Table 2. A spatial modulation can be observed along d_1 (cross-section AA'). Within this modulation the peaks are grouped in pairs: each pair repeats with period $\ell_1 = 4.5\pm0.2$ μm.

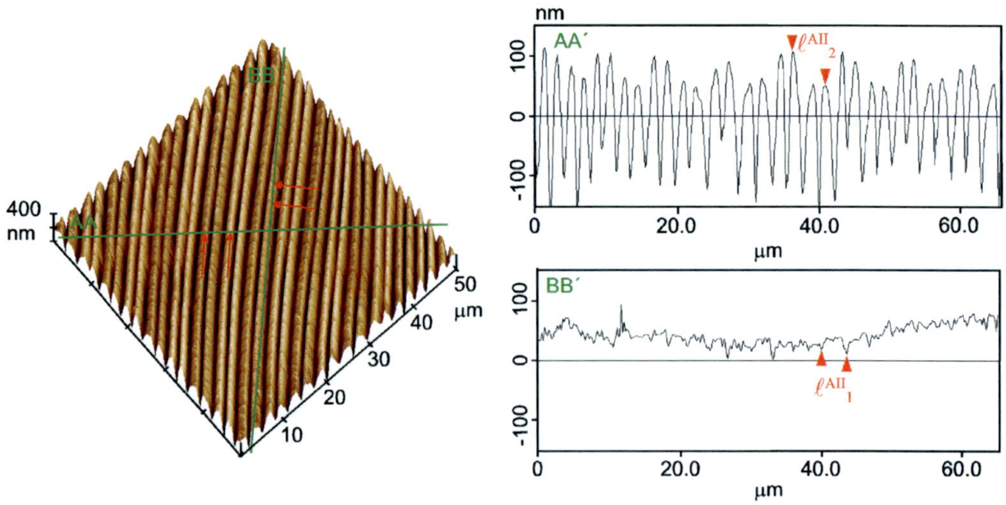

Figure 14. AFM 3D topography image (50x50 μm^2 scan) of the free surface of sample IIB, 60% PU / 40% PBDO, obtained by subjecting sample IIA to stress along d_1. The roughness parameters are listed in Table 2. Spatial periodicities can be observed along d_2 (cross section AA') and along d_1 (cross section BB'); the latter reveals the imprint of state A. See the text for details.

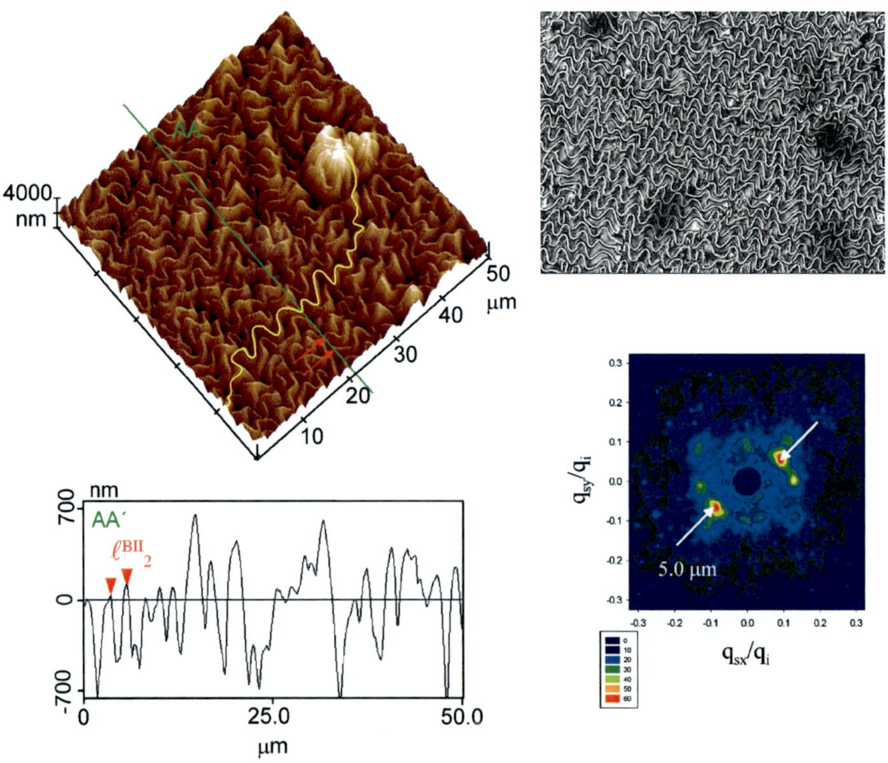

Figure 15. Sample III-2B, 60% PU / 40% PBDO, which was obtained by extracting sample IIB in toluene. (a) AFM 3D topography image (50x50 μm^2 scan) of the free surface: the roughness parameters are listed in Table II; (b) POM image; (c) SALS pattern. See the text for details.

The peak-to-valley height in the gap between peaks of different heights is $h_{11} = 220 \pm 70$ nm and the peak-to-valley heights in the gap between peaks of similar heights are $h_{12} = 120 \pm 30$ nm and $h_{13} = 200 \pm 50$ nm for the lower (darker) and higher (brighter) peaks, respectively. The peak-to-peak height difference between consecutive peaks of different heights is constant and equal to $h_{14} = 40 \pm 20$ nm. The profile along d_2 (cross section BB') reveals the imprint of the precursor of sample IIB, sample IIA, with characteristic period $\ell_2 = 2.4 \pm 0.3$ μm and peak-to-valley height $h_2 = 15 \pm 5$ nm.

Sample IIB was then extracted in toluene, yielding sample III-2B. It was imaged by AFM, see figure 15a, revealing a rougher surface than either sample IIB (its precursor), III-20 or III-2A, see Table 2. The wave vector of the corrugations varies along d_1 (figure 15a, cross section AA'). The peak-to-peak periodicity $\ell_1 = 2.0 \pm 0.5$ μm is of the same order of magnitude as the period imprinted on the sample after UV irradiation (i.e., on sample II0). The solid wavy line is one of a number of sinusoid-like features, also observed by POM, see figure 15b, and which tend to have wave vectors along d_2 and a peak-to-peak distance around 5 μm, as found from the SALS pattern and marked by the arrows in figure 15c.

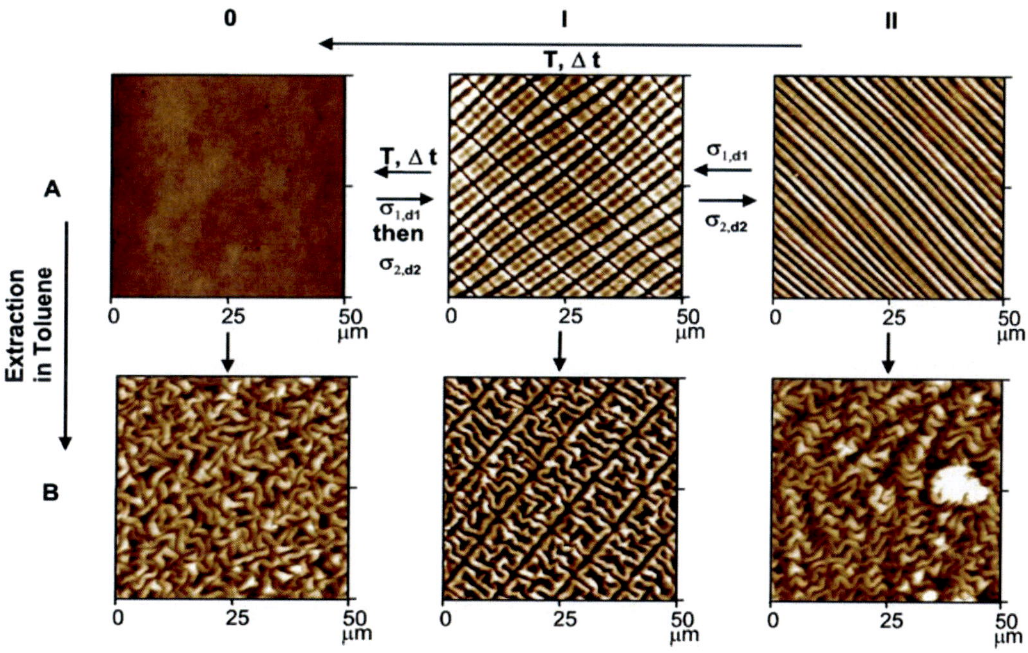

Figure 16. Summary of AFM 3D topography images of the free surface of the 60% PU / 40% PBDO elastomer film. Type II samples are samples that have been exposed to UV radiation ($\lambda = 254$ nm) for two days. Type III-2 samples originate from type II samples after extraction in toluene. A sample is said to be in state 0 if it has not been subjected to mechanical treatment. Sample IIA (a type II sample in state A) is obtained by subjecting sample II0 (a type II sample in state 0) to sequential extensions along d_1, upon applying the stress $\sigma_1 = 0.22$ MPa, and along d_2, upon applying the stress $\sigma_2 = 0.19$ MPa. Sample IIB (a type II sample in state B) is obtained from Sample IIA (a type II sample in state A) after extension along d_2 upon applying the stress σ_2. Samples of type II can be switched between states A and B by applying/removing σ_1 along d_1 (A→B) and σ_2 along d_2 (B→A), respectively. States A and B can be erased, giving state 0, by raising the temperature above 65° C, or by waiting at room temperature for a time that depends on the rate of extension.

Sample III-2B has more anisotropic tensile properties than samples III-20 or III-2A, see Table 1, which may be a consequence of the fact that its precursor, sample IIB, has a high degree of order induced by the mechanical field. However, these are the lowest moduli of all the samples we prepared: it appears that deformation *and* solvent extraction have a softening effect.

Finally, figure 16 summarises our experimental procedures and findings, as well as the notation for naming samples. The very important fact should be highlighted that a sample of type II (i.e., UV-irradiated) can be reversible switched between states A and B by the application of a stress σ_1 along d_1 (A to B) or σ_2 along d_2 (B to A). Alternatively, states IIA or IIB can be erased, giving state II0, by raising the sample temperature above 65°C. At room temperature, these states decay with a time constant that is a function of the rate of extension which was used to produce them; in all cases state II0 is recovered after at most a few months. However, no such reversible switching is possible between states III-2A and III-2B: swelling in toluene freezes the imprinted patterns permanently on the film surfaces. Currently we are not able to explain the 1.3 μm and 11 μm periodicities found in samples III-20 and III-2A, respectively, as the experimental techniques that we have used so far do not yield enough information on them. We are looking into the possibility of elucidating their nature by means of transmission electron microscopy (TEM), as done in, e.g., [43].

4.2.3. Effect of UV Irradiation Times

In order to investigate the effect of different UV irradiation times on elastomer film morphologies, and thus establish how the same can be controllably produced, samples were exposed for 23, 25, 26, 31 and 49 h, then extracted in toluene and dried [35]. AFM 3D topographies, POM images and SALS patterns are shown in figure 17a, 17b and 17c, respectively.

The samples start to wrinkle after 23 h, and two regimes can be identified: before and after 26 h. Between 23 h and 26 h, the typical spacing between structur is fairly contstant at ~1.3 μm and there is only slight coarsening. For exposures longer than 26 h, there is dramatic coarsening of the surface textures, with the corresponding increase in their characteristic lengthscale to ~5 μm, as is consistently borne out by AFM and SALS. This correlates with an increase in roughness and in the peak-to-valley distance of the wrinkles, see Table 3.

The abruptness of the crossover between the two regimes suggests that a minimum crosslink density is necessary to have fully developed wrinkles.

4.3. Other Compositions

Urethane/urea elastomers were also prepared with weight ratios 75% PU / 25% PBDO [39], 70% PU / 30% PBDO [40], 50% PU / 50% PBDO, and 25% PU / 75% PBDO [39]. All films were isotropic and translucent prior to the application of any shear stress; the 75% PU / 25% PBDO, 50% PU / 50% PBDO samples did not exhibit any shear-induced band formation, the 70% PU / 30% PBDO sample only marginally, with very small features, whereas the 75% PU / 25% PBDO intriguingly exhibited a *third* periodicity (besides stripes and bands), oblique to the direction of applied stress.

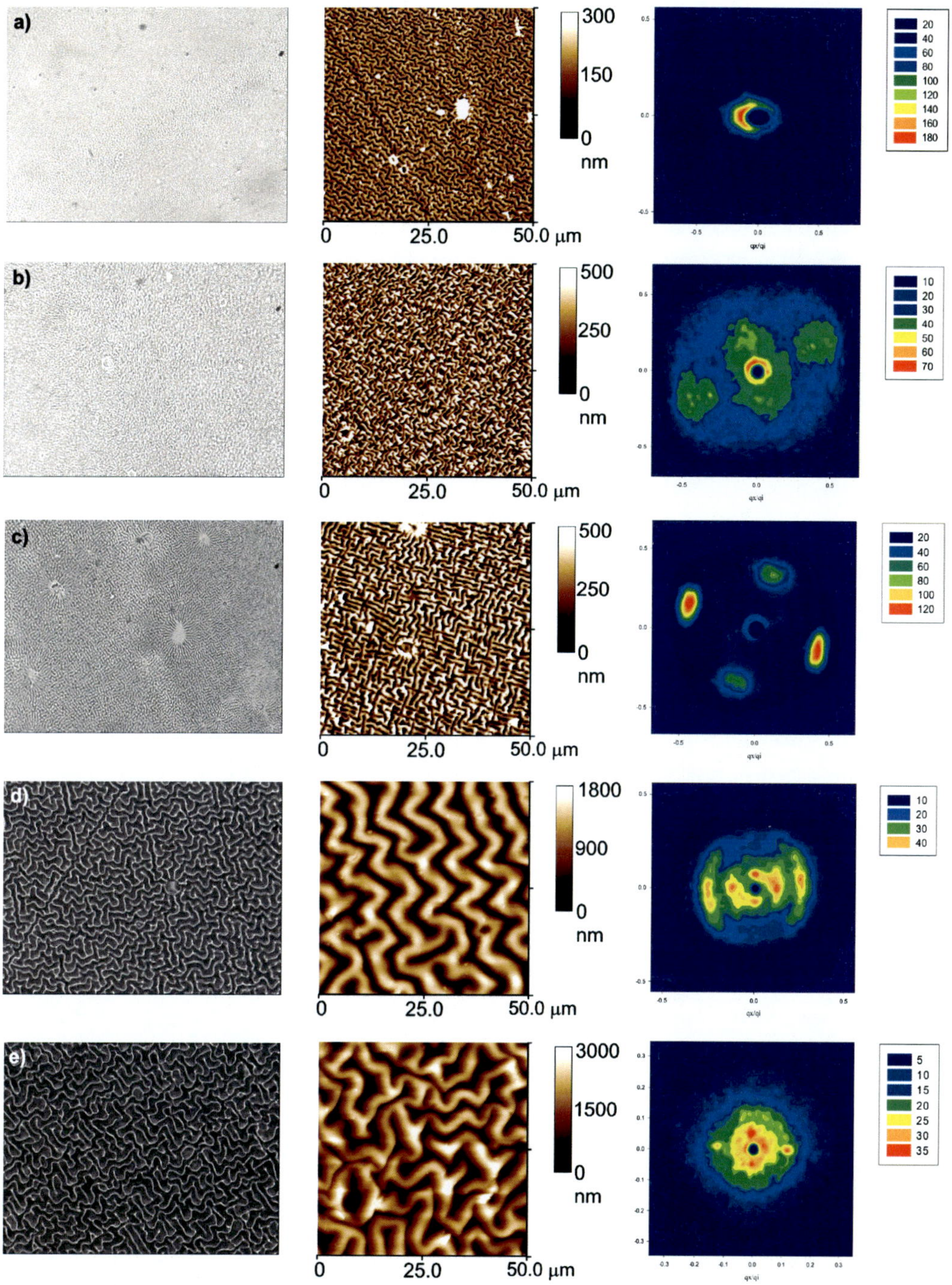

Figure 17. 60% PU / 40% PBDO sample after UV exposures of varying duration (as given), followed by extraction by toluene. (a) AFM 3D topographies; (b) POM images; (c) SALS patterns.

This suggests that there is a critical PBDO content below which no mechanico-optical effect is observed.

Table 3. Roughness parameters as determined by AFM for the urethane/urea 60% PU / 40% PBDO films in figure 16

Exposure time (h)	Root mean square roughness R_q (nm).	Peak-to-valley distance (nm)	Characteristic lengthscale (μm)
23	62	193	1.36
25	130	383	1.36
26	142.88	381	1.28
31	507.38	1550	5.46
49	623.54	2039	5.12

5. Modelling

In this section we discuss a first attempt at modelling the mechanical behaviour of the 60% PU / 40% PBDO elastomer films, for which we have a more extensive collection of data. We start by noting that the observed corrugations ("wrinkles") have amplitudes (of the order of tens or hundreds of nm) much smaller than the film thickness (~60-100 μm). Furthermore, the film surface not exposed to UV did not exhibit any features. This suggests that the effect of UV irradiation is to modify the properties of a relatively thin layer on the film surface. This layer will likely have a higher density of cross-links and therefore be stiffer, which is consistent with the data in Table 1: UV-irradiated sample II0 indeed has higher elastic moduli than unirradiated sample I0. The change is, however, not large, 4-16% depending on direction, which hints at a surface layer thickness (provided the same can be defined unambiguously) of a few percent of the total film thickness.

In a recent paper, Cerda and Mahadevan [44] addressed the wrinkling of a thin, stiff elastic layer sitting atop a thicker, softer substrate. They found that two different types of wrinkling can occur. First, when the system is stretched by clamping its boundaries, it is not able to contract laterally close to the clamps, as it is required to do to conserve volume. This sets up a biaxial state of stress that is compressive away from the clamped boundaries, which leads to buckling with wave vector perpendicular to the direction of stretching. Second, when the natural dimensions of the thin surface layer are greater than those of the underlying substrate, then the thin surface layer will buckle. In both cases buckling prevails because a thin elastic plate finds it cheaper to bend rather than compress [45]. The wavelength of the resulting corrugation is determined by the competition between the bending stiffness of the thin film on top (which penalises short wavelengths) and the energy of stretching the substrate below (which penalises long wavelengths); from [46], $\lambda \propto t\left(E_l/E_b\right)^{1/3}$, where E_l is the Young's modulus of the surface layer of thickness t, and E_b that of the substrate (bulk).

On the basis of the above we propose the following model for explaining our results.

- The stripes reported in earlier papers [30,31,37,40] are just Cerda and Mahavdevan's "wrinkles in a plastic sheet": they are parallel to the direction of applied stress and have small amplitude.
- On stretching the elastomer, the surface layer, being stiffer, deforms plastically by more than the substrate. So on removing the stress, the natural dimensions of the surface layer and of the substrate are now mismatched, hence the bands. A similar effect was recently reported by Efimenko et al. for a PDMS film [46]. This is also consistent with the fact that one can switch reversibly between samples IIA and IIB, and that the surface textures of these samples can be erased by either heating them to 65°C, or waiting for long enough at room temperature, which is suggestive of recovery due to viscoelastic flow of a plastically-deformed state. One further supporting argument is the fact that many of the periodic surface patterns (e.g., that of sample IIB) are characterised by more than one wavelength: indeed such a buckling "cascade" has been reported by other authors [46,47] in cases where a plastically deformed thin sheet has to relieve a compressive stress. As the sheet buckles its effective thickness changes, so a new wavelength of instability is selected.

We carried out simulations of this model using the Finite Element package ABAQUS [48]. We assumed the surface layer to comprise about 10% of the film thickness and estimated its elastic modulus from the data of Table 1 for samples I0 and II0, approximating our system as a composite of layer + substrate under same-strain conditions (see figure 18a).

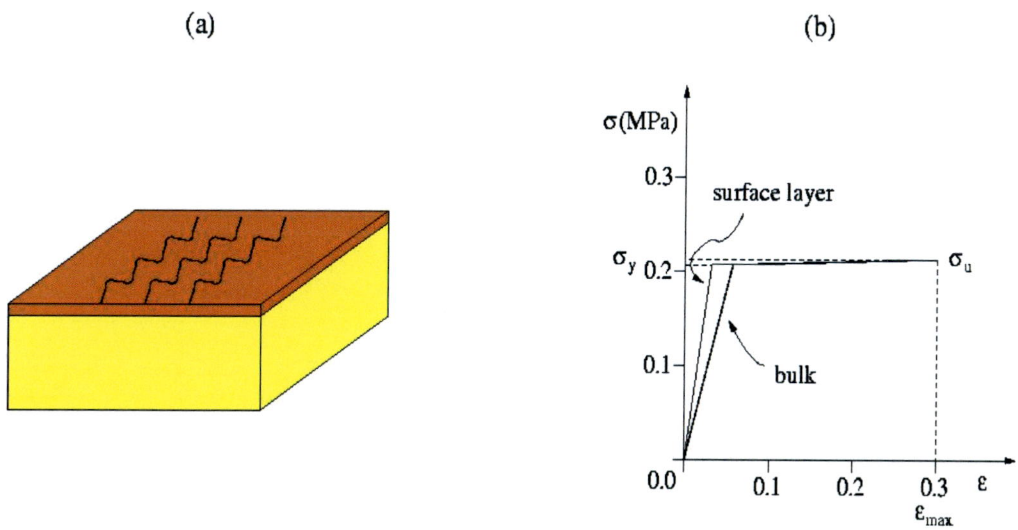

Figure 18. (a) Schematic representation of the `upper crust´ model: a thin, stiff layer atop a thick, softer substrate. (b) Stress-strain curves for the thin layer and for the substrate (bulk). Technical reasons require that the plastic region of the curves should have a small, finite slope; this is shown exaggerated for clarity.

Table 4. Parameters used in the Fine Elements simulation of the urethane/urea 60% PU / 40% PBDO films. Bulk Young's modulus are from sample I0, see Table 1. We have neglected the elastic anisotropy at this stage

	Surface layer	Bulk
Young's modulus (MPa)	5.92	4.22
Poisson's ratio	0.49	0.49
Yield stress (MPa)	0.220	0.220
Ultimate stress (MPa)	0.221	0.221
Maximum total strain	0.30	0.30

The simplest plastic behaviour was assumed: a nearly flat stress-strain curve past the yield stress, which was taken to be the same for both surface and bulk material (see figure 18b and Table 4 for the parameters used in the simulations). The films have two planes of mirror symmetry, along and perpendicular to the direction of stretching, hence only one quarter of the domain was modelled. Appropriate boundary conditions were applied in the two symmetry planes and rigid body motion was prevented by additionally fully constraining the lower node common to those planes. In addition, nodes pertaining to a surface where stress is applied are constrained, during application of stress, so as not to allow translation in the respective plane. This mimics the fact that a film is stretched by clamping two of its edges, which therefore are not free to deform, leading to non-uniform strains/stresses. Because the surface material is stiffer than the bulk, it will enter the plastic regime earlier and upon unloading (for the same final elongation, see figure 18b) will retain a larger plastic strain.

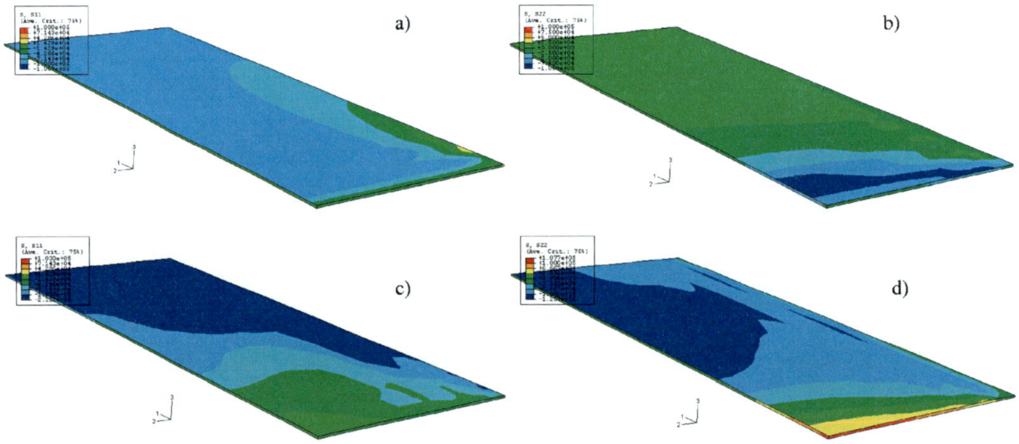

Figure 19. Distribution of residual stresses σ_1 [(a), (c)] and σ_2 [(b), (d)] in our model elastomer film. (a) and (b): after application and release of stress $\sigma_1 = 0.88$ MPa along d_1 (parallel to the long edge of the samples) to the undeformed elastomer; (c) and (d): after application and release of stress $\sigma_2 = 0.19$ MPa along d_2 (parallel to the short edge of the samples) to the elastomer previously stressed along d_1. Only one quarter of the film is shown: the top edge and the near edge are in the planes of symmetry, see the text for details.

Figure 19 shows the distribution of stresses σ_1 and σ_2 in the elastomer film after stress was applied sequentially along $\mathbf{d_1}$ (figure 19a, 19b) and $\mathbf{d_2}$ (figure 19c, 19d). There is a clear differential response along the layer thickness: upon release of the applied stress, the surface layer is in compression, whereas the bulk is mostly under traction. Compression of the surface layer is conducive to the occurrence of undulations associated with elastic instabilities. It remains to extract the wavelengths of these surface undulations. This work is in progress and will be reported elsewhere.

Our modelling is, of course, at a very preliminary stage. Indeed, we do not know at present whether there is a sharp transition between the surface layer and the substrate and, if so, what is the thickness of the former, which we have treated as a free parameter. We have only proposed a simple mechanism whereby bands *may* form, and shown that it is plausible – we have not determined their equilibrium amplitude(s) or wavelength(s). Nor have we explained why the bands disappear after a few months; this, as well as the kinetics of their formation, will very likely require a model for an elastic film on a viscoelastic substrate, such as that recently proposed by Huang and Suo [49,50]. At the moment, we do not have enough data to allow a detailed comparison with this theory, which would require knowledge of: (i) the band wavelengths and amplitudes as a function of the applied strains, for which we have only a few data points; (ii) the thickness of the surface layer; and (iii) the time evolution of the band wavelengths and amplitudes over the very short interval (less than 1 second) in which they form, as well as over the very long interval (several months) in which they disappear. We have also completely neglected the elastic anisotropy of the samples, which might explain, on the basis of the simple theory of [46], why the bands have different wavelengths along $\mathbf{d_1}$ and $\mathbf{d_2}$. Finally, the role of swelling in freezing/erasing surface textures remains poorly understood.

6. Applications

6.1. Soft Microlithography

Recently there has been much interest in soft lithography methods for the realisation of advancd organic flexible devices patterned on the micro- and nano-scale [51]. For this prototypical application, a 120 μm urethane/urea elastomer film of weight ratio 60% PU / 40% PBDO was prepared as described in section 2, except that exposure to UV radiation was for two days, through a silk paper mask with the capital letter E (font size ??pt) printed on in black ink [41]. The letter E was thus written into the film surface as a region of lower crosslinking density; however it is invisible to the naked eye and needs to be developed, which can be done in one of two ways:

1) By applying, and then removing, a mechanical shear stress perpendicular to the direction of the casting shear; or
2) By extracting the sample in toluene and drying it afterwards.

Figure 20 summarises the procedure, where we use the same notation as in section 4.2.

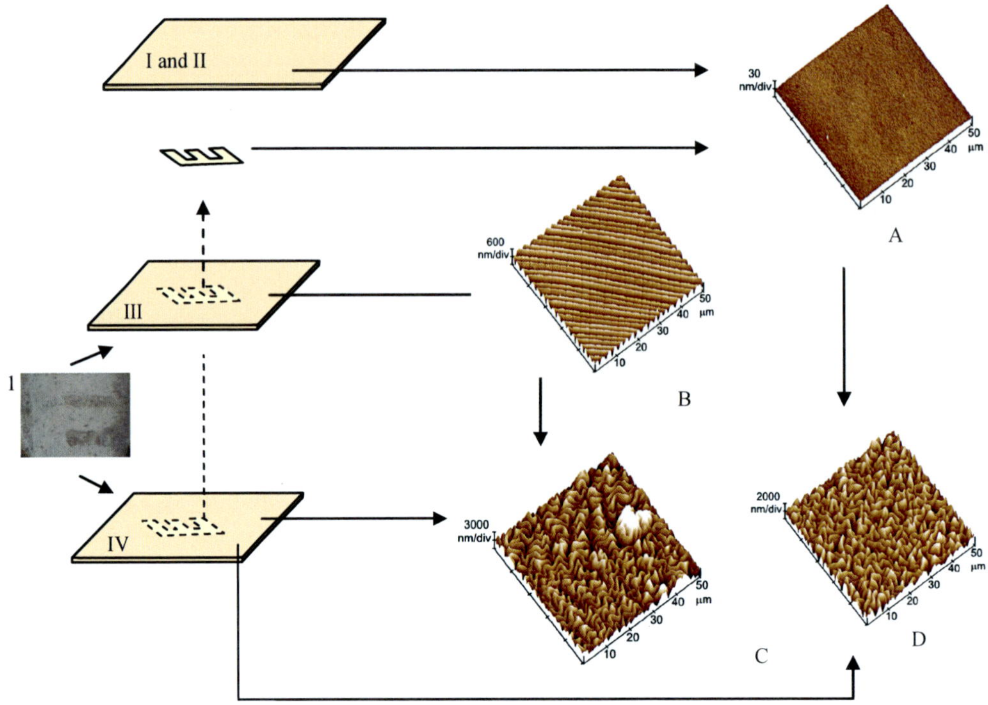

Figure 20 goes here.

(This is figures 2 and 3 of [41] combined.)

Figure 20. Schematics of the microlithography surface modification method. Sample I0 is an elastomeric film as cast; sample II0 is sample I0 after exposure to UV radiation through the paper mask, which thus bears the imprinted image. This is invisible but can be developed by applying a mechanical stress along d_2 (perpendicular to the direction of casting shear) and then removing it, thereby producing sample IIC; or by extraction in toluene, leading to sample III-20. Sample IIC can be converted back to sample II0 (i.e., the image can be erased) by raising the temperature to 60°C.

The shape and size of the micropatterns can be modulated depending on the processing conditions. The imprinted image displays a dynamics behaviour: it can be read, erased, or left for at least months on the elastomer film surface, invisible unless a mechanical field is applied, or the film is immersed in an appropriate solvent.

6.2. Substrates for Cell Adhesion

One of the major problems associated with biomedical materials used in implants or indwelling devices concerns the existence of biomaterial.centred infections from coagulase-negative staphylococci [52]. Biomaterial-centred infections start from bacterial adhesion [53] that occurs due to specific interactions between cell-surface structures and specific molecular groups of the biomaterial, and by intermolecular forces. After the initial adhesion, a second process of bacterial cell-cell adhesion results in a biofilm that is very difficult to eliminate, because bacteria inside it are protected from phagocytosis and antibiotics [54]. We have studied the ability of each type of urethane/urea elastomer to be colonised by *S. epidermidis*, in order to evaluate their possible application in the manufacture of medical indwelling devices [42].

The bacterial strain was provided by Gerald B. Pier (Harvard Medical School, Boston, MA, USA). Trypticase soy broth (TSB) and trypticase soy agar (TSA) plates were prepared according to the manufacturer's instructions. The strain was grown for 24±2 h at 37°C in a shaker rotating at 130 rpm in 15 ml TSB, using as inocula bacteria grown on TSA plates not older than 2 days. Then 50 µl of each cell suspension was transferred to 30 ml fresh TSB, which was incubated for 18±2 h at 37°C and 130 rpm. After being harvested by centrifugation (for 5 min at 9000g and 4°C), cells were washed twice and resuspended in saline solution (0.9% NaCl in distilled water) at a concentration of approximately 109 cells/ml, determined by its optical density at 640 nm. These cell suspensions were used in the subsequent adhesion assays.

Five different types of urethane/urea elastomer films were prepared as described in section 2, in a 60% PU / 40% PBDO weight ratio; these were thoroughly characterised in section 4.2. In the remainder of this section they are denoted "materials" "materials" and numbered 1 to 5; in parentheses we give their names in the notation of section 4.2, where available: material 1 is a non-modified elastomer (sample I0); material 2 is an elastomer extracted in toluene as-cast (sample III-10); material 3 is an elastomer irradiated and then extracted in toluene (sample III-20); material 4 is a elastomer that was UV-irradiated, stretched along the casting shear direction, and then extracted in toluene; finally material 5 is an elastomer that was UV-irradiated, stretched along the casting shear direction, then along the direction perpendicular to the casting shear, and lastly extracted in toluene (sample III-2A).

Film samples were placed in 6-well tissue-culture plates containing 3.5 ml of a suspension of $2x10^9$ cells/ml in saline solution, Initial adhesion to each substrate was allowed to proceed for 2 h at 37°C in a shaker rotating at 130 rpm. Negative controls were obtained by placing squares of elastomer film in a saline solution without bacterial cells. The squares were then gently transferred to 100-ml glass beakers containing distilled water, and were allowed to rest for approximately 10 s, after which they were transferred to a different 100-ml glass containing distilled water, followed by a third transfer 10 s later. These washing steps were carefully performed in order to remove only the cells that were suspended in the liquid and to minimise cell detachment from the film surface [55]. The films with adhered cells were dried at 37°C. All experiments were done in triplicate and repeated four times.

Before image collection, the films were stained with a 0.2% safranin solution or with a 0.01% DAPI (4'-6-diamidino-2-phenylindole) solution, for better image contrast. Direct bacterial counts were done using an epifluorescent microscope coupled to a 3 CCD video camera that acquires images with 820x580-pixel resolution at a magnification of 1000x. With this magnification, 1 cm^2 is equivalent to $1.198x10^4$ captured images (as determined by a Neubauer chamber). Cells were counted using automated enumeration software. Dried samples of films with adhered bacteria were stuck on metal holders with double-sided adhesive tape and gold-coated in a vacuum evaporator. Observations were performed at 10-15 kV with a Leica S360 scanning electron microscope (SEM).

SEM images [42] confirm that the substrates are rough as previously reported [37,38] and that *S. epidermidis* adhered preferentially to the "valleys" on the corrugated surfaces of materials 3-5. This was to be expected, since cells at these locations are more effectively sheltered from shear forces [56]. Fluorescence micrographs [42] show greater bacterial adhesion on the mechanically-modified elastomers (materials 4 and 5). Figure 21 gives the number of *S. epidermidis* cells adhered to each elastomer film; these results were obtained from a number of different experiments and standard deviations are larger for the materials with more irregular surfaces, especially 4 and 5, possibly also because these may exhibit more pronounced uncontrollable variations in surface topography. The number of adhered cells is found to increase from smooth (materials 1 and 2) to periodically modulated (materials 4 and 5) surfaces, in accordance with expectations. However, material 3 does not conform to this trend, which may be due to micro-phase separation of hydrophobic (adhesion-enhancing) and hydrophilic (adhesion-) elastomer blocks.

These results show that the surface roughness is not the only significant mechanism involved in the adhesion process and suggest that micro-phase separation, characteristic of block copolymer elastomers, may play a significant role. The ability to modify the surface properties of our systems by acting on easily accessible experimental parameters thus establishes their usefulness as model surfaces for bacterial adhesion studies.

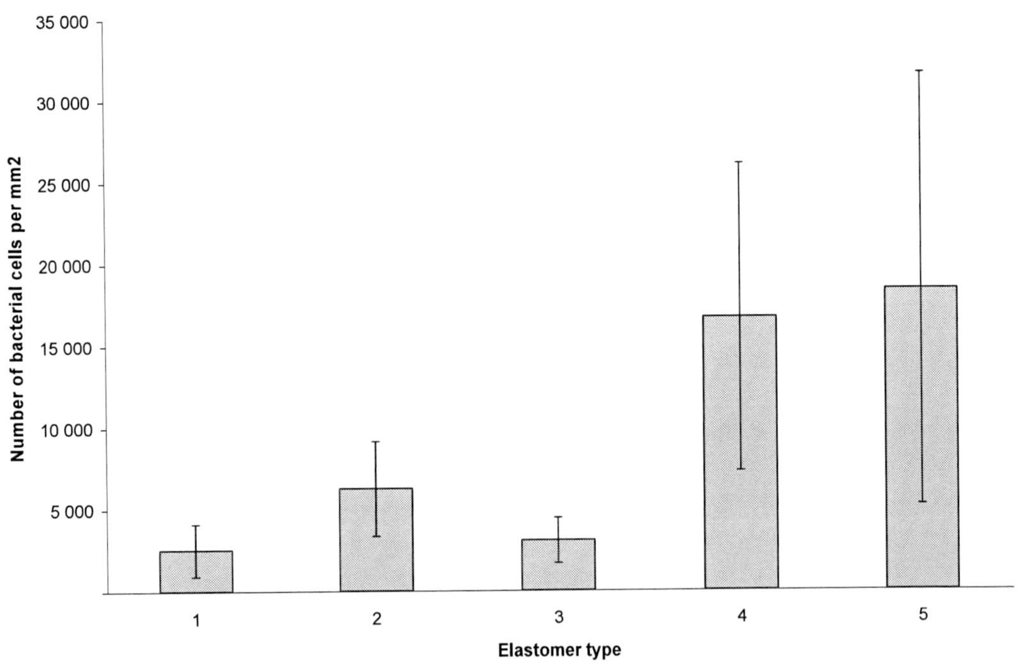

Figure 21. Adhesion of *S. epidermidis* on modified urethane/urea elastomer films. The bars are standard deviations at the 95% confidence level.

7. Conclusions

In this chapter we describe an elastomeric material with different functionalities resulting from the textures generated by UV irradiation of one of its surfaces. The synthesis is straightforward and uses current chemicals and methods; no sophisticated deposition or emulsification steps are necessary. A tuneable topographic system can be obtained from a single formulation of a urethane/urea elastomer upon varying the UV exposure time, the temperature, applying a mechanical field and/or performing solvent extraction. The topographies, with amplitudes in the tens of nm range and periods of a few μm, are remarkably defect-free over distances of the order of a few mm, and can be switched between themselves reversibly, without any cracks appearing. If left unstressed, an imprinted pattern has a lifetime of a few months; it can be erased by heating. We have proposed a very simple model to account for band formation; more detailed aspects of the system, however, still await a proper theoretical treatment. Finally, we have discussed in some detail two potential applications: image recording and retrieval, and "non-stick" substrates for medical implants.

A number of questions remain open. It will be crucial to be able to correlate composition with the degree of phase separation (bulk, microphase, or mixed) in the structures, which then determines their mechanical properties [43]. Our simple mechanical model will have to be extended, possibly to include anisotropy and viscoelasticity. And the detailed mechanism whereby swelling alters textures will also need to be elucidated.

A related system which appears promising is a ferrogel, obtained by doping the elastomer with ferromagnetic nanoparticles [57,58]. These have been shown to elongate or contract under the action of a magnetic field gradient [59], which property may lead to many different applications, from soft actuators and micromanipulators to artificial muscles or drug carriers [29,60]. Recently, one of us and co-workers have shown that Fe_3O_4 grains can be efficiently incorporated into urethane/urea elastomers of appropriate compositions [61,62]. The doped elastomers then exhibit enhanced strain birefringence relative to the undoped samples. This work is ongoing.

Acknowledgements

The research described in this chapter was done in collaboration with J. Azeredo, A. M. Deus, J. G. Fonseca, L. V. Melo, R. Oliveira, M. N. de Pinho, M. Simões, P. Teixeira, A. C. Trindade and C.-T. Zhao. We gratefully acknowledge financial support from Projects PRAXIS XXI no. 3/3.1/CEG/2620/95, POCTI/32597/CTM/2000-FEDER, POCTI/CTM/38330/2001-FEDER, POCTI/ CTM/42456/2001-FEDER, POCI/CTM/ 56382/ 2004-FEDER, and from the Portuguese Science Foundation (FCT) through Pluriannual Contracts with CENIMAT and ICEMS. We thank C. Carfagna for helpful discussions, and M. Warner for hospitality and many insightful discussions.

References

[1] C. Koulic and R. Jérôme, Macromolecules 37, 888 (2004).

[2] T. L. Morkved, M. Lu, A. M. Urbas, E. E. Ehrichs, H. M. Jaeger, P. Mansky and T. P. Russel, *Science* 931, 273 (1996).

[3] M. Camacho-Lopez, H. Finkelmann, P. Palffy-Muhoray and M. Shelley, *Nature* Materials 3, 307 (2004).

[4] G. Evmenenko, C. J. Yu, S. Kewalramani and P. Dutta, *Langmuir* 20, 1698 (2004).

[5] N. Bowden, S. Brittain, A. G. Evans, J. W. Hutchinson and G. M. Whitesides, *Nature* 393, 146 (1998).

[6] N. Stutzmann, T.A. Tervoort, K. Bastiaansen and P. Smith, *Nature* 407, 613 (2000).

[7] M. Watanabe, H. Shirai and T. Hirai, *J. Appl. Phys.* 92, 4631 (2002).

[8] S. R. Quake and A. Scherer, *Science* 290, 1536 (2000).

[9] N. A. Peppas and R. Langer, *Science* 263, 1715 (1994).

[10] W. T. S. Huck, N. Bowden, P. Onck, T. Pardoen, J. W. Hutchinson and G. M. Whitesides, *Langmuir* 16, 3497 (2000).

[11] S. P. Lacour, S. Wagner, Z. Huang and Z. Suo, Appl. Phys. Lett. 82, 2404 (2003).

[12] P. J. Yoo and H. H. Lee, *Phys. Rev. Lett.* 91, 154502 (2003).

[13] P. J. Yoo, K.Y. Suh, H. Kang and H. H. Lee, *Phys. Rev. Lett.* 93, 034301 (2004).

[14] J. S. Sharp and R. A. L. Jones, *Phys. Rev.* E 66, 011801 (2002).

[15] J. S. Sharp, D. Vader, J.A. Forrest, M. I. Smith, M. Khomenko and K. Dalnoki-Veress, *Eur. Phys. J.* E 19, 423 (2006).

[16] C. Jiang, S. Markutsya and V. V. Tsukruk, *Adv. Mater.* 16, 157 (2004).

[17] C. Jiang, S. Markutsya, Y. Pikus and V. V. Tsukruk, *Nature Materials* 3, 721 (2004).

[18] See, e.g., I. W. Hamley, *The Physics of Block Copolymers* (Oxford University Press, Oxford, 1998).

[19] M. Hasegawa, T. Ikawa, M. Tsuchimori, O. Watanabe and Y. Kawata, Macromolecules 34, 7471 (2001).

[20] C. Park, J. Yoon and E. L.Thomas, *Polymer* 44, 6725 (2003).

[21] E. A. Moschou, S. F. Peteu, L. G. Bachas, M. J. Madou and S. Daunert, *Chemistry of Materials* 16, 2499 (2004).

[22] M. Park, C. Harrison, P. M. Chaikin, R. A. Register and D. H. Adamson, *Science* 276, 1401 (1997).

[23] L. Leibler, *Macromolecules* 13, 1602 (1980).

[24] A. L. Gamboa, E. J. M. Filipe and P. Brogueira, *Nano Letters* 2, 1083 (2002).

[25] T. Kaneko, K. Yamaoka, Y. Osada and J. P. Gong, Macromolecules 37, 5385 (2004).

[26] J. Fu, B. Luan, X. Yu, Y. Cong, J. Li, C. Pan, Y. Han, Y. Yang and B. Li, *Macromolecules* 37, 976 (2004).

[27] C. Hepburu, in *Polyurethane Elastomers*, 2nd ed. (Elsevier Applied Science, New York, 1992).

[28] J. N. C. VanBogart, A. Lilaonitkul and S. L. Cooper, in *Multiphase Polymers,* ed. by G. M. Ester, *Advances in Chemistry Series* (American Chemical Society, Washington DC, 1979), p. 176.

[29] C.-T. Zhao and M. N. de Pinho, *Polymer* 40, 6089 (1999).

[30] M. H. Godinho, J. L. Figueirinhas, C.-T. Zhao and M. N. de Pinho, *Macromolecules* 33, 7675 (2000).
[31] A. C. Trindade, M. H. Godinho and J. L. Figueirinhas, *Polymer* 45, 5551 (2004).
[32] M. Warner and E. M. Terentjev, *Liquid Crystal Elastomers* (Clarendon Press, Oxford, 2003).
[33] M. H. Godinho, L. V. Melo and P. Brogueira, Mat. Sci. Eng. C-Bio S, 23, 919 (2003).
[34] M. H. Godinho, A. C. Trindade, J. L. Figueirinhas, L. V. Melo and P. Brogueira, *Biomol. Eng.* 24, 97 (2007).
[35] A. C. Trindade, M. H. Godinho, M Simões and J. L. Figueirinhas, *Mol. Cryst. Liq. Cryst.* 412, 93[1703] (2004).
[36] M. H. Godinho, L. V. Melo and P. Brogueira, Mater. *Sci. Eng.* C 23, 919 (2003).
[37] M. H. Godinho, A. C. Trindade, J. L. Figueirinhas, L. V. Melo, P. Brogueira, A. M. Deus and P. I. C. Teixeira, *Eur. Phys.* J E 21, 319 (2006).
[38] M. H. Godinho, J. L. Figueirinhas, C.-T. Zhao and m. N. de Pinho, *Mol. Cryst. Liq. Cryst.* 365, 447 (2001).
[39] M. H. Godinho, A.C. Trindade, J. L. Figueirinhas, D. Vidal, L. V. Melo and P. Brogueira, *Synthetic Metals* 147, 209 (2004).
[40] M. H. Godinho, A. C. Trindade, J. L. Figueirinhas, L. V. Melo and P. *Brogueira, Mol. Cryst. Liq. Cryst.* 437, 53[1297] (2005).
[41] P. Teixeira, A. C. Trindade, M. H. Godinho, J. Azeredo, R. Oliveira and J. G. Fonseca, *J. Biomater. Sci. Polymer Edn.* 17, 239 (2006).
[42] D. P. Queiroz, M. C. Gonçalves and M. N. Pinho, *J. Appl. Polym. Sci.* 103, 315 (2007).
[43] E. Cerda and L. Mahadevan, *Phys. Rev. Lett.* 90, 074302 (2003).
[44] L.D. Landau and L.M. Lifshitz, *Theory of Elasticity*, 3rd ed. (Pergamon Press, New York, 1986).
[45] K. Efimenko, M. Rackaitis, E. Manias, A. Vaziri, L. Mahadevan and J. Genzer, *Nature Materials* 4, 293 (2005).
[46] E. Sharon, B. Roman, M. Marder, G.-S. Shin and H. L. Swinney, *Nature* 419, 579 (2002).
[47] http://www.abaqus.com or http://www.hks.com.
[48] R Huang and Z. Suo, Int. *J. Solids and Structures* 39, 1791 (2002).
[49] R. Huang, *J. Mech. Physics Solids* 53, 63 (2005).
[50] Y. N. Xia and G. M. Whitesides, Angewandte Chemie-*Int. Ed.* 37, 551 (1998).
[51] M. E. Rupp and G. D. Archer, *Clin. Infect. Dis.* 19, 231 (1994).
[52] Y. H. An and R. J. Friedman, *Handbook of Bacterial Adhesion: Principles, Methods and Applications* (Humana Press, Totowa NJ, 1991).
[53] B. D. Hoyle and J. W. Costerton, *Prog. Drug Res.* 37, 91 (1991).
[54] N. Cerca, G. Pier, R. Oliveira and J. Azeredo, *Res. Microbiol.* 55, 755 (2004).
[55] M. Quirynen and C. M. L. Bollen, *J. Clin. Periodontol.* 22, 1 (1995).
[56] M. Zrinyi, L. Barsi and A. Büki, *J. Chem. Phys.* 104, 8750 (1996).
[57] V. T. Lebedev, Gy. Torok, L. Cser, A. L. Buyanov, L. G. Revelskaya, D. N. Orlova and A. I. Sibilev, *J. Magn. Magn. Mater.* 201, 136 (1999).
[58] M. Zrinyi, L. Barsi and A. Büki, Polym. Gels Networks 5, 415 (1997).

[59] S. Bohlius, H. R. Brand and H. Pleiner, *Phys. Rev.* E 70, 061411 (2004).
[60] A. M. Figueiredo Neto, M. H. Godinho, T. Toth-Katona and P. Palffy-Muhoray, *Brazilian J. Phys.* 35, 184 (2005).
[61] C. Sena, M. H. Godinho and A. M. Figueiredo Neto, *J. Appl. Phys.* 102, 073524 (2007).

In: Biomimetic and Supramolecular Systems Research ISBN: 978-1-60456-405-1
Editor: Arturo H. Lima

Chapter VI

Biomimetic Apatite Coatings – A Review

Mei Wei[4] and Haibo Qu
Materials Science and Engineering Program, Department of Chemical, Materials, and Biomolecular Engineering, University of Connecticut, USA

Apatite is the main inorganic component of natural bone. It is both bioactive and osteoconductive. Apatite has been used extensively as a coating on the surface of bioinert materials, especially metals, to enhance the bioactivity of these materials. There are many coating techniques available for applying apatite onto different substrates, including plasma spraying [1], chemical vapor deposition (CVD) [2], sputtering coating [3], sol-gel coating [4], electrochemical deposition [5], electrophoresis [6], and biomimetic coating [7]. Among these, plasma spraying is the most widely used technique for applying apatite coating on metal substrates. However, the high temperature employed in the coating process can result in decomposition and reduced crystallinity of apatite coating, and in situ incorporation of drugs and proteins becomes impossible with this coating technique. Although the substrate is remained relatively cool (approximately 200°C) during plasma spraying, the temperature applied is still too high for most polymeric substrates.

In the 1980s, Professor Kokubo from the Kyoto University and his colleagues proposed a biomimetic method to form a bone-like apatite layer at a physiological temperature and pH using simulated body fluid (SBF) [8]. SBF was initially used to assess the bioactivity of different biomaterials. It was found that bioactive materials could induce apatite formation, but bioinert materials could not [9]. Later, the biomimetic method was more widely used as a coating technique to apply bone-like apatite onto different biomaterials. However, a coating could only be formed on the surface of bioactive materials, such as silica gel [10], titania gel [11], tantalum [12], and alkaline-treated titanium and its alloys [13]. It is generally believed that certain surface functional groups are responsible for the apatite formation on these materials. A recent study has evaluated the effect of the four most promising surface

[4] Corresponding author: Dr. Mei Wei, Tel: (860)486-9253, Fax: (860)486-4745; Email: m.wei@ims.uconn.edu.

functional groups, including -OH, $-SO_3H$, $-PO_4H_2$ and –COOH, on apatite formation, and found that carboxyl end groups yield more apatite deposition than any other functional groups [14]. In order to form an apatite coating on relatively inert materials, a nucleation aid, normally $CaO-SiO_2$-based glass, has been used. First, the substrate is placed in contact with glass particles in SBF thus forming apatite nuclei on its surface. The substrate is then transferred to another SBF with relatively higher calcium and phosphate concentrations (e.g. 1.5×SBF) to allow the further growth from the apatite nuclei in situ. Using this approach, bone-like apatite has been successfully deposited onto the surface of polymers, alumina and 316 stainless steel to name but a few materials [15-17]. An alternative approach to form apatite coating on inert substrates is the utilization of concentrated SBF alone. Not only does apatite form on the surface of inert materials, but also its formation rate is significantly improved by using concentrated SBF. In recent years, biomimetic apatite coating has been applied to the surface of bioinert tissue-engineered scaffolds to enhance the osteogenic potential of the material [18, 19]. Homogenous apatite coatings have been obtained on different types of scaffolds, such as titanium foam [18] and poly(D,L-lactic-co-glycolic acid) (PLGA) scaffold [19]. These coatings have been found to support osteoblast attachment and proliferation, which are important properties of bone regenerative materials.

Calcium and Phosphate Effects

Original SBF (1×SBF) has relatively low calcium and phosphate ion concentrations, 2.5 mM and 1.0 mM, respectively. It normally takes 7-28 days to form an apatite coating with a thickness of 5~30 μm [7, 20, 21]. To make the coating method more cost-effective, an SBF with higher ion concentrations, such as 1.5× [22, 23], 2× [24], 2.5× [25], 5× [26, 27] and even 10× [28], has been employed to expedite the coating process. Table 1 summarizes typical biomimetic conditions used to expedite the apatite coating formation.

The apatite coating growth rate has not been substantially improved by using 1.5×SBF. It still takes 14 days to deposit a homogeneous coating on substrates [22]. However, the apatite coating forming rate has increased substantially by using 2.5×SBF. A 2-3 μm apatite coating is formed within 3 days of soaking [25]. The apatite coating formation rate can be further enhanced by using 5×SBF or even higher calcium- and phosphate-containing SBF.

Table 1. Typical biomimetic coating conditions (concentration, temperature and pH) used to expedite the apatite coating formation [30]. (The effect of temperature and initial pH on biomimetic apatite coating. Copyright (Journal of Biomedical Materials Research Part B: Applied Biomaterials)) Reprinted with permission of John Wiley & Sons, Inc.

Author/Group	SBF	Temperature	pH	Reference
Tas *et al*	10×	22°C	6.25	[28]
Li	2.5×	45°C	"neutral"	[25]
Habibovic *et al*	5×	50°C	5.7	[29]
Lin *et al*	1.5×	80°C	7.4	[17]

By using 5×SBF, an apatite coating over 30 μm is formed within 24 hours [29]. For 10×SBF, an apatite coating of a thickness up to 60 μm is formed within 6 hours [28]. It has been reported that there is no difference in bonding strength for coatings formed in 10× and 1.5×SBF [28]. However, careful study also revealed that increasing concentration of SBF may result in higher Ca/P ratio and carbonate content in the apatite [24]. Therefore, it may be necessary to adjust the initial Ca/P ratio and other ions concentrations in a highly concentrated SBF to render an apatite coating which resembles the chemical composition of natural bone.

Besides Ca^{2+} and PO_4^{3-}, the effect of other ions in SBF, such as Mg^{2+}, HCO_3^-, Na^+, and Cl^-, on apatite formation has also been explored. It has been revealed that these ions significantly affect the apatite formation and its crystal growth. In addition, the protein content in SBF, the pH of the solution and the operating temperature also play important roles in controlling the quality of apatite coatings.

Magnesium Effect

Magnesium is an essential cation in biological systems. The typical concentration of magnesium ions in human blood plasma is within the range of 0.7-0.9 mM [31, 32]. Magnesium is also one of the most abundant trace ions in the human hard tissues, such as dentine, enamel and bone. It is generally accepted that bone apatite has a trace of (0.6%) magnesium substitution [33].

It is believed that magnesium ions inhibit the formation and growth of calcium phosphate phases, especially apatite [34]. It is also well accepted that magnesium ions can absorb onto apatite nuclei and thereby retard apatite growth [31]. Figure 1 illustrates that the growth rate of different calcium phosphate phases (dicalcium phosphate dehydrate (DCPD), octacalcium phosphate (OCP) and HA) changes as magnesium concentration increases [34]. The growth rate of HA is suppressed more when magnesium concentration increases compared with other calcium phosphate phases. Figure 2 suggests that not only does Mg^{2+} retard apatite formation but also slows down apatite growth rate [35].

Although magnesium can inhibit apatite growth, it is an important component for apatite coating formation. Attempts have been made to prepare apatite coatings using magnesium-free 5×SBF [36]. It was found that no apatite coating was formed on the titanium substrate. Instead, only apatite precipitation in SBF solution was obtained. X-ray photoelectron spectroscopy (XPS) study on the apatite coating formed in 5×SBF with magnesium ions indicates that the Mg^{2+} content at the interface between the apatite coating and titanium substrate is much higher than any other sites [36]. This result suggests that the absorption of Mg^{2+} onto substrates is important for apatite coating initiation and its bonding to the substrate. Also, during the apatite coating growth, the Mg^{2+} can incorporate into the apatite structure to form poorly crystallized apatite. X-ray diffraction results show that when the Mg^{2+} concentration increases, the apatite phase is more amorphous. These studies have collectively suggested that not only does the Mg^{2+} slow down the apatite growth, but it also helps the initiation of the apatite nucleation on substrates. Due to the effect of Mg^{2+} on apatite nucleation, Mg^{2+}-free calcium phosphate solution (SCP) was used to produce apatite coating on apatite pre-coated substrates [37].

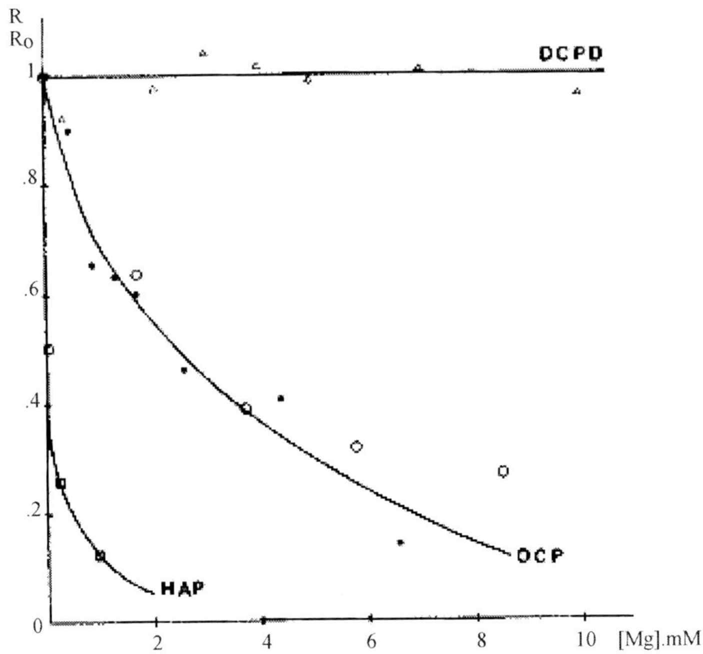

Figure 1. Normalize the growth rate of DCPD, OCP and apatite in the presence of magnesium ions. R and R_o are the growth rates in the presence or absence of magnesium ions, respectively. Reprinted with permission from [34]. Copyright (1985) American Chemical Society.

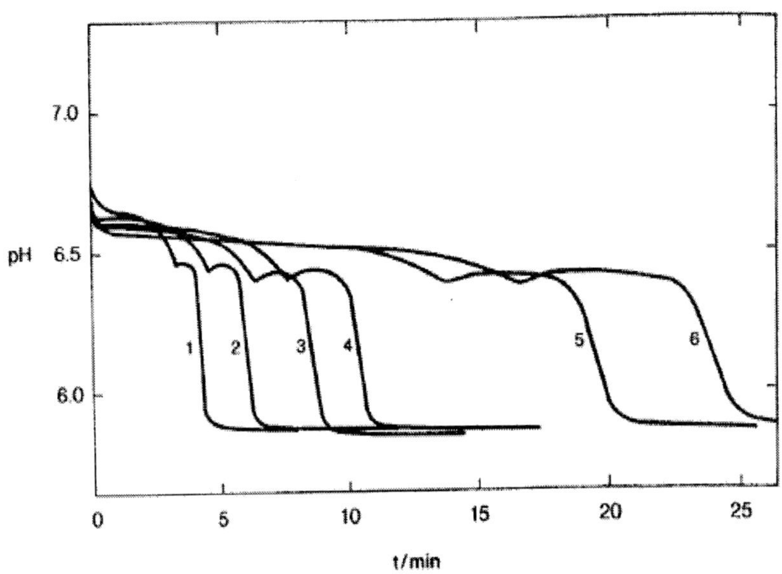

Figure 2. pH curve against time for calcium phosphate precipitates with different Mg/Ca ratios. The Mg/Ca ratio is (1) 0, (2) 0.04, (3) 0.08, (4) 0.16, (5) 0.20, and (6) 0.24, respectively [35]. Reproduced by courtesy of Elsevier limited.

Bicarbonate Effect

It has been reported that the average sized adult produces about 200 ml of carbon dioxide as the end-product of metabolism every minute at rest [38]. According to Henry's law, some of the CO_2 is dissolved in the body fluid to form carbonic acid. This carbonic acid can further dissociate into bicarbonate ions and hydrogen ions. These reactions also provide the body fluid with a buffering capacity.

$$CO_2 + H_2O \longleftrightarrow H_2CO_3 \longleftrightarrow HCO_3^- + H^+ \quad (1)$$

A precise analysis reveals that there are 1.2 mM CO_2, 0.005 mM H_2CO_3 and 23 mM bicarbonate ions (HCO_3^-) in the human body fluid [38]. At biological conditions, carbonate ions can be easily incorporated into the apatite lattice structure. They either substitute phosphate groups to form type-B carbonated apatite or hydroxyl groups to form type-A carbonated apatite. Although the relative amount of carbonation in apatite varies with age, the carbonate content is generally about 3–8 wt% in human bone mineral [39].

The presence of bicarbonate ions suppresses OCP and DCPD growth and favors the type-B carbonate apatite formation [26]. Studies have shown that DCPD precipitation can be obtained from 5×SBF in the absence of bicarbonate, and that very tiny calcium phosphate particles have been found to deposit on the titanium substrate (Figures 3) [26]. When bicarbonate is present in the 5×SBF, a think carbonated-apatite coating with a low crystallinity is formed (Figure 4).

The carbonate in SBF contributes to the formation of carbonated apatite – a defective apatite structure. A high HCO_3^- content in SBF can effectively slow down apatite growth, but it in turn renders a dense and strong apatite coating [26]. Table 2 shows that the density and bonding strength of carbonated apatite coating increase with the HCO_3^- concentration [40]. Bonding strength tests have revealed that the apatite coating tends to fail at the coating/substrate interface instead of within the apatite coating as bicarbonate content increases. Thus, it can be concluded that the bonding strength of apatite coating is improved by increasing the HCO_3^- /calcium ratio in the SBF solution at the expense of growth rate. It is worth noting that the HCO_3^-/calcium ratio in human blood plasma is 10.8, which is about 4 times higher than the highest HCO_3^-/calcium ratio used in the study above (2.76) [40]. Thus, the apatite formed in natural bone should be even denser than the coatings attained in the above study [40].

Taking the above study further, one can envisage using biocarbonate to tailor the coating density/morphology. It is known that the bicarbonate in SBF naturally decomposes into CO_2 and OH^-, as shown in reaction (2) [41].

$$HCO_3^- \rightarrow CO_2 + OH^- \quad (2)$$

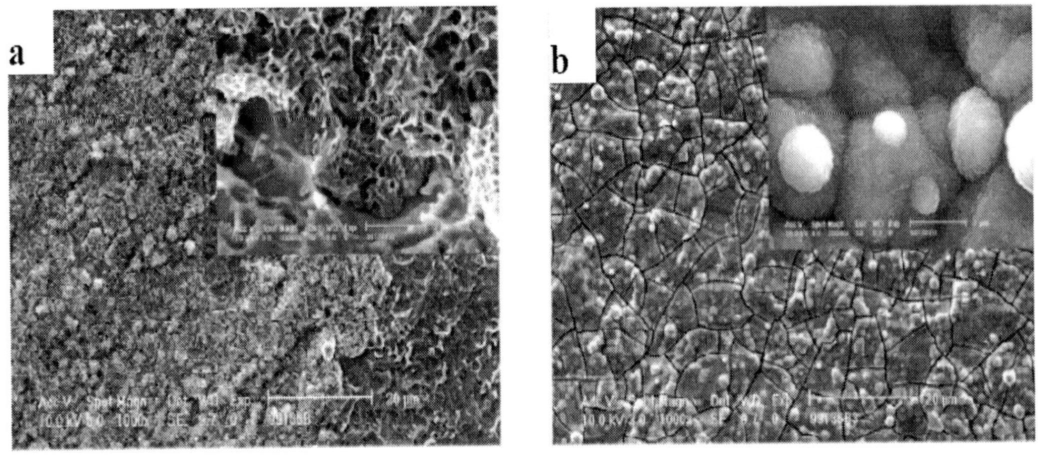

Figure 3. SEM images of (a) the coating formed on metal substrates in the absence of bicarbonate ions, and (b) coating formed on metal substrates with the presence of bicarbonate ions. Only tiny calcium phosphate particles deposit on the titanium substrate in the absence of bicarbonate ions while dense coating was formed in the SBF with the presence of bicarbonate ions [26]. Reproduced by courtesy of Elsevier limited.

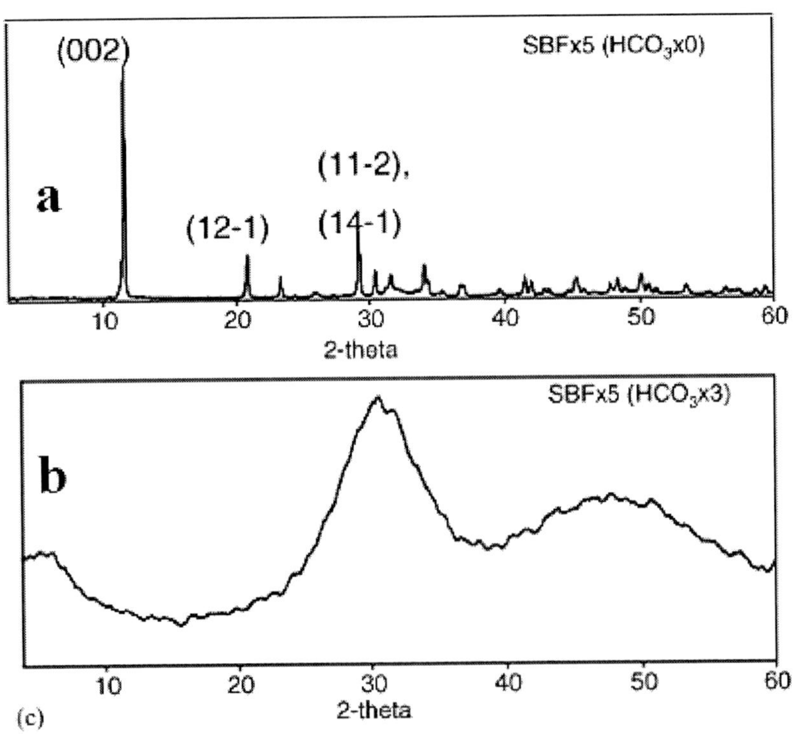

Figure 4. XRD patterns of (a) the precipitate formed in the SBF without bicarbonate ions and (b) the coating formed on metal substrates with bicarbonate ions. A highly crystallized DCPD phase with sharp peaks was obtained in an SBF in the absence of bicarbonate ions. The coating formed in the presence of bicarbonate ions has a low crystallinity [26]. Reproduced by courtesy of Elsevier limited.

Table 2. Final HCO_3^- content, density and bonding strength of different apatite coatings formed in 5×SBF. It can be found that the density and bonding strength of apatite coating increase as the final HCO_3^- content in the SBF increases [40]

	Coating A	Coating B	Coating C
Final HCO_3^- concentration (mM)	8.2	11.5	33.5
Density (% of apatite theoretical density)	67%	80%	85%
Bonding strength (MPa)	8.52	10.36	17.68

Therefore, in an open SBF system, the HCO_3^- concentration decreases as soaking time increases. This leads to the formation of a dense coating adjacent to the substrate and a porous coating on the surface. As bicarbonate ions gradually decompose and bicarbonate concentration gradually decreases in SBF, an apatite coating with a gradient structure is achieved (unpublished results). Figure 5 shows a gradient apatite coating formed on the surface of a polymeric substrate. The coating next to the substrate surface demonstrates the highest density due to the high bicarbonate concentration in the initial coating stage, while the surface of the apatite coating is porous due to the low bicarbonate concentration in SBF in the later stage of coating formation. Thus, it is expected that the dense coating adjacent to the substrate provides a strong bond to the substrate, while the porous surface of the coating can facilitate cell attachment and bone formation in vivo.

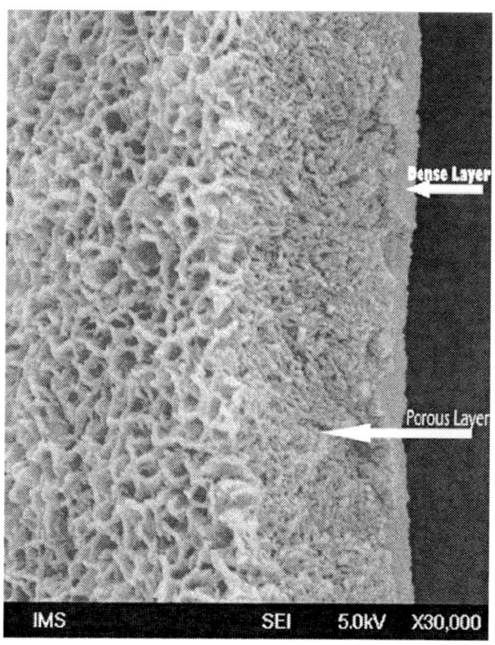

Figure 5. A gradient apatite coating formed on a polymeric substrate. The coating nearest the substrate surface has a greater density due to the high bicarbonate concentration at the initial stage of the coating. The surface of the apatite coating is less dense because of the low bicarbonate concentration in SBF at a later coating formation stage.

Ionic Strength Effect

In aqueous solutions with high ionic strengths, the average number of water molecules surrounding ions is low. With less water molecules surrounding the ions, the average size of hydrated ions (ions with surrounding water molecules) is smaller. During the crystal growth, the hydrated ions move from solution onto substrate surfaces. They are first absorbed onto the substrate surface or crystal nuclei. The size of the hydrated ions with less water molecules is similar to the size of the ions in the crystal lattice. The absorption step is easier for the hydrated ions with less water molecules surrounded (smaller hydrated size). After absorption, the hydrated ions dehydrate and reorganize themselves into crystal structures. Again, it is more favorable for hydrated ions with less water molecules surrounded to dehydrate and reorganize. Therefore the adsorption and incorporation of hydrated ions onto substrate surfaces as well as their dehydration may be facilitated when the crystal growth process takes place in solutions with a higher ionic strength as less water molecules are surrounded the ions [42, 43].

In an SBF solution, Na^+ and Cl^- have the highest ion concentrations and therefore their ionic strengths have a significant impact on apatite coating formation. Barrere *et al* carefully investigated the effect of SBF ionic strength on apatite coating formation by adjusting the content of NaCl in SBF [26]. Three SBF solutions, 5 times SBF (SBF×5) with no NaCl (NaCl×0), 3 times NaCl (NaCl×3) and 5 times NaCl (NaCl×5), was employed to study the effect of ionic strength on apatite coating formation. Homogenous apatite precipitation in the SBF solution occurred in all three solutions. However, in SBF×5 (NaCl×0), apatite precipitation commenced immediately after the completion of preparing SBF, while there was at least a 4 hour delay before precipitation occurred in the SBF with both NaCl×3 and NaCl×5. The early apatite precipitation dramatically lowers the calcium and phosphate supersatuation in SBF, resulting in less calcium and phosphate ions available for apatite coating formation. As a result, the coating formed in SBF×5 (NaCl×0) was much thinner than those formed in SBF×5 (NaCl×3) and SBF×5 (NaCl×5). The substrate underneath the apatite coating formed in SBF×5 (NaCl×0) was still visible after 4 hours of soaking in the coating solution. The correlation between ionic strength (NaCl contents), precipitation formation and coating thickness in 5×SBF is summarized in Table 3. This suggests that high ionic strength delays apatite precipitation in the SBF solution and in turn favors the apatite coating formation. In other words, the homogenous apatite nucleation (both on the substrate surface and in the SBF solution) can readily occur at a low ionic strength, while heterogeneous apatite coating (only at the substrate surface) is more favorable at a high ionic strength.

Protein Effect

Apatite coating can be an effective delivery system due to its low degradation rate and good biocompatibility. In most of the studies using apatite as a carrier material, proteins, such as bone morphogenetic protein (BMP) and transforming growth factor-beta (TGF-β), are simply dropped onto the surface of the apatite coating or scaffold. In these cases, the majority of the protein dissociates from the carrier within the first 48 hours in a so-called "burst" release, which is not effective for bone regeneration *in vivo*.

Table 3. Kinetics of precipitation and coating formation in 5×SBF with various NaCl contents (ionic strengths) [26]

	Precipitation Time	Coating thickness
SBF×5 (NaCl ×0)	0 hours	Very thin (substrate sill visible)
SBF×5 (NaCl×3)	4.5 hours	Thick (substrate invisible)
SBF×5 (NaCl×5)	4.5hours	Thick (substrate invisible)

Researchers have also incorporated various proteins into the apatite coating by simply adding them into SBF solution [37, 44-46]. During the coating formation, the protein molecules are co-precipitated with apatite coating and thus probably incorporated into the apatite structure. Proteins are generally believed to act as apatite forming inhibitors [47, 48]. The apatite coating thickness decreases as the protein concentration in SBF increases. The apatite crystal morphology also changes with protein incorporation. Before protein incorporation, the crystal plates of apatite coating are flat and straight, but they becomes curved and bent after co-precipitated with proteins (Figure 6) [37, 45].

Sustained BMP-2 release has been achieved by co-precipitating BMP with apatite using a biomimetic method [49, 50]. It has been reported that the co-precipitated BMP-2 demonstrated a sustained release profile in a 5-week follow-up study period in an ectopic rat model, and substantial amount of bone formation was observed [50]. By contrast, no bone tissue was found for the group with BMP-2 directly dropped onto the surface of the biomimetic apatite coating. Thus, the approach of co-precipitation of proteins into the biomimetic coating structure has demonstrated a great potential, which can contribute significantly to the osteointegration of dental and orthopedic implants in vivo.

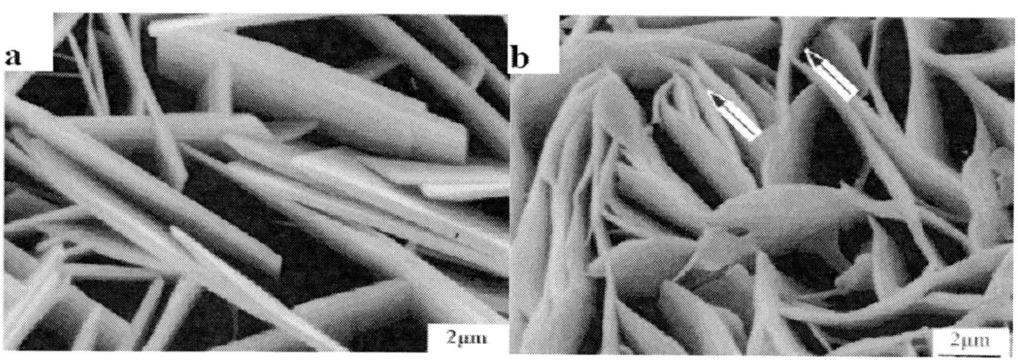

Figure 6. SEM images of apatite coating formed in an SBF a) without and b) with protein (BSA). The apatite crystal plates formed in the SBF without protein are flat and straight, while those formed in the SBF with protein are curved and bent [45]. (Incorporation of bovine serum albumin in calcium phosphate coating on titanium. Vol. 46, No.2, 1999, 245-252, Copyright (1999 and Journal of Biomedical Materials Research)) Reprinted with permission of John Wiley & Sons, Inc.

pH and Temperature Effects

It has also been discovered that the coating conditions, such as pH and temperature, are very important in controlling the apatite coating quality. It has been reported that when calcium and phosphate concentrations are high in SBF, DCPD is most likely to form due to its highest nucleation rates among different Ca-P phases [51]. In order to achieve apatite rather than other Ca-P phases (i.e. DCPD or OCP), the pH and temperature of SBF have to be carefully selected. Simply using body temperature and pH may not produce an optimal apatite coating. It has been reported that the morphology, thickness, composition and crystallinity of the biomimetic apatite coating can be tailored by manipulating the initial pH and temperature of the SBF [30]. During the apatite precipitation or coating growth, the pH of the SBF solution decreases owing to the consumption of OH^- in the solution according to reaction (3) [36, 52].

$$5Ca^{2+} + 3HPO_4^{2-} + 4OH^- \rightarrow Ca_5(PO_4)_3OH + 3H_2O \quad (3)$$

According to reaction (2), the bicarbonate content in SBF decreases due to its natural decomposition. In our recent study, we used 3×SBF to study the effect of temperature and pH on biomimetic apatite coating formation [53, 30]. It has been found that the higher the temperature, the faster the bicarbonate ions decompose. After 12 hours of soaking in SBF, about 90, 70, and 23% of total inorganic carbon remained in SBF when the temperature was held at 24, 40, and 60°C, respectively [30]. Similarly, the pH of the SBF also increases with the temperature based on reaction (2) [30].

It has been found that the apatite coating only forms when the pH of the SBF falls within an appropriate pH range [30]. This pH range shifts downward as the temperature increases. SBF solutions with three different initial pHs were prepared in our previous study [30]. The pH of each SBF can be brought up to the apatite formation range via bicarbonate ion decomposition. Once the pH of SBF reaches the apatite formation range, the pH-time curve starts to bend and the degree of bending is determined by the apatite formation rate (Figures 7) [30]. When the pH value in SBF exceeds the apatite formation range, apatite nucleation not only occurs on the substrate surface but also in the SBF solution. As a result, a sharp decrease of pH was observed for SBF solutions with a high initial pH value at all temperatures studied (24, 40, and 60°C). This is due to the high OH^- consumption rate resulting from both the apatite coating formation on the substrate surface and the apatite precipitation in the SBF solution. Only a thin coating is generated at a high temperature as many of the calcium ions in the SBF have contributed to the precipitation formation in the solution.

Due to the difference in initial pH values, it takes different amounts of bicarbonate ions to decompose to reach the apatite formation pH range. Figure 7 shows that at 40°C it takes approximately 5, 9, and 13 hours to reach the critical pH range for the SBF with a high, medium and low initial pH, respectively [30]. As a result, the remaining bicarbonate ion concentration in the SBF increases with the initial pH of the solution. It is known that the bicarbonate ion concentration has a great influence on the surface morphology of apatite coatings [26]. The higher the bicarbonate ion concentration in the SBF, the denser the resulting coating is. The FESEM images (Figure 8) of coatings produced at 40°C reveal that the apatite coating formed in an SBF with a low initial pH is more porous than the coating formed in an SBF with a high or medium initial pH.

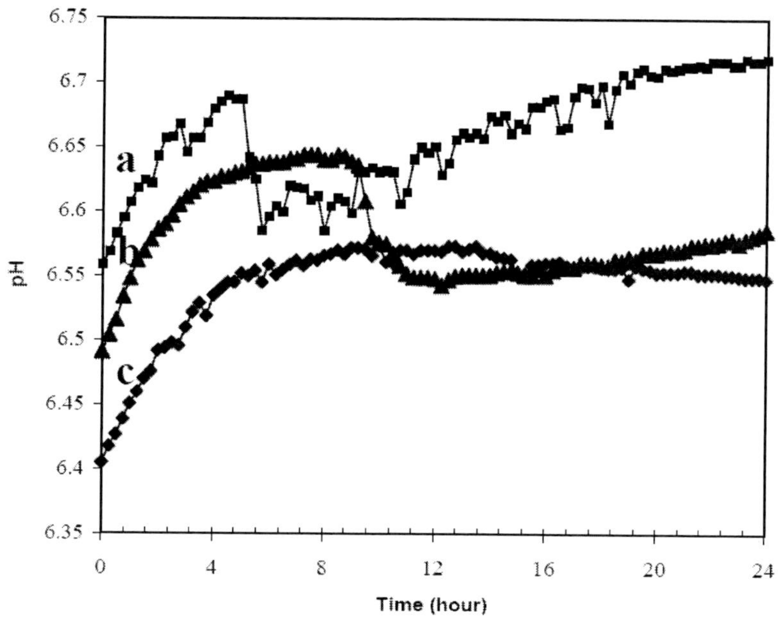

Figure 7. pH versus immersion time of SBF solution with a (a) high initial pH, (b) medium initial pH and (c) low initial pH at 40°C. The SBF with a high initial pH value startes to form colloidal precipitates in the solution after 4.5 hr of soaking. The rest two solutions remain stable and clear throughout the whole experiment. The pH of all SBF solutions increases at an initial rate of 0.028~0.032 pH unit/hr. It then deviates based on different apatite formation scenarios [30]. (The effect of temperature and initial pH on biomimetic apatite coating. Copyright (Journal of Biomedical Materials Research Part B: Applied Biomaterials)) Reprinted with permission of John Wiley & Sons, Inc.

At a high temperature (60°C), apatite coatings with plate-shaped crystals are formed in SBFs with both high and medium initial pHs (Figure 9), while ball-shaped apatite crystals are developed in an SBF with a low initial pH [30]. The FTIR analysis of the apatite coating formed at 60°C demonstrates that coatings formed in SBF at different initial pH values have different CO_3^{2-} contents [30]. Carbonate (CO_3^{2-}) groups were observed for apatite coating formed at 60°C in the SBF with a high initial pH, but they were not detected in the coatings formed in the SBF with either a medium or a low initial pH at 60°C. This is mainly due to that the SBF with a higher initial pH requires a shorter time to reach the apatite formation pH range. When the apatite coating is formed, there is sufficient CO_3^{2-} retained in the SBF solution resulting in the formation of a CO_3^{2-}-rich coating. In comparison, the SBF with a lower initial pH takes a longer time to reach the critical pH range. Most of the CO_3^{2-} in the SBF solution may have decomposed and released in the form of CO_2 before the apatite coating forms. In addition, the fast bicarbonate ion decomposition rate at 60°C also results in less or no CO_3^{2-} content in the apatite coating formed at the later stage of the coating process.

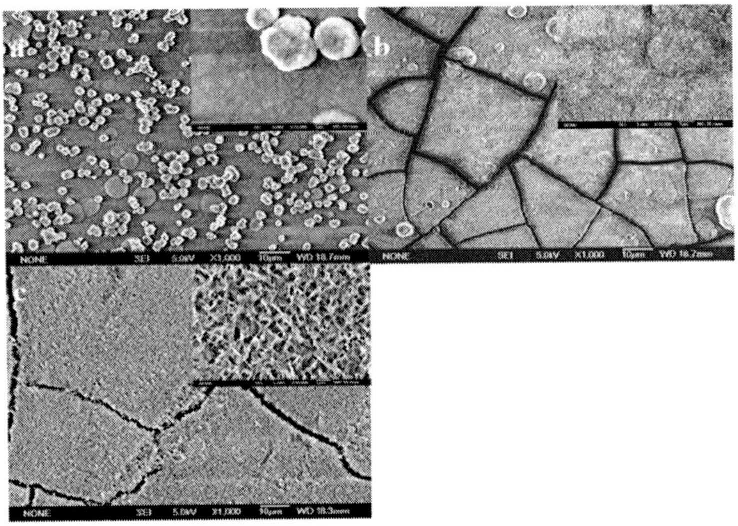

Figure 8. FESEM images (at magnifications 1000× and 10000×) of apatite coating formed at SBF at a (a) high initial pH, (b) medium initial pH and (c) low initial pH at 40°C. The apatite coating formed in the SBF with a medium or high initial pH is dense, while that formed in the SBF with a low initial pH is porous [30].(The effect of temperature and initial pH on biomimetic apatite coating. Copyright (Journal of Biomedical Materials Research Part B: Applied Biomaterials)) Reprinted with permission of John Wiley & Sons, Inc.

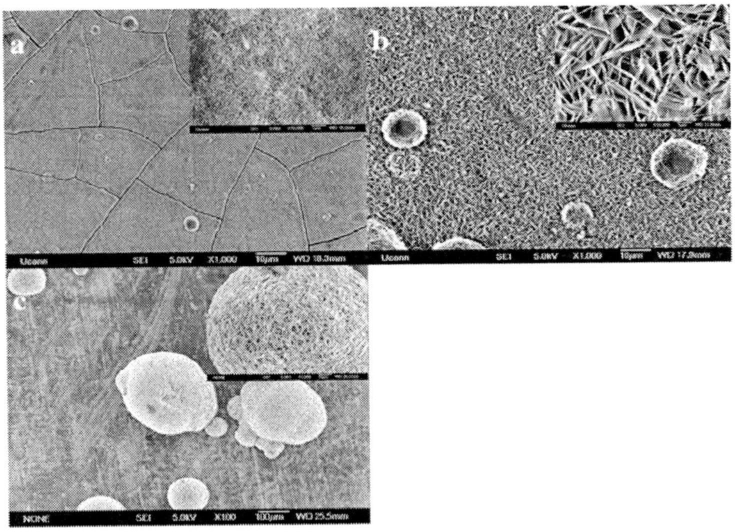

Figure 9. FESEM images (at magnifications 1000× and 10000×) of apatite coating formed in an SBF with a (a) high initial pH, (b) medium initial pH and (c) low initial pH at 60°C. Apatite coating is uniformly deposited on substrates in SBF with both medium and high initial pHs. The apatite coating formed in SBF with a high initial pH is dense but with cracks, while the coating formed in the SBF with a medium initial pH was porous and rough composing of ball-shaped particles around 10 μm. By contrast, instead of forming continuous coating, ball-shaped apatite particles with a diameter in a range 50-200 μm are deposited on the substrate in the SBF with a low initial pH [30]. (The effect of temperature and initial pH on biomimetic apatite coating. Copyright (Journal of Biomedical Materials Research Part B: Applied Biomaterials)) Reprinted with permission of John Wiley & Sons, Inc.

In summary, the biomimetic coating technique has been developed rapidly in the last two decades. It is no longer limited to produce the coating at body temperature and pH. Varied coating conditions, including altered temperatures, compositions, initial pHs, ion concentrations and buffer solutions, have been attempted and a better understanding of the coating system has been gained. A multi-functional apatite coating with a fast-forming rate, tailored surface morphology, strong bonding strength to the substrate, capability of delivering proteins, and good osteoconductivity and osteoinductivity has been developed. Such an apatite coating will be suitable for a wider range of biomedical applications.

Acknowledgement

The authors would like to thank the supports from National Science Foundation (DMI 0500269 and BES 0503315) and Connecticut Innovations under the Yankee Ingenuity Technology Competition.

References

[1] Ong JL, Appleford M, Oh S, Yang Y, Chen WH, Bumgardner JD, et al. The characterization and development of bioactive hydroxyapatite coatings. *JOM Journal of the Minerals, Metals and Materials Society* 2006;58(7):67-69.

[2] Liu X, Chu PK, Ding C. Formation of apatite on hydrogenated amorphous silicon (a-Si:H) film deposited by plasma-enhanced chemical vapor deposition. *Materials Chemistry and Physics* 2007;101(1):124-128.

[3] Ozeki K, Aoki H, Fukui Y. Dissolution behavior and in vitro evaluation of sputtered hydroxyapatite films subject to a low temperature hydrothermal treatment. *Journal of Biomedical Materials Research Part A* 2006;76A(3):605-613.

[4] Ben-Nissan B, Choi AH. Sol-gel production of bioactive nanocoatings for medical applications. Part 1: an introduction. *Nanomedicine* 2006;1(3):311-319.

[5] Peng P, Kumar S, Voelcker NH, Szili E, Smart RSC, Griesser HJ. Thin calcium phosphate coatings on titanium by electrochemical deposition in modified simulated body fluid. *Journal of Biomedical Materials Research Part A* 2006;76A(2):347-355.

[6] Wei M, Ruys AJ, Milthorpe BK, Sorrell CC. Precipitation of hydroxyapatite nanoparticles: Effects of precipitation method on electrophoretic deposition. *Journal of Materials Science: Materials in Medicine* 2005;16(4):319-324.

[7] Wei M, Uchida M, Kim H-M, Kokubo T, Nakamura T. Apatite-forming ability of CaO-containing titania. *Biomaterials* 2002;23(1):167-172.

[8] Kokubo T, Hayashi T, Sakka S, Kitsugi T, Yamamuro T. Bonding Between Bioactive Glasses, Glass-Ceramics or Ceramics in a Simulated Body Fluid. *Journal of the Ceramic Society of Japan* 1987;95(8):785-791.

[9] Li P, Ohtsuki C, Kokubo T, Nakanishi K, Soga N, Groot Kd. The role of hydrated silica, titania, and alumina in inducing apatite on implants. *Journal of Biomedical Materials Research* 1994;28(1):7-15.

[10] Li P, Ohtsuki C, Kokubo T, Nakanishi K, Soga N, Nakamura T, et al. Apatite Formation Induced by Silica Gel in a Simulated Body Fluid. *Journal of the American Ceramic Society* 1992;75(8):2094-2097.

[11] Li P, Kangasniemi I, Groot K, Kokubo T. Bonelike Hydroxyapatite Induction by a Gel-Derived Titania on a Titanium Substrate. *Journal of the American Ceramic Society* 1994;77(5):1307-1312.

[12] Miyazaki T, Kim HM, Kokubo T, Ohtsuki C, Kato H, Nakamura T. Enhancement of bonding strength by graded structure at interface between apatite layer and bioactive tantalum metal. *Journal of Materials Science: Materials in Medicine* 2002;13(7):651-655.

[13] Wei M, Kim HM, Kokubo T, Evans JH. Optimising the bioactivity of alkaline-treated titanium alloy. *Materials Science and Engineering*: C 2002;20(1-2):125-134.

[14] Liu DP, Majewski P, O'Neill BK, Ngothai Y, Colby CB. The optimal SAM surface functional group for producing a biomimetic HA coating on Ti. *Journal of Biomedical Materials Research Part A* 2006;77A(4):763-772.

[15] Tanahashi M, Yao T, Kokubo T, Minoda M, Miyamoto T, Nakamura T, et al. Apatite Coating on Organic Polymers by a Biomimetic Process. *Journal of the American Ceramic Society* 1994;77(11):2805-2808.

[16] Abe Y, Kokubo T, Yamamuro T. Apatite coating on ceramics, metals and polymers utilizing a biological process. *Journal of Materials Science: Materials in Medicine* 1990;1(4):233-238.

[17] Lin F-H, Hsu Y-S, Lin S-H, Sun J-S. The effect of Ca/P concentration and temperature of simulated body fluid on the growth of hydroxyapatite coating on alkali-treated 316L stainless steel. *Biomaterials* 2002;23(19):4029-4038.

[18] Jalota S, Bhaduri SB, Tas AC. Osteoblast proliferation on neat and apatite-like calcium phosphate-coated titanium foam scaffolds. *Materials Science and Engineering: C* 2007;27(3):432-440.

[19] Li J, Beaussart A, Chen Y, Mak FT. Transfer of apatite coating from porogens to scaffolds: Uniform apatite coating within porous poly(DL-lactic-co-glycolic acid) scaffold in vitro. *Journal of Biomedical Materials Research* 2007;80A(1):226-233.

[20] Li P, Ducheyne P. Quasi-biological apatite film induced by titanium in a simulated body fluid. *Journal of Biomedical Materials Research* 1998;41(3):341-348.

[21] Gou Z, Chang J, Zhai W. Preparation and characterization of novel bioactive dicalcium silicate ceramics. *Journal of the European Ceramic Society* 2005;25(9):1507-1514.

[22] Takeuchi A, Ohtsuki C, Miyazaki T, Tanaka H, Yamazaki M, Tanihara M. Deposition of bone-like apatite on silk fiber in a solution that mimics extracellular fluid. *Journal of Biomedical Materials Research Part A* 2003;65A(2):283-289.

[23] Bigi A, Boanini E, Bracci B, Facchini A, Panzavolta S, Segatti F, et al. Nanocrystalline hydroxyapatite coatings on titanium: a new fast biomimetic method. *Biomaterials* 2005;26(19):4085-4089.

[24] Dorozhkina EI, Dorozhkin SV. Structure and properties of the precipitates formed from condensed solutions of the revised simulated body fluid. *Journal of Biomedical Materials Research Part A* 2003;67A(2):578-581.

[25] Li P. Biomimetic nano-apatite coating capable of promoting bone ingrowth. *Journal of Biomedical Materials Research Part A* 2003;66A(1):79-85.

[26] Barrere F, van Blitterswijk CA, de Groot K, Layrolle P. Influence of ionic strength and carbonate on the Ca-P coating formation from SBFx5 solution. *Biomaterials* 2002;23(9):1921-1930.

[27] Chou Y-F, Chiou W-A, Xu Y, Dunn JCY, Wu BM. The effect of pH on the structural evolution of accelerated biomimetic apatite. *Biomaterials* 2004;25(22):5323-5331.

[28] Tas AC, Bhaduri SB. Rapid coating of Ti6Al4V at room temperature with a calcium phosphate solution similar to 10 × simulated body fluid. *J. Mater Res.* 2004;19(9):2743.

[29] Habibovic P, Barrere F, Blitterswijk CA, Groot K, Layrolle P. Biomimetic Hydroxyapatite Coating on Metal Implants. *Journal of the American Ceramic Society* 2002;85(3):517-522.

[30] Qu H, Wei M. The effect of temperature and initial pH on biomimetic apatite coating. *Journal of Biomedical Materials Research Part B: Applied Biomaterials,* accepted.

[31] TenHuisen KS, Brown PW. Effects of magnesium on the formation of calcium-deficient hydroxyapatite from $CaHPO_4 \cdot 2H_2O$ and $Ca_4(PO_4)_2O$. *Journal of Biomedical Materials Research* 1997;36(3):306-314.

[32] Rosolova H, Mayer O, Reaven G. Effect of Variations in Plasma Magnesium Concentration on Resistance to Insulin-Mediated Glucose Disposal in Nondiabetic Subjects 1. *Endocrine Soc*, 1997. p. 3783-3785.

[33] Mayer I, Schlam R, Featherstone JDB. Magnesium-containing carbonate apatites. *Journal of Inorganic Biochemistry* 1997;66(1):1-6.

[34] Salimi MH, Heughebaert JC, Nancollas GH. Crystal growth of calcium phosphates in the presence of magnesium ions. *Langmuir* 1985;1(1):119-122.

[35] Kibalczyc W, Christoffersen J, Christoffersen MR, Zielenkiewicz W. The effect of magnesium ions on the precipitation of calcium phosphates. *Journal of crystal growth* 1990;106(2-3):355-366.

[36] Barrere F, van Blitterswijk CA, de Groot K, Layrolle P. Nucleation of biomimetic Ca-P coatings on Ti6Al4V from a SBFx5 solution: influence of magnesium. *Biomaterials* 2002;23(10):2211-2220.

[37] Liu Y, Layrolle P, Bruijn Jd, Blitterswijk Cv, Groot Kd. Biomimetic coprecipitation of calcium phosphate and bovine serum albumin on titanium alloy. *Journal of Biomedical Materials Research* 2001;57(3):327-335.

[38] Kruse JA. Calculation of plasma bicarbonate concentration versus measurement of serum CO_2 content. pK' revisited. *Clinical Intensive Care* 1995;6(1):15-15.

[39] Landi E, Tampieri A, Celotti G, Langenati R, Sandri M, Sprio S. Nucleation of biomimetic apatite in synthetic body fluids: Dense and porous scaffold development. *Biomaterials* 2005;26(16):2835-2845.

[40] Qu H, Wei M. Improvement of bonding strength between biomimetic apatite coating and substrate. *Journal of Biomedical Materials Research Part B: Applied Biomaterials* 2008;84B(2):436-443.
[41] Wen HB, Wijn JRd, Cui FZ, Groot Kd. Preparation of calcium phosphate coatings on titanium implant materials by simple chemistry. *Journal of Biomedical Materials Research* 1998;41(2):227-236.
[42] Zuddas P, Mucci A. Kinetics of Calcite Precipitation from Seawater: II. The Influence of the Ionic Strength. *Geochimica et Cosmochimica Acta* 1998;62(5):757-766.
[43] Lasaga AC. Fundamental approaches in describing mineral dissolution and precipitation rates. *Reviews in Mineralogy and Geochemistry* 1995;31(1):23-86.
[44] Wen HB, Moradian-Oldak J, Fincham AG. Modulation of apatite crystal growth on Bioglass by recombinant amelogenin. *Biomaterials* 1999;20(18):1717-1725.
[45] Wen HB, de Wijn JR, van Blitterswijk CA, de Groot K. Incorporation of bovine serum albumin in calcium phosphate coating on titanium. *Journal of Biomedical Materials Research* 1999;46(2):245-252.
[46] Uchida M, Oyane A, Kim HM, Kokubo T, Ito A. Biomimetic Coating of Laminin-Apatite Composite on Titanium Metal and Its Excellent Cell-Adhesive Properties. *Advanced Materials* 2004;16(13):1071-1074.
[47] Martin RI, Brown PW. Formation of hydroxyapatite in serum. Journal of Materials Science: *Materials in Medicine* 1994;5(2):96-102.
[48] Radin S, Ducheyne P. Effect of serum proteins on solution-induced surface transformations of bioactive ceramics. *Journal of Biomedical Materials Research* 1996;30(3):273-279.
[49] Liu Y, de Groot K, Hunziker EB. BMP-2 liberated from biomimetic implant coatings induces and sustains direct ossification in an ectopic rat model. *Bone* 2005;36(5):745-757.
[50] Liu Y, Hunziker EB, van de Vaal C, de Groot K. Biomimetic coatings vs. collagen sponges as a carrier for BMP-2: a comparison of the osteogenic responses triggered in vivo using an ectopic rat model. *Key Engineering Materials* 2004;254:619-622.
[51] Lu X, Leng Y. Theoretical analysis of calcium phosphate precipitation in simulated body fluid. *Biomaterials* 2005;26(10):1097-1108.
[52] Barrere F, Layrolle P, van Blitterswijk CA, de Groot K. Biomimetic calcium phosphate coatings on Ti6AI4V: a crystal growth study of octacalcium phosphate and inhibition by Mg^{2+} and HCO_3^-. *Bone* 1999;25(2 Suppl):107S-111S.
[53] Qu H, Wei M. The effect of initial pH on morphology of biomimetic apatite coating. *Key Engineering Materials* 2007;330-332:757-760.

In: Biomimetic and Supramolecular Systems Research
Editor: Arturo H. Lima
ISBN: 978-1-60456-405-1

Chapter VII

Mimicking Biomineralization Processes: How Far can We Go ?

Thibaud Coradin and Nadine Nassif
UPMC Univ Paris 06 and CNRS, UMR 7574, Chimie de la Matière Condensée de Paris, F-75005, Paris, France

Abstract

The ability of living organisms to control the structure and morphology of biominerals has attracted a lot of attention over the past 20 years. In parallel to biological studies devoted to *in vivo* processes, so-called "biomimetic" approaches have been widely developed. These approaches aim at reproducing *in vitro* the biomineral morphology and/or structure in biologically-relevant conditions. First steps in this direction have extensively relied on the identification of biomolecules extracted from living organisms and their evaluation in mineralization experiments in diluted solutions. Further insights on biomineralization processes were gained by using biomimetic synthetic molecules. Recently, more accurate model systems that take into account confinement effects and question the role of mineral precursors have been developed. However, in many instances, such biomimetic approaches have so far failed to accurately reproduce the complexity of natural materials. This raises the question whether *in vitro* models will allow the complete elucidation of biomineralization processes.

This commentary discusses the recent achievements, actual limitations and possible perspectives of such biomimetic approaches. After a general introduction aiming at presenting the main concepts involved in current biomineralization mimetic studies, we will illustrate the state-of-art of this very active field of research through the examples of some popular models in this field. We will show that, despite the differences that exist in the associated mineral phases and in the conditions of their formation, common trends in biomimetic strategies and resulting materials can be identified. Finally, we will try to identify to what extent *in vitro* experiments may allow the elucidation of the *in vivo* mechanisms of biomineralization. In parallel, we will also emphasize that an accurate replication of the biomineral features may not be required to obtain synthetic materials useful for future applications.

Introduction

The frontiers between living systems and minerals have a long history of controversies and debates. Extensive efforts of survey and classification undertaken by naturalists revealed striking resemblances between mineral growth patterns and biological forms, as illustrated by the works of S. Leduc and D.W. Thomson [1]. However, the development of molecular biology led to a strict DNA-based (or cell-based) division between living and inert matter. As a result, being mainly extra-cellular components, minerals are absent from most current biology textbooks. Nevertheless, there are two domains of science where minerals and living systems remain strongly linked: medicine and geology. Indeed, the human body elaborates physiological (*e.g.* bone) and pathological (*e.g.* kidney stones) inorganic materials and their process of formation needs to be studied to develop therapeutic tools [2]. In geological science, where the mineral phase is considered as a signature of the former presence of a living system, shells are used for soil dating as well as for climate change monitoring [3]. Moreover, it was recently proposed that the contribution of biological activities to mineral deposition is under-estimated, leading to the development of a novel scientific discipline termed "geomicrobiology" [4].

Therefore, although the division between these two phenomena is not always straightforward, it is useful to establish a distinction between geological mineral deposition, due to chemical and/or physical perturbations, and biomineralization, for which the mineral phase formation is triggered by the presence of a living organism [5]. The term "presence" has to be taken into account carefully because it is currently admitted that two types of biomineralization processes, respectively termed bio(logically)-induced [6] and bio(logically)-controlled [7], should be distinguished. In bio-controlled mineralization, as performed by crustaceans, vertebrates, mollusks and some unicellular algae, mineral formation involves a specific metabolic activity of the living organism and is regulated by a complex macromolecular system termed the organic matrix (Figure 1). In contrast, bio-induced mineralization, corresponds to an incidental mineral deposition following a modification of the chemical conditions (pH, ionic strength, composition) of a given media due to the biological activity of a living system not specialized in mineralization.

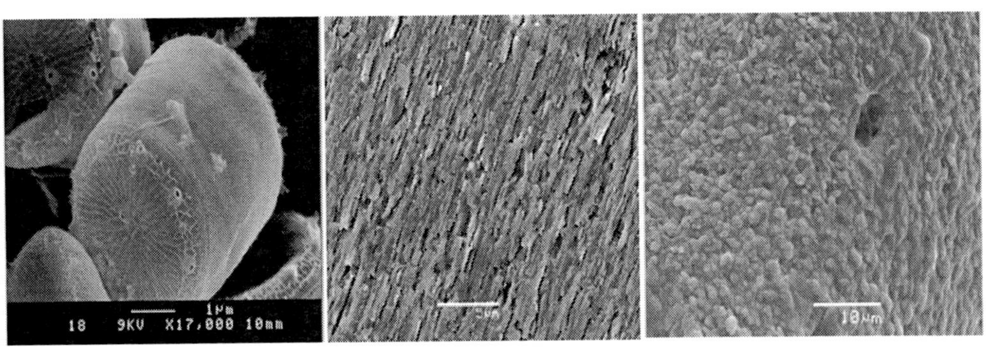

Figure 1. Examples of biomineral structures: Scanning electron micrographs of (left) the shell of *Thalassiosira pseudonana* diatom (Courtesy B. Tesson, Université de Nantes), (middle) the nacreous layer of *Nautilus Pompilius* mollusk shell (Courtesy of A. G. Checa, Universidad de Granada) (right) the compact bone structure of a veal.

For instance, bone formation is bio-controlled by the specific activity of osteoblast cells [8] whereas kidney stone precipitation is bio-induced by the supersaturation of urine fluid with calcium salts, especially calcium oxalate [9].

Once again, the distinction between bio-induced and bio-controlled mineralization may not be easy. In both cases, cells are responsible for the first steps of mineralization due to their ability to synthesize and export chemical agents that trigger mineral deposition. However, physico-chemical processes also seem to play a major role on the progress of the mineralization reaction, regulating not only the organic macromolecular system reactivity but also the inorganic precursor charge and solubility and hence mineral growth. Therefore, the course of the biomineralization reaction, and therefore the features of the deposited material, strongly depends on the relative importance of both biological and physico-chemical processes. In many cases, the implication of a specific biological activity in the formation of a mineral phase is easily suggested after examining the structure, composition and morphology of the deposited material. As an example, the precipitation of $CaCO_3$ from calcium and carbonate ions aqueous solutions at room temperature without the use of organic additives leads to calcium carbonate crystals in its thermodynamically-stable allomorph, *i.e.* calcite, presenting rhombohedral habits. Hence, the presence of $CaCO_3$ hexagonal platelets exhibiting the aragonite structure in the nacreous layer of some mollusks indicates that these crystals are formed under a biological control [10]. Another and maybe more specific feature of bio-controlled minerals is that they are in fact not purely inorganic phases but composite materials, containing part of the organic matter used by the living organism to control the mineral formation. Noticeably, this aspect was long debated for systems containing low fraction of organics (< 1 wt%) but recent developments in analytical methods lead to suggest that it is a common feature of bio-controlled minerals [11].

As stated earlier, the understanding of these biomineralization processes is important for a wide range of research domains related to biological science. In addition, bio-controlled minerals often exhibit specific properties in terms of morphological complexity, mechanical stability or self-repairing ability that still have no equivalent in man-made synthetic materials [12]. Hence, a recent trend in material science is to get inspiration from biomineralization processes to design novel materials [13]. However, a pre-requisite for the elaboration of such "bio-inspired" materials is to get information on the strategies used by living organism to control the formation of mineral phases.

The most intuitive and straightforward approach to understand biomineralization processes is to study the mineralizing living system itself and/or characterize the biomaterial itself. This is mainly achieved *via* microscopy observations (*e.g.* fowl and fish bone [14,15], abalone nacre [16,17], crabs [18] and inverterbrates [19,20]) as well as using structural and chemical characterizations (X-ray diffraction (*e.g.* human bone [21]), microprobe analysis (brachipod shell [22]), magnetic resonance (*e.g.* horse bone [23]),...), on both *in situ* sampled or *in laboratory* grown organisms. In the latter case, it is also possible to modify the growth conditions to evaluate its impact on the mineralization process [24].

Alternatively, huge efforts have been made to identify the main cellular components involved in the mineral deposition reaction *in vivo* and to reproduce in the most accurate manner the conditions of this reaction *in vitro*. This is the scope of this commentary to present the principles and most recent achievements of such "biomimetic" approaches in the field of bio-controlled mineralization. It is shown that pioneer works in this area, focusing on the extraction of putative mineralizing molecules from living organisms and evaluation of their effect on mineral precipitation, have only led to partial, if not non-relevant, information on biomineralization mechanisms. This lack of relevance originates from the omission of essential criteria in the experimental set-up design. In contrast, recent developments that have taken into account the role of confinement of the reaction as well as have re-considered the influence of the nature of mineral precursor provide more accurate data on the conditions of the *in vivo* reaction. As a conclusion, a critical re-evaluation of the principles and goals of biomimetic approaches in biomineralization is presented and some perspectives for further developments in this field are provided.

I. Principles of Biomineralization and Biomimetism

I.1. Bio-Controlled Mineralization

The key actors involved in controlled biomineralization have already been well-identified by the pioneer works of H. Lowenstam and S. Weiner [5] as well as S. Mann [7]. Starting from a more general level, the environment in which organisms are living first determines the availability and nature of inorganic precursors and the temperature at which the mineralization will proceed. Other parameters can directly influence the cell activity (organic content (nutrients), light intensity/oxygen availability (for photosynthesis/respiration processes),...). Inorganic precursors can also originate from the resorption of a former biomineral, as observed for cuticle during crab moult cycle and bone upon osteoclastic activity.

Second the living organism itself, that can be uni- or pluri-cellular, is primarily responsible for the production of the organic components involved in the mineralization control. This production occurs under genetic control and is not only strictly monitored from a chemical point of view (nature, composition and structure of the molecules) but also from a kinetics point of view (matter fluxes). Some living systems can also control the inorganic precursors availability since mineral species in such organisms transit through the cells before reaching the mineralization site [25]. This implies that the mineral species involved in the mineral deposition process may differ from the precursors that were up-taken by the living organisms from the environment. For instance, an ion-by-ion growth mechanism has long been assumed to be responsible for biomineral formation [26]. However, recent reports indicate that amorphous particles may be used by living organisms both for ion storage and transport and may therefore constitute the form of precursor that are directly involved in the biomineral deposition process [27,28].

An important feature of this mineralization site, which can be intra- or extra-cellular, is that it is a spatially-delineated compartment, inside an intracellular vesicle in uni-cellular systems or through an organic matrixin pluri-cellular (Figure 2). In most cases, this organic matrix is confined behind a semi-permeable barrier, such as a lipid membrane and/or an epithelium. This compartmentalization allows the control of the organic and inorganic matter fluxes inside the mineralization site. This barrier may also act as a mold to determine the morphology of the growing mineral at the macroscale. When present, the organic matrix consists of a macromolecular network constraining the spatial expansion of the growing material at a lower scale. Not only does this network influence the mineralization process in many ways but its preservation within the mineral network confers higher mechanical properties to the resulting material thanks to its composite structure. Finally, the mineralization sites also include small organic molecules, mostly described as highly charged macromolecules, that are found intimately associated to the mineral phase. They would interact directly with the inorganic precursors and/or the growing mineral phase and act as structure-directing agents, or templates, to control the crystallographic structure and the morphology of the biomineral. In some cases, specific enzymes, such as alkaline phosphatase or carbonic anhydrase, capable of modifying the structure of the organic molecules, the nature of inorganic precursors or the pH conditions are also found at the mineralization site [29].

In a way, the structural and morphological complexity of biominerals reflects the diversity of scales and strategies used by living systems to control the mineral deposition process. It is therefore not surprising that after many years of investigation focusing on the biology of whole organisms, several aspects of biomineralization processes are still poorly understood. With the increased popularity of reductionist approaches in biology, it has been suggested that instead of studying the mineralization process at the living organism level, fruitful information should be accessible by focusing on phenomena occurring at lower scales, i.e. at a lower degree of complexity. This concept led to the development of biomimetic approaches that aim at mimicking *in vitro* the *in vivo* processes in the most accurate but simple manner.

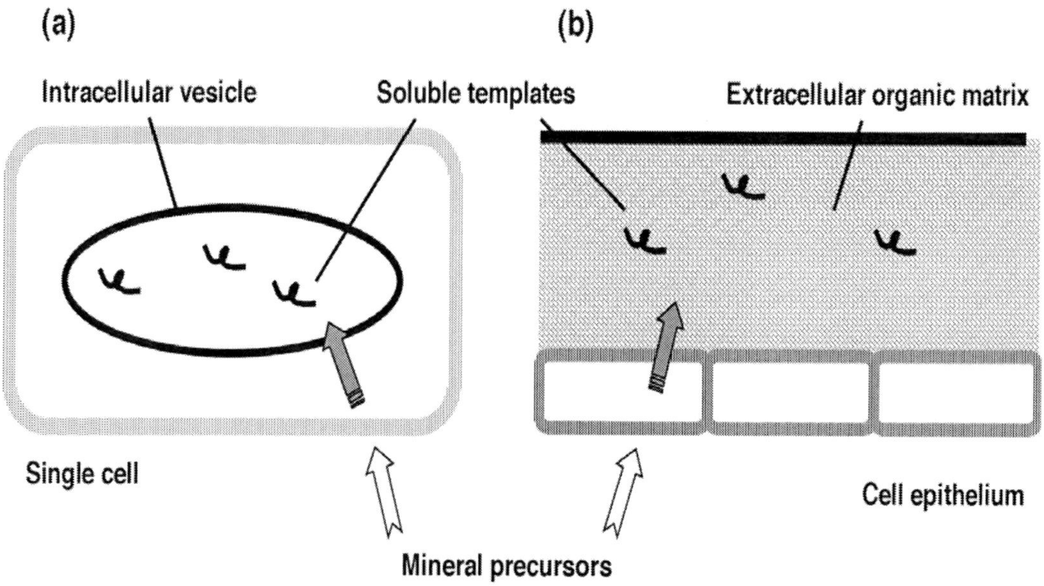

Figure 2. Schematic illustration of mineralization sites in (a) intra-cellular vesicles, as for diatoms and (b) extracellular matrix, as for gastropods.

I.2. Biomimetic Approaches

Biomimetic approaches may be described as synthetic/engineering procedures that closely resemble certain stages of biological processes [30]. Such approaches have been developed not only in material science but also in organic chemistry, where synthetic procedures mimic the biochemical reactions responsible for the formation of a given product in natural systems [31]. Noticeably, it is possible to get inspiration from living organisms to synthesize non-natural products, for instance to use cell-extracted macromolecules to control the growth of non-biogenic materials [32]. Alternatively, synthetic procedures that have nothing to do with biomineralization processes may allow the elaboration of materials resembling biominerals [33]. Both approaches may more suitably considered as "bio-inspired" although the distinction between biomimetism and bioinspiration is sometimes debatable.

As biomimetic approaches aim at re-creating *in vitro* the conditions of *in vivo* phenomena, one may suggest that model experiments should incorporate as much actors involved in the biomineralization processes as possible. However, as mentioned earlier, it might be useful in a first step to design simple systems allowing the study of a limited number of elements and reaction conditions. However, the question arises whether such simplified approaches, that focus on a specific element or step of the biomineralization process, can nevertheless provide information relevant to the whole mineralization system. To illustrate this point, the next section is dedicated to the main achievements and obvious limitations of such simple biomimetic approaches.

II. Molecular Templates in Biomineralization Mimicry

II.1. In Vitro Evaluation of Cell-Extracted Molecules

A traditional approach in biomineralization studies first starts with the extraction of the organic matrix from the composite material by mineral dissolution. As mentioned earlier, the extract usually consists of small organic molecules and a macromolecular network, termed the soluble and insoluble fraction, respectively. At this stage, one has to notice that such an attribution is sometimes arbitrary as it strongly depends on the extraction procedures, including pH, ionic strength or denaturing effects. Moreover, the chemical nature or organization of the macromolecules, and therefore their solubility, may change with time and different forms of the same molecule may coexist within the biomaterial [34].

A second step of the experiment consists in the identification of the extract content. Indeed, the organic matrix is a complex mixture so that, in many cases, the so-called insoluble fraction is discarded. The remaining soluble fraction is analyzed by chromatography and/or mass spectrometry and, in most cases, one of its component is purified and characterized in more details (amino acid sequence, structure,..). This component is then added to a solution containing mineral precursors to evaluate its effect on mineralization. However, although it is a common knowledge in biochemistry, the fact that protein activity depends on its concentration and environment has often been neglected in the field of biomineralization. In fact, typical studies explore a large range of mineral/macromolecule ratio but in concentrations that may strongly differ from the *in vivo* conditions.

Following such procedures, it was found that poly-anionic, calcium-binding proteins have a significant influence on calcium carbonate and calcium phosphate formation whereas positively-charged macromolecules can activate silica formation [35]. These results can be explained on the basis of physico-chemical concepts such as electrostatic interactions or epitaxial growth. However, following examples show that their contribution to the understanding of the biological processes may have been over-rated.

The studies devoted to silica formation by diatoms nicely illustrate the possibilities and limitations of these approaches. Over the last few years, Kröger and Sumper extracted and purified a set of proteins, termed silaffins, as well as several poly-amines macromolecules [36]. They show that silaffins present specific post-translational modifications, allowing these proteins to activate silica formation [37]. Noticeably, it was shown that the extraction procedure can partially degrade the natural proteins so that their *in vivo* structure may not be fully identified [38]. Moreover, these proteins, as well as poly-amines, exhibit self-assembly propertiesthat may guide the silica network patterning [39]. In fact, *in vitro* precipitation experiments led to particulate and/or porous morphologies that bear some resemblance with diatom shells at the nano-to-microscale [36]. Interestingly, most complex morphologies were obtained using silaffin/poly-amines mixtures rather than purified single templates [40]. However, at the macroscale, the silica framework obtained *in vitro* is quite far from the natural hierarchically organized silica structures in diatoms. Interestingly, the overcharged structure of such natural extracted protein seems to be ideal to interact, concentrate and then stabilize locally the precursors but not act as a mould to structure the mineral. This statement is in agreement with biological studies suggesting that the shell morphology may be controlled by the shape of the specific intra-cellular vesicle in which silica deposition occurs [41].

In the case of bone formation, a clear distinction has been long established between collageneous and non-collageneous proteins forming the organic matrix. Whereas the former are considered to be involved in the macromolecular network formation, the large number of additional, "soluble" proteins makes their role difficult to study. Flade *et al.* therefore focused on the most abundant protein in the tissue, a specific γ-carboxyglutamic acid protein called osteocalcin [42]. Indeed, this protein exhibits expected calcium binding properties. Moreover, spacings between the Gla residues of osteocalcin were found to be very similar to the inter-atomic distances of calcium sites of hydroxyapatite 0001) planes. However, the activity of this protein was only studied in contact with calcium phosphate phases that are not present in natural bone [43] Moreover, these experiments were performed in a diluted media that does not take into account the influence of the confinement associated to the collagen network that exists in bone.

In calcium carbonate shells, different groups have demonstrated that soluble components play an important role in determining the nature of the deposited $CaCO_3$ polymorph. Belcher *et al.* suggested that calcite or aragonite could be selected by the sole presence of extracted acidic macromolecules [44]. However these conclusions are to be taken carefully as similar crystal morphologies could be obtained without any additives [45]. Additionally, Falini *et al.* showed that the soluble proteins must be associated with the macromolecular network, made of β-chitin-silk fibroin assembly, to fully exert their templating activity [46]. Moreover, a recent study of molluscan shell proteins found in nacre proves that many proteins could be less acidic than previously thought [34], suggesting that other amino acid sequences may play a major role in the morphological control of calcium carbonate biominerals.

These examples suggest that the traditional approach relying on the extraction, purification and mineralization evaluation of cellular components in dilute solutions can only partially contribute to the understanding of biological strategies involved in the controlled elaboration of materials. However, they constitute a useful basis to design model systems allowing the study of the bio-mineral interface at the molecular level.

II.2. Synthetic Templates

From a physico-chemical point of view, a protein is a rather complex molecule whose ability to interact with a mineral precursor depends on its amino acid composition and distribution as well as on its structure. Thus, in order to design model systems, synthetic molecules have been used that present some common key properties with the natural templates.

Thus, synthetic anionic polymers selected for their strong interactions with calcium were used to mimic acidic proteins found in mollusk shells and bone. In diluted solution, they were found to induce the formation and stabilization of amorphous phases as mentioned above. [47-50]. Other studies showed that these polymers could interact with specific crystal faces and therefore guide the crystal growth [51,52] Similarly, many studies have been devoted to the interactions between poly-lysine, a model for diatom-extracted cationic macromolecules, and silica precursors (Figure 3) [53,54]. These studies were useful in identifying the influence of polymer charge density, chain length and concentration as well as the effect of pH or stirring conditions on the formation of silica.

A simplification of the polymer composition has a strong impact on its structure. One way to overcome this problem is to design spatially-organized model systems. This was achieved using 2D synthetic templates such as Langmuir or self-assembled monolayers (SAMs) [55,56]. In the latter case, a precise control of the template spatial distribution is nowadays achievable using soft lithography techniques [57]. It is also possible to combine both strategies by using patterned SAMs and additional charged organic molecules in solution. The main advantage of such strategies is that they take into account the adsorption processes that may occur within the macromolecular networks. Nevertheless, it is an open question whether they provide unambiguous informations about the relation between the structure of a polymeric template and a developing mineral phase. For instance, similar calcite truncated morphologies were obtained *(i)* in the presence of carboxylate-terminated SAMs in the presence of magnesium [58], or COOH-terminated regions of patterned self-assembled monolayers of alkane thiols in the presence of L-asp and D-asp [59]; *(ii)* without monolayers but in the presence of magnesium and malic acid [60] or carbonic acid [61]; *(iii)* without monolayer or magnesium ions but in the presence of organic additives such as

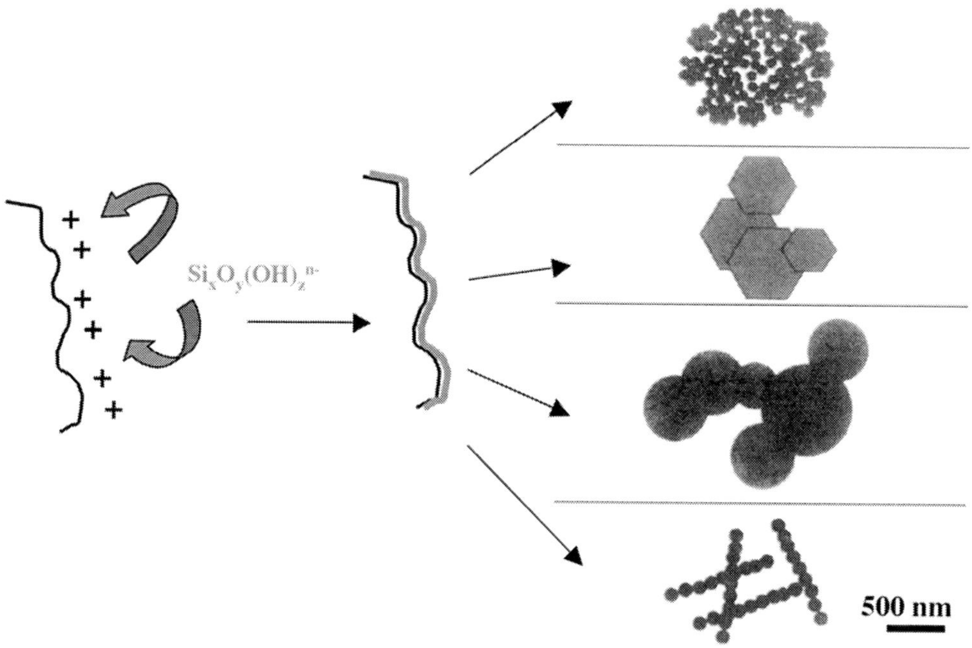

Figure 3. Schematic illustration of the activation of silica condensation by cationic polymers mimicking diatom silaffins and resulting *in vitro* silica particle morphologies (Adapted from [53]).

desulfated dermatane sulphate proteoglycan [62] or sodium salt of poly(L-isocyanoalanyl-D-alanine) [63]. If these structures are identical, it suggests that some elements present in the precipitation media could be redundant or that the main factor that drives the general crystallization scheme is underestimated or not identified.

Nevertheless, in some cases, the use of synthetic templates may bring fruitful information about the reactivity of the cell-extracted molecules. For instance, it was shown that the silica-precipitating activity of diatom-extracted silaffins and poly-amines was highly dependent on the presence of phosphate ions [38]. Using poly-(allylamine hydrochloride) as a model polymer, it was possible to study in details the influence of these anions on the self-assembly properties of poly-amines and hence to explain the behavior of cell-extracted templates [64].

This example illustrates the strong interdependency that exists between these two approaches. On the one hand, the design of more accurate model systems relies on a better knowledge of the *in vivo* processes, including a more extended identification of the cellular components involved in the biomineralization reaction, and their possible interplay. On the other hand, the elucidation of physico-chemical parameters involved in this reaction may help in determining the respective role of these components and, hence, contribute to a better understanding of the *in vivo* process. However, many of the examples provided above indicate that the biological relevance of such model systems is to be examined carefully due to the omission or under-estimation of key parameters in the experimental set-up. In the following section, recent attempts to address these limitations are presented.

III. Recent Biomimetic Approaches in Biomineralization

III.1. Confinement Effects

Until recently, most biomimetic studies were performed using soluble molecules in diluted solutions. This is quite surprising as biomineral formation always occurs in a confined media. Indeed, one of the key features in biomineralization is the capacity of the cells to control the process by localizing it in a specific compartment. This partitioning of the space allows a better control and regulation of the physico-chemical parameters. Therefore the mineral phase precipitation can occur independently of the environmental conditions or at least being less sensitive to these external conditions.

In general four different types of confined environments have been identified: phospholipids vesicles, protein cages, cellular assemblies and extracellular matrices. The so-called magnetosomes [65], that direct the formation of iron oxide nanoparticles in magnetotactic bacteria, and ferritin [66], an iron storage protein, can be considered as paradigms for the two first types of environment. In both cases, the compartment boundaries control the diffusion of Fe^{II} ions inside the vesicle and limit the size of the growing particles (*diffusion* and *steric* effects). In the case of ferritin, this nucleation step is activated by interaction of iron ions with some of the amino acids present at the surface of the protein cavity (*surface* effects). In mollusk exoskeletons, the organic matrix, consisting of polysaccharide and proteins layered network, surrounds the aragonite crystals. It thus serves as template for the crystal platelets growth, acts as a mold limiting the *c*-axial growth of the mineral and thus defines the platelet thickness whereas the growth of the crystal lateral faces is stopped by the crystals in contact [67]. Once again, this growth process is controlled *via* ion diffusion and interlayer spatial confinement. Nucleation also appears controlled by interactions of calcium ions with layer surface groups.

As silica formation in diatoms occurs in a vesicle, it appears clearly that synthetic vesicular systems should constitute suitable models to mimic biosilicification. In this context, silica formation within so-called "onion phases" was recently described [68]. Biomimetic systems were investigated using sodium silicate solutions as silica precursors and the natural lipid L-α-phosphatidylcholine as the main component of the lyotropic phase. In acidic conditions, multilamellar particles in the 100-500 nm diameter range were obtained (Figure 4a). In contrast to previous works on similar systems performed in non-biomimetic conditions, the silica network is not condensed but consists of un-connected silica nanoparticles forming a monolayer in the interlamellar space of the vesicle. These colloids are *ca.* 40 Å in diameter, to be compared with the *ca.* 50 Å wall-to-wall distance of the surfactant network. This therefore suggests that the growth of silica is limited by the dimensions of the reactor, *i.e.* the interlamellar distance, as expected. However, the fact that these particles are not aggregated is more difficult to understand as acidic conditions correspond to a low positive charge surface of silica that should favor inter-particle interactions. The possibility for an influence of the surfactant bilayer surface on silica condensation cannot be put aside. In fact, it was observed that the interlamellar distance decreases upon silica condensation, suggesting the possibility for direct silica/surfactant interactions and therefore particle nucleation on the bilayer surface. Therefore, this example illustrates two consequences of confinement: a steric effect limiting the size of the growing particle and a probable surface effect that influences nanoparticule nucleation. Similar results were more recently obtained using the channels of nanoporous membranes as confined media for silica formation (Figure 4b) [69]. It was observed that the silica particle size depends on the channel dimension, due to the modification of the diffusion behavior of mineral species. Moreover, the introduction of a model polyamine (poly-lysine) within the membrane pores also led to a change in the silica network structure [70].

Similarly, an interesting synthetic route was proposed that involves a polycarbonate porous membrane to precipitate calcium carbonate using amorphous precursors. Through this method, it was possible to constrain the morphology of single crystals in the absence of any additional soluble proteins [71] Interestingly, this membrane not only plays the role of a confined media but its chemical nature, poly-carbonate, makes it suitable to mimic calcium-binding environments. In a step further, it was possible to elaborate polymer replica of sea urchin porous skeleton and use them as mold to form calcium carbonate materials with complex morphologies [72].

In bone, several identified non-collageneous proteins are currently studied to understand their influence in hydroxyapatite formation. An important work performed by Hunter and co-workers has interestingly proved that the activity of these proteins clearly depends on confinement conditions [73]. The effect of polyanionic phosphoproteins, bone sialoprotein (BSP) and osteopontin, on hydroxyapatite crystal formation was determined in the presence of an agarose gel in order to mimic the *in vivo* condensed extracellular matrix consisting of type I collagen. This approach revealed that, within the gel, BSP favour crystal nucleation rather than inhibition.

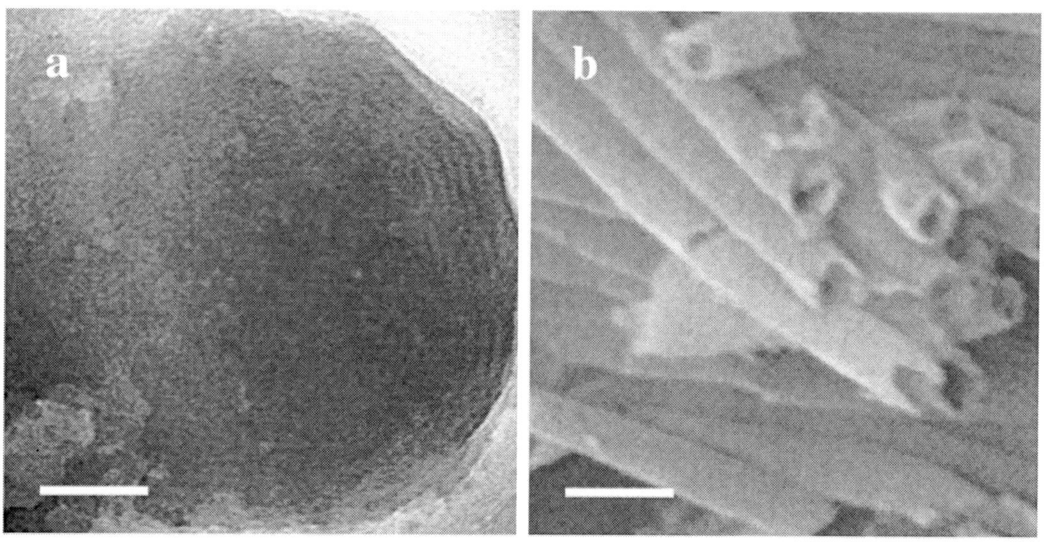

Figure 4. Growth of silica in confined environments: (a) transmission electron micrograph of hybrid phospholipids/silica vesicles (scale bar = 10 nm) [68], and (b) scanning electron micrograph of silica tubes formed in poly-carbonate porous membranes (scale bar = 500 nm) [69].

A similar result had been obtained for surface-immobilized BSP by Linde *et al.* [74] whereas Addadi and co-workers did not detect any effect of diluted solutions of BSP on the hydroxyapatite formation [75]. These data demonstrate that biomimetic experiments performed in diluted conditions may not only be biologically non-relevant but may even bring un-correct information.

These approaches represent significant advances in biomimetic model design. Nevertheless, such systems may be improved further. In many cases, confinement conditions can be reproduced but with macromolecular networks that differ chemically and/or structurally from the natural organic matrix. For instance, in the field of bone biomineralization, many studies were performed using the major structural protein, type I collagen as a template. However, the three-dimensional arrangement of collagen fibrils in compact bone forms a regularly ordered network that is not easy to reproduce *in vitro* [76,77]. One interesting alternative involves the use of demineralised matrices [50,78,79]. However, efforts still have to be made to reconstitute this supramolecular arrangement as it should help to better understand the contribution of the different actors involved in the mineralization process (ionic strength, organic matrix organization, protein activities and cellular control).

III.2. Role of the Mineral Precursors

Another key aspect of biomimetic approaches that has long been underestimated is related to the nature of inorganic precursors as well as the conditions of their incorporation to the model systems. In fact, it is often difficult to establish the more suitable source and concentration of mineral species due to the lack of information available from the living system itself or the difficulty in accurately reproducing *in vivo* conditions.

The specific case of silica formation in diatoms and sponges deserves to be studied in more details. It is known that these organisms uptake soluble silica in the form of silicic acid $Si(OH)_4$ [80]. This silicic acid is then transported through the cytoplasm until it reaches the specific vesicle where silica formation occurs. However, the high concentration of soluble silica found within these cells suggest that silicon is not present as silicic acid but in some stabilized form that can be either colloidal silica or some silicon complex. In parallel, the synthesis of silica by chemists usually involve silicon alkoxide $Si(OR)_4$ precursors [81], that are not soluble in water and release ROH alcohol molecule upon hydrolysis, so that it is very unlikely that they are present within diatom cells. However, as biomimetic studies of silicification were mainly triggered by the material science community, silicon alkoxides are the most widely used precursors in this field, whereas aqueous silicates, silicon complexes and colloidal silica have been much less studied [82].

However, the selection of a given silica source is not genuine because each precursor has a different reactivity and therefore its interactions with mineralization templates can be very different. For instance, it was found that silica could be associated to collagen within certain marine sponges [83]. Therefore, the influence of several silica precursors on the self-assembly process of collagen triple-helices was studied. These investigations showed that diluted aqueous silicate solutions or hydrolyzed silicon complexes were able to favor the self-assembly process by bridging triple-helices via hydrogen bond (Figure 5a) [84]. In contrast, concentrated silicate solution and colloidal silica hinder the fibrillogenesis due to electrostatic interactions leading to triple helices precipitation (Figure 5b). Finally, silicon alkoxides either in solution or in the vapour phase, do not perturb the collagen behavior [85,86]. In the same time, the presence of collagen do not significantly influence the rate of silica formation but appear to affect the hydrolysis of silicon complexes and silicon alkoxides [84]. Thus, diverging results may be obtained depending on the nature of the precursors, which makes the biomimetic relevance of these experiments debatable, especially in the absence of clear information about the natural form of silica source at the mineralization site.

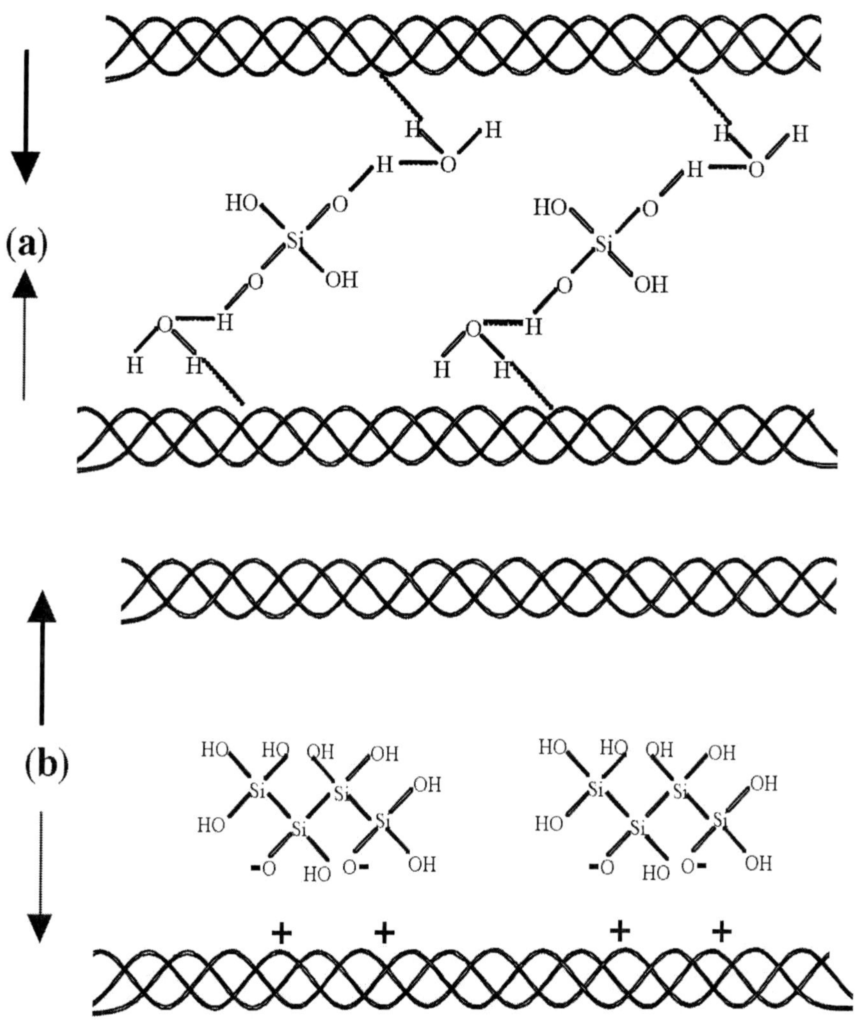

Figure 5. Influence of silica precursors on collagen fibrillogenesis: (a) diluted silicate solutions favor triple helix self-assembly through hydrogen bonds and (b) silicates hinder triple helix self-assembly through electrostatic interactions [84].

Another aspect of this problem is related to the influence of the experimental set-up on the mineralization process. A significant example was recently described by Gehrke *et al.* for $CaCO_3$ crystallization [45]. Using the vapour diffusion method, based on the thermal decomposition of ammonium carbonate for the slow generation of CO_2, it was found that experimental conditions should be chosen carefully, as they could generate several simultaneous reaction pathways. Remarkably rosetta-shaped mesostructures were obtained without any additives, whereas a related morphology was previously obtained in a $CaCl_2$-NH_4HCO_3 mixed solution and attributed to the presence of proteins extracted from sea shells [44]. In this case, an open question is if these structures are indeed identical, which would mean that the proteins are redundant and that the general crystallization scheme is driven by the ammonium ions. Thus, it shows that reaction conditions can drastically influence the crystallization process leading to specific crystal shapes without any additives but only in the presence of small ions that are associated to the mineral precursors [87]. The experimental conditions may also modify the mineral precursor/additive interactions. Thus, the same polymer can induce many different $CaCO_3$ morphologies, by simply altering experimental conditions [48,88]. In a recent publication, experiments performed by Kulak et al. even showed that it is possible to induce a continuous structural evolution of calcium carbonate particles by changing the ratio of calcium/polymer [89].

On the one hand, these recent developments help to progress in the understanding of the influence of physico-chemical and structural conditions on biomineralization processes. On the other hand, they question many of the conclusions that were drawn from previous simpler model approaches. Indeed, as the complexity of biomimetic systems increases, the accuracy of each component should also be optimized to maintain a suitable level of relevance for the whole system. Chemists then face a number of limitations that originate both from the lack of information concerning the *in vivo* processes and from the technical challenge associated with the *in vitro* mimicry of biological conditions. The question therefore arises whether the design of suitable model systems that would allow a full understanding of biomineralization processes is possible in a next future.

IV. Perspectives

As mentioned earlier, most biomimetic studies have been performed following a reductionnist perspective, meaning that the mineral formation proceeds from physico-chemical interactions of chemical species in specific experimental conditions. In other words, they consider living organisms as “living factories” that produce these chemical species and control these conditions. On this basis, a first assumption is that the whole biomineralization process can be reproduced *in vitro* if suitable reagents are placed in suitable environments. Indeed, these approaches do not underestimate the complexity of the process but this complexity is accommodated through a second assumption stating that some parameters are more determining than the others. Therefore, in principle, a minimum set of criteria should be gathered to accurately mimic the whole process. The key point is then to identify these criteria and to be able to reproduce them *in vitro*.

As demonstrated above, these two requirements are not easily reached. Ideally, the identification of suitable conditions would rely on a step-by-step approach starting from a simple system and increasing its complexity. However, more elaborated biomimetic systems may contradict simpler ones, as shown previously for the divergent behavior of bone proteins in solution and in confined environment. Moreover, the design of experimental set-ups that allows the study of more complex systems may be technically demanding. For instance, up to now, no methods are available to study in details the growth of silica in vesicles that would mimic diatom specific compartment.

Indeed, a way to enhance the biological relevance of biomimetic systems would involve the introduction of living cells within the model system. But the question then arises about the minimum level of complexity that should be achieved *in vitro* to reproduce the *in vivo* process. From a biological point of view, it is commonly admitted that living systems do not waste energy, meaning that most element involved in the biomineralization process has its specific role. Therefore, a perfect replication of the biomineral formation reaction *in vitro* without using the specialized cells appears impossible. To our point of view, the elucidation of biomineralization mechanisms therefore relies on a more extensive study of the living organism itself, particularly *via* genetic approaches, or at least by the association of the specialized cells in the presence of the reproduced organic matrix in an appropriate ionic/nutritive medium. However, it should be mentioned that the introduction of cells in such biomimetic systems also implies an increased implication of biologists to this field of research. Further cell-extraction experiments may also provide fruitful information but efforts should be made to evaluate the role of identified molecules within the living organism and not *in vitro*.

In parallel, it is worth asking to what extent we need to copy natural mineral structures. Biomineralization studies stand at the frontier between at least four scientific domains: biology, medicine, geology and materials science. In the first case, the ultimate goal is to identify the biological processes of mineralization, in the more general context of the study of the biosphere. Medicine is also mainly concerned by the whole organism rather than by the mineral itself and therefore directly benefit from the biological data for therapeutic purposes. In geology, both the biomineralizing system and the deposited material are of importance as this field is interested in ancient and contemporary mineral formation. Finally, in many cases, biomimetic approaches in material science focus on the biomineral itself, as it may act as a model to design bio-inspired functional materials. As biomineralization is intrinsically a trans-disciplinary field of research, the distinction between "bio-inspired" and "bio-mimetic" should be use carefully to elaborate a coherent interdisciplinary interactions. However, it suggests that the required extent of understanding, and hence the level of complexity that biomimetic systems should achieve, may differ significantly from one field to another.

In those cases where the mineral itself, rather the full biomineralization process, is of interest, the biological relevance of the chemical procedures may not be the key aspect to consider. For instance, silica morphologies that have nothing in common with natural structures have been obtained using silicon alkoxides and "biomimetic" templates (peptides, poly-amines) [53]. In parallel, "biomimetic" silica structures resembling radiolaria shells have been recently achieved *via* synthetic methodologies and in experimental conditions that have nothing to do with the *in vivo* process [90]. It is our opinion that it not possible to draw any conclusion about biological processes from these studies, so that the term "biomimetic" is here un-adequate.

In contrast, we feel that there is room for more general investigations of the interface between biological systems and minerals, that may have strong implications for the understanding of biomineralization processes. In fact, the reactivity of mineral precursors in biological conditions as well as the influence of these precursors on bio-molecule and cell behavior are still poorly known. As an illustration, diatom cells were recently immobilized in silica gels to evaluate their behavior in a mineralized environment [91]. These experiments revealed that diatoms have the ability to dissolve the silica network, a possibility that was never reported before and that implies that these cells have developed a strategy to recover silicate precursors required for their shell formation from condensed mineral phases. Thus, although these experiments were not initially developed in a "biomimetic" perspective, they finally provide novel insights about the biomineralization process performed by these organisms.

Acknowledgments

We thank Dr M.M. Giraud-Guille for her critical reading and helpful comment of the manuscript.

References

[1] Thompson, D. W. *On Growth and Form*; Cambridge University Press: Cambridge, 1917.

[2] *Biomineralization: Medical Aspects of Solubility* ; Königsberger, E.; Königsberger, L.; Eds.; J. Wiley and Sons : Chichester, 2006.

[3] Montanez, I. P. *Proc. Natl. Acad. Sci. USA* 2002, *99*, 15852-15854.

[4] Newman, D. K.; Banfield, J. F. *Science* 2002, *296*, 1071-1077.

[5] Lowenstam, H. A.; Weiner, S. *On Biomineralization*; Oxford University Press: New York, 1989.

[6] Lowenstam, H. A. *Science* 1981, *211*, 1126-1131.

[7] Mann, S. *Struct. Bonding* 1983, *54*, 125-174.

[8] Glimcher, M. J.; In *Metabolic Bone Disease and Clinically Related Disorders* ; Avioli, L. V.; Krane, S. M.; Eds.; Academic Press: San Diego, 1998, pp 23-50.

[9] Ryall, R. L.; Fleming, D. E.; Grover, P. K.; Chauvet, M.; Dean, C. J.; Marshall, V. R. *Mol. Urol.* 2000, *4*, 391-402.

[10] Thompson, J. B.; Poloczi, J. T.; Kindt, J. H.; Michenfelder, M.; Smith, B. L.; Stucky, G. D.; Morse, D. E.; Hansma, P. K. *Biophys. J.* 2000, *79*, 3307-3312.

[11] *Biomineralization : from Paleontology to Materials Science* ; Arias, J. L. ; Fernandez, M. S. ; Eds.; Editorial Universitaria : Santiago de Chile, 2007.

[12] *Handbook of Biomineralization* ; Baeuerlein, E.; Ed.; Wiley-VCH : Weinheim, 2007; Vol. 1.

[13] *Handbook of Biomineralization* ; Baeuerlein, E.; Behrens, P.; Epple, M.; Eds.; Wiley-VCH : Weinheim, 2007; Vol. 2.
[14] Fitton-Jackson, S. *Proc. R. Soc. London Ser. B* 1957, *146*, 270-280.
[15] Chen, J.; Burger, C.; Krishnan, C. V.; Chu, B.; Hsiao, B. S.; Glimcher, M. J. *Macromol. Chem. Phys*. 2005, *206*, 43-51.
[16] Rousseau, M.; Lopez, E.; Stempflé, P.; Brendlé, M.; Franke, L.; Guette, A.; Naslain, R.; Bourrat, X. *Biomaterials* 2005, *26*, 6254-6262.
[17] Nassif, N.; Pinna, N.; Gehrke, N.; Antonietti, M.; Jäger, C.; Cölfen, H. *Proc. Natl. Acad. Sci. U.S.A*. 2005, *102*, 12653-12655.
[18] Giraud-Guille, M. M.; Bouligand, Y. In *Chitin World*, Karnicki, Z.S.; Ed.; Wirtschaftsverlag, 1995, pp. 136–144.
[19] Aizenberg, J.; Lambert, G.; Weiner, S.; Addadi, L. *J. Am. Chem. Soc*. 2002, *124*, 32-39.
[20] Aizenberg, J.; Weaver, J. C.; Thanawala, M. S.; Sundar, V. C.; Morse, D. E.; Fratzl, P. *Science* 2005, *309*, 275-278.
[21] Gupta, H. S.; Seto, J.; Wagermaier, W.; Zaslansky, P.; Boesecke, P.; Fratzl, P. *Proc. Natl. Acad. Sci. U.S.A*. 2006, *103*, 17741-17746.
[22] England, J.; Cusack, M.; Lee, M. R. *Lethaia* 2007, *40*, 2-10.
[23] Jaeger, C.; Groom, N. S.; Bowe, E. A.; Horner, A.; Davis, M. E. Murray, R. C.; Duer, M. J. *Chem. Mater*. 2005, *17*, 3059-3061.
[24] Merz, M. U. E. ; Zankl, H. *Facies* 1993, *29*, 75-80.
[25] Garrone, R. ; Simpson, T. L. Pottu-Boumendil, J. ; In *Silicon and Siliceous Structures in Biological Systems* ; Simpson, T. L.; Volcani, B. E.; Eds.; Springer Verlag, New York, 1981.
[26] Addadi, L.; Weiner, S. *Angew. Chem. Int. Ed. Engl*. 1992, *21*, 153-169.
[27] Watabe, N.; Meenakshi, V. R.; Blackwelder, P. L.; Kurtz, E. M.; Dunkelberger, D. G.; In *Mechanisms of Biomineralization in the Invertebrates and Plants* ; Watabe, N.; Wilbur, K. M.; Ed.; University South Carolina Press: Columbia, 1976, pp. 283-308.
[28] Politi, Y.; Arad, T.; Klein, E.; Weiner, S.; Addadi, L. *Science* 2004, *306*, 1161-1164.
[29] Giraud-Guille, M. M. *Cell Tissue Res*. 1984, *236*, 413-420.
[30] Sanchez, C.; Arribart, H.; Giraud-Guille, M. M. *Nat. Mater*. 2005, *4*, 277-288.
[31] Yoder, R. A. ; Johnston, J. N. *Chem. Rev*. 2005, *105*, 4730-4756.
[32] Sumerel, J. L. ; Yang, W. ; Kisailus, D. ; Weaver, J. C. ; Choi, J. H. ; Morse, D. E. *Chem. Mater*. 2003, *15*, 4804-4809.
[33] Sun, Q.; Vrieling, E. G.; van Santen, R. A.; Sommerdijk, N. A. J. M. *Curr. Opin. Solid State Mater. Sci*. 2004, *8*, 111-120
[34] Marin, F.; Luquet, G. *C. R. Palevol* 2004, *3*, 469-492.
[35] Mann, S.; B*iomineralization. Principles and Concepts in Bioinorganic Materials Chemistry* ; Oxford University Press :Oxford, 2001.
[36] Sumper, M.; Kröger, N. *J. Mater. Chem*. 2004, *14*, 2059-2065
[37] Kröger, N.; Deutzmann, R.; Sumper, M. *Science* 1999, *286*, 11229-1132.
[38] Kröger, N.; Lorenz, S.; Brunner, E.; Sumper, M. *Science* 2002, *298*, 584-586
[39] Sumper, M. *Science* 2002, *295*, 2430-2432.
[40] Kröger, N.; Deutzmann, R.; Bergsdorf, C.; Sumper, M. *Proc. Natl Acad. Sci. USA* 2000, *97*, 14133-14138.

[41] Hildebrand, M. *Prog. Org. Coat.* 2003, *47*, 256-264.

[42] Flade, K.; Lau, C.; Mertig, M.; Pompe, W. *Chem. Mater.* 2001, *13*, 3596-3602.

[43] Glimcher, M. J. *Rev. Miner. Geochem.* 2006, *64*, 223-282.

[44] Belcher, A. M.; Wu, X. H.; Christensen, R. J.; Hansma, P. K.; Stucky, G. D.; Morse, D. E. *Nature* 1996, *381*, 56-58.

[45] Gehrke, N.; Cölfen, H.; Pinna, N.; Antonietti, M.; Nassif, N. *Cryst. Growth Des.* 2005, *5*, 1317-1319.

[46] Falini, G.; Albeck, S.; Weiner, S.; Addadi, L. *Science* 1996, *271*, 67-69.

[47] Guillemet, B.; Faatz, M.; Gröhn, F.; Wegner, G.; Gnanou, Y. *Langmuir* 2006, *22*, 1875-1879.

[48] Nassif, N.; Gehrke, N.; Pinna, N.; Shirshova, N.; Tauer, K.; Antonietti, M.; Cölfen, H. *Angew. Chem. Int. Ed.* 2005, *117*, 6004-6009.

[49] Olszta, M. J.; Douglas, E. P.; Gower, L. B. *Calcif. Tissue Int.* 2003, *72*, 583-591.

[50] Gehrke, N.; Nassif, N.; Pinna, N.; Antonietti, M.; Gupta, H. S.; Cölfen, H.*Chem. Mater.* 2005, *17*, 6514-6516.

[51] Marentette, J. M.; Norwig, J.; Stöckelmann, E.; Meyer, W. H.; Wegner, G. *Adv. Mater.* 1997, *9*, 647-650.

[52] Gower, L. B.; Odom, D. J. *J. Cryst. Growth* 2000, *210*, 719-734.

[53] Lopez, P. J.; Gautier, C.; Livage, J.; Coradin, T. *Curr. Nanosci.* 2005, *1*, 73-83.

[54] Patwardhan, S. V.; Clarson, C. J.; Perry, C. C. *Chem. Commun.* 1113-1119

[55] Mann, S.; Heywood, B. R.; Rajam, S.; Birchall, J. D. *Nature* 1988, *334*, 692-695.

[56] Bunker, B. C.; Rieke, P. C.; Tarasevich, B. J.; Campbell, A. A.; Fryxell, G. E.; Graff, G. L.; Song, L.; Liu, J.; Virden, J. W.; McVay, G. L. *Science*, 1994, *264*, 48-55.

[57] Aizenberg, J. *Adv. Mater.* 2004, *16*, 1295-1302.

[58] Han, Y.J.; Aizenberg, J. *J. Am. Chem. Soc.* 2003, *125*, 4032-4033.

[59] Orme, C. A.; Noy, A.; Wierzbicki, A.; McBride, M. T.; Grantham, M.; Teng, H. H.; Dove, P. M.; DeYoreo, J. J. *Nature* 2001, *411*, 775-779.

[60] Meldrum, F. C.; Hyde, S. T. *J. Cryst. Growth* 2001, *231*, 544-558.

[61] Kralj, D.; Kontrec, J.; Brecevic, L.; Falini, G.; Nöthig-Laslo, V. *Chem. Eur. J.* 2004, *10*, 1647-1656.

[62] Arias, J. I.; Jure, C.; Wiff, J. P.; Fernandez, M. S.; Fuenzalida, V.; Arias, J. L. *Mat. Res. Soc. Symp. Proc.* 2002, *711*, 243-248.

[63] Donners, J. J. J. M.; Nolte, R. J. M.; Sommerdijk, N. A. J. M. *J. Am. Chem. Soc.* 2002, *124*, 9700-9701.

[64] Brunner, E.; Lutz, K.; Sumper, M. *Phys.Chem.Chem.Phys.* 2004, *64*, 854-859.

[65] Schuler, D.; Frankel, R. B. *Appl. Microbiol. Biotechnol.* 1999, *52*, 464-473.

[66] Theil, E. C. *Ann. Rev. Biochem.* 1987, *56*, 289-315.

[67] Belcher, A. M.; Gooch, E. E. In *Biomineralization* ; Bäuerlein, E.; Ed, Wiley-VCH: Weinheim, Germany, 2000, pp. 221-249.

[68] El-Rassy, H.; Belamie, E.; Livage, J.; Coradin, T. *Langmuir* 2005, *21*, 8584-8586.

[69] Gautier, C.; Lopez, P. J.; Hemadi, M.; Livage, J.; Coradin, T. *Langmuir* 2006, *22*, 9082-9084.

[70] Gautier, C.; Lopez, P. J.; Livage, J.; Coradin, T. *J. Colloid Interf. Sci.* 2007, *309*, 44-48.

[71] Loste, E.; Meldrum, F. C. *Chem. Commun.* 2001, 901-902.

[72] Loste, E.; Park, R. J.; Warren, J.; Meldrum, F. C. *Adv. Funct. Mater*. 2004, *14*, 1211-1220.

[73] Hunter, G. K.; Goldberg, H. A. *Proc. Natl. Acad. Sci. U. S. A*. 1993, *90*, 8562-8565.

[74] Linde, A.; Lussi, A.; Crenshaw, M. A. *Calcif. Tissue Int*. 1989, *44*, 286-295.

[75] Addadi, L.; Weiner, S. *Proc. Natl. Acad. Sci. U. S. A*. 1985, *82*, 4110-4114.

[76] Mosser, G.; Helary, C.; Eglin, D.; Giraud-Guille, M. M. *Micron* 2005, *36*, 602-608,

[77] Mosser, G.; Anglo, A.; Helary, C.; Bouligand, Y.; Giraud-Guille, M. M. *Matrix Biol*. 2006, *25*, 3-13.

[78] Chen, J.; Burger, C.; Krishnan, C. V.; Chu, B.; Hsiao, B. S.; Glimcher, M. J. *Macromol. Chem. Phys*. 2005, *206*, 43-51,

[79] Watabe, N.; Bernhardt, A.M.; Kingsley, R. J.; Wilbur, K. M. *Trans. Am. Microsc. Soc*. 1986, *105*, 311-318.

[80] Coradin, T.; Lopez, P. J.; Gautier, C.; Livage, J. *C. R. Palevol* 2004, *3*, 443-452.

[81] Brinker, C. J.; Scherrer, G. Sol-gel Science, *The Physics and Chemistry of Sol-Gel Processing* ; Academic, Boston, MA, 1990.

[82] Coradin, T.; Livage, J. *Acc. Chem. Res*. 2007, *40*, 819-826.

[83] Müller, W. E. G. ; Krasko, A. ; Le Pennec, G. ; Schröder, H. C. *Microsc. Res. Tech*. 2003, *62*, 368-377.

[84] Eglin, D.; Shafran, K. L.; Livage, J.; Coradin, T.; Perry, C. C. *J. Mater. Chem*. 2006, *16*, 4220-4230.

[85] Eglin, D.; Mosser, G.; Giraud-Guille, M.M.; Livage, J.; Coradin, T. *Soft Matter* 2005, *1*, 29-33.

[86] Heinemann, S.; Ehrlich, H.; Knieb, C.; Hanke, T. *Int. J. Mat. Res*. 2007, *98*, 603-608.

[87] Pastero, L.; Costa, E.; Bruno, M.; Rubbo, M.; Sgualdino, G.; Aquilano, D. *Cryst. Growth Des*. 2004, *4*, 485-490.

[88] Wang, T. X.; Antonietti, M.; Cölfen, H. *Chem. Eur. J*. 2006, *12*, 5722-5730.

[89] Kulak, A. N.; Iddon, P.; Li, Y.; Armes, S. P.; Cölfen, H.; Paris, O.; Wilson, R. M.; Meldrum, F. C. *J. Am. Chem. Soc*. 2007, *129*, 3729-3736.

[90] Wang, J.; Xiao, Q.; Zhou, H.; Sun, P.; Li, B.; Ding, D.; Chen, T. *J. Phys. Chem. C* 2007, *111*, 16544-16548.

[91] Gautier, C.; Coradin, T.; Livage, J.; Lopez, P. J. *Chem. Commun*. 2006, 4611-4613.

In: Biomimetic and Supramolecular Systems Research
Editor: Arturo H. Lima
ISBN: 978-1-60456-405-1

Chapter VIII

Molecular Engineering of Phenyleneethynylenes towards Specific Molecular and Supramolecular Organization for Optoelectronics

Ivana Moggio[5] and Eduardo Arias
Centro de Investigación en Química Aplicada
Boulevard Enrique Reyna 140, 25100, Saltillo, Coahuila, Mexico

Abstract

Phenyleneethynylenes (PEs) are a promising class of conjugated materials for optoelectronic devices due to their high electronic delocalization and strong emission properties. Depending on the type of optoelectronic application, specific molecular and supramolecular organization is required. For electroluminescent diodes, it is desirable to deposit the emitting layer with the phenyleneethynylene molecules oriented with the conjugated backbone, responsible of the electroluminescent properties, in the same direction of the electrons-holes injection from the electrodes. On the other hand, when the final application is a solar cell, it is preferred that the photoconductive part of the molecule is exposed to the incident light as much as possible. These requirements must also match other physicochemical characteristics that are necessary for any optoelectronic device, such as solubility, thermal and chemical stability, among others. With this idea in mind, in this chapter, we report on the molecular engineering of phenyleneethynylenes by changing the chemical structure either in the main conjugated backbone or with proper lateral substituents. Different molecular architectures: totally rod-like, semirigid, dendrimers or functionalized nanoparticles can be synthesized and specific supramolecular organization can be induced by varying the film deposition technique.

[5] Tel. 0052 844 438 98 30. Fax: 0052 844 438 98 39. e-mail: imoggio@ciqa.mx, earias@ciqa.mx.

Introduction

Organic optoelectronics has emerged in the last decades as an interesting technology being an intermediate between electronics and photonics. In order to be applied in an optoelectronic device such as a light emitting diode (LED), solar cell, smart window etc., a material must satisfy the following requirements: i) processability in homogeneous thin films (usually around 100 nm), ii) thermal and chemical stability in order to have a good life time in the device, iii) low cost of production. Some of these parameters can be obtained through a proper molecular design. Although, additional issues related to the specific device configuration can also be addressed. For instance, it is well established that the driving voltage of an LED depends on the emitting layer thickness that, in turn, can be controlled with proper experimental conditions of the film preparation technique. Other important factors are associated with the working functions of the electrodes that are related to the energetic difference between their Fermi level and the HOMO or LUMO values of the emitting material. In this respect, different metals can be proposed as cathodes in single layer organic light emitting diodes (oLEDs) or, multilayer devices can be fabricated adding charge-injecting layers sandwiching the emitting film and the electrodes in order to decrease these energetic barriers and enhance the stability of the emitting layer [1]. Small organic molecules can be used as optoelectronic materials and can usually be deposited by vacuum evaporation. This allows us to obtain highly homogenous films with uniform thickness and with charge mobility, usually one order of magnitude higher than those achievable by solution process methods [2]. Nevertheless, this technique can be applied only to low molecular weight materials and the use of high vacuum implies expensive equipment, which increases the production costs of the device. Since the discovery of electrical conductivity in doped poly(acetylene) [3], conjugated polymers have been extensively studied for many optoelectronic applications such as light emitting diodes [4], solar cells [5], light emitting electrochemical cells [6], smart windows [7], sensors [8] and biosensors [9]. Conjugated polymers, being semiconductor materials, present, with respect to small organic molecules, the advantages of good processability, low operating voltages, fast response time and color tunability through chemical modulation of the HOMO-LUMO values. Nowadays, a large variety of conjugated macromolecules can be obtained, mainly classified in the three most studied groups: poly(p-phenylenevynilidene)s (PPV), poly(p-phenylene)s (PPP) and poly(phenyeleneethynylene)s (PPE), (Figure 1). The molecular shape can be rigid rod-like, dendrimers, star-like, ladder type etc. [10]. These different structures act as building blocks for specific organic architectures with defined optoelectronic characteristics. Usually, the process to pass from the molecule to the organic bulk material is driven by supramolecular forces, and the intra- and inter-chain π–π interactions that occur as a consequence (or origin) of this assembly affect the device performance [11]. Under this approach, we report, in this chapter, on the chemical modulation of the optical and optoelectronic properties of families of phenyleneethynylenes. The first section presents the chemical modulation of the intrinsic optical properties. The following paragraphs deal with the molecular and supramolecular organization in bulk and thin films and its effect on the optoelectronic properties. Concluding remarks on nano-organization in nanoparticles of PPE and polymer brushes are made in the final part.

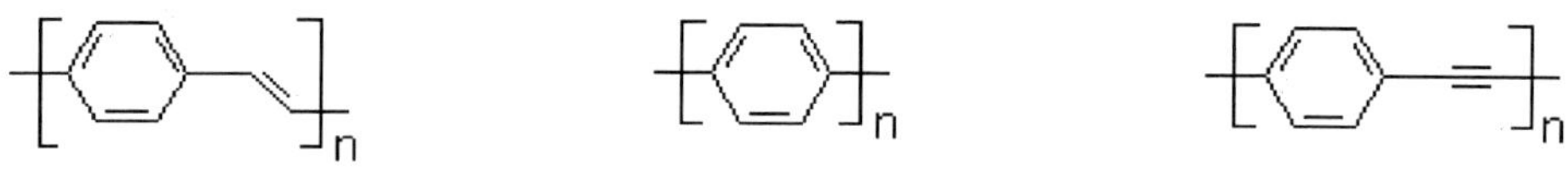

Figure 1. Chemical structure of the three most studied groups of conjugated polymers.

Molecular Design. Structure and Optical Properties in Solution

We have focused our interest in aryleneethynylenes and especially phenyeleneethynylenes (PEs) because they present an orthogonal π-orbital of the triple bond that is not involved in the rigid conjugated backbone. The triple bond increases the thermal stability with respect to PPV derivatives for which it has been reported a susceptibility to be oxidized [12] and impedes the formation of *cis-trans* isomers. Moreover, high quantum yields have been reported as well as interesting non linear optical properties and sensoring capacity [13]. In Figure 2, a collection of some representative PE macromolecules designed and synthesized in our group for different optoelectronic applications, is reported.

Family n. 1. Cholesterol PE oligomers and polymer

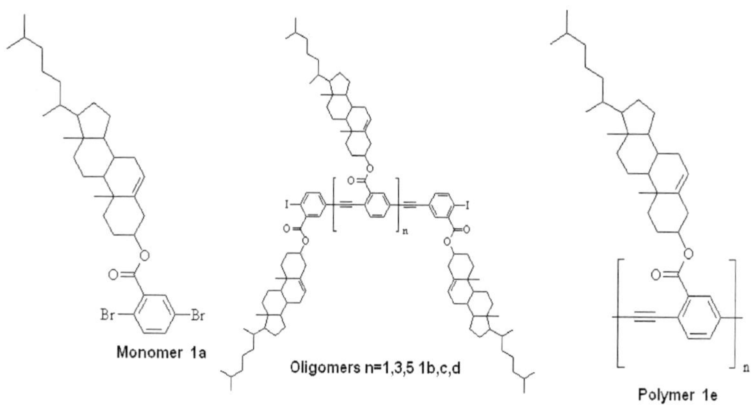

Family n. 2. Alcoxy PE oligomers and polymers

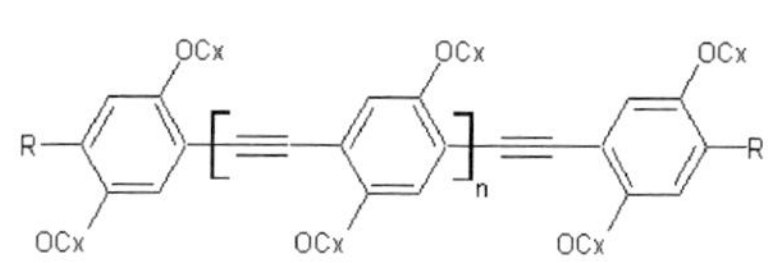

n	x	R	compound
1	12	H	2a
1	12	Br	2b
1	12	I	2c
1	4	H	2d
1	4	Br	2e
3	12	H	2f
3	4	H	2g
3	4	Br	2h
5	12	H	2i
polymer	12		2l
polymer	4		2m

Family n. 3. Sequenced PE copolymers

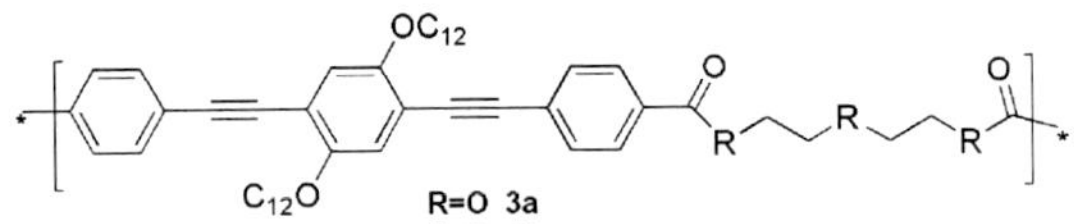

R=O 3a
R=S 3b

Family n. 4. Arylene ethynylene copolymers

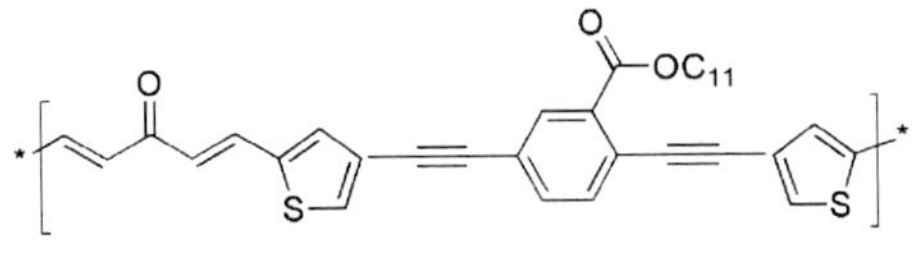

4a

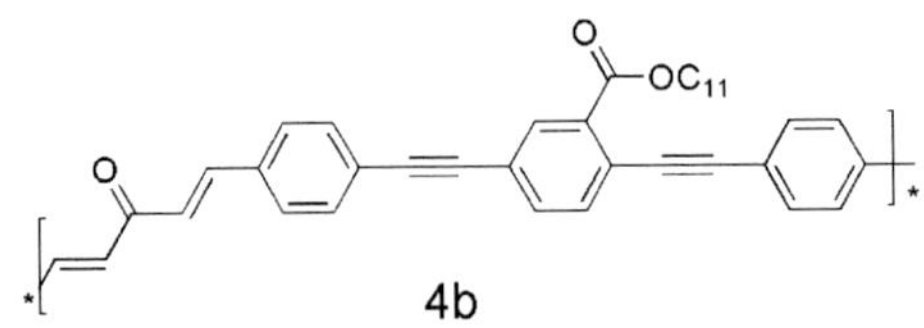

4b

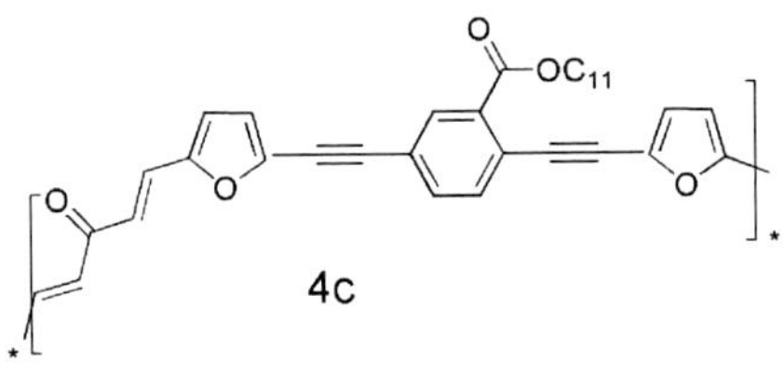

4c

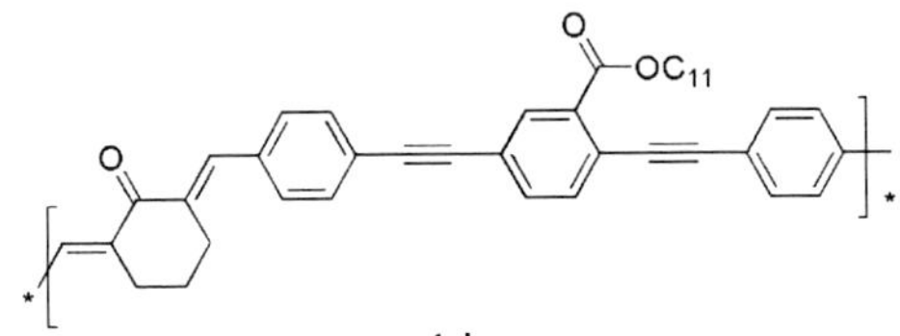

4d

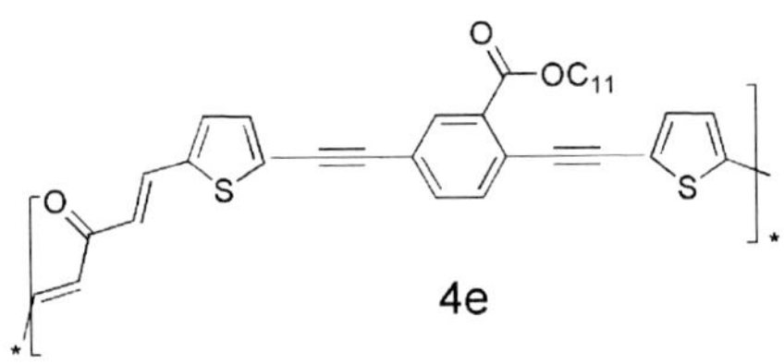

4e

Family n. 5. PE oligomers and polymers bearing diacetylenic moieties as central group.

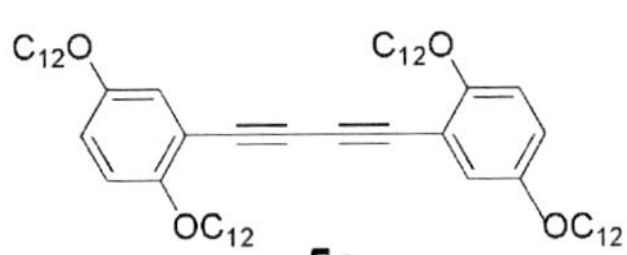

5a

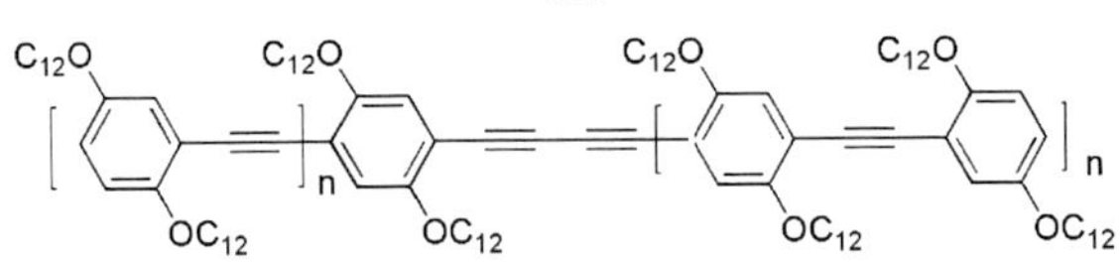

n=1,2,3 5b,c,d

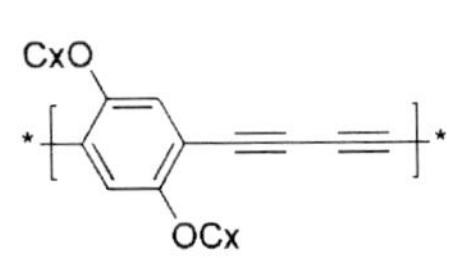

x=4,8,12 5e,f,g

Family n. 6. Fuleropyridine PE oligomers and their aldehyde precursors.

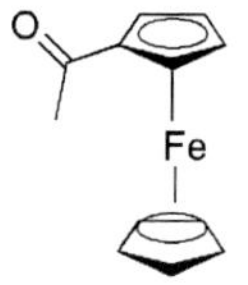

6a

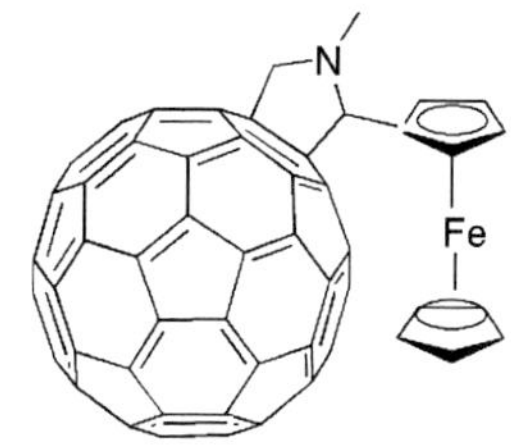

6b

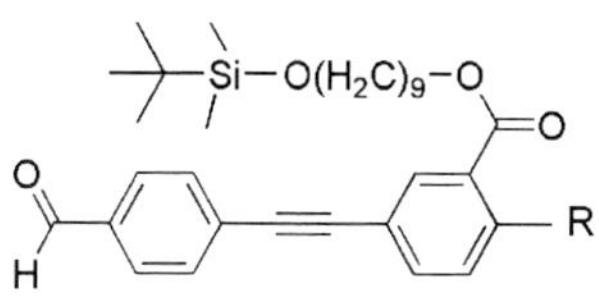

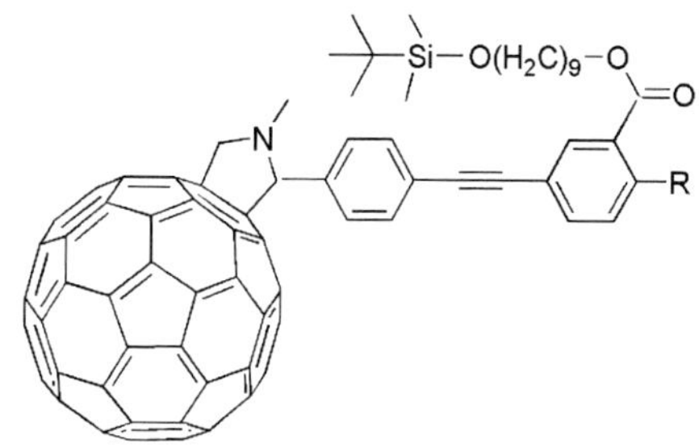

R= 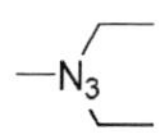

6c

R= 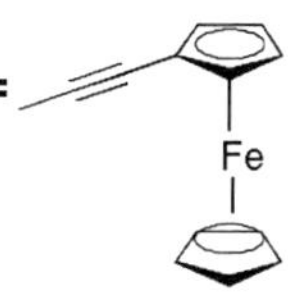

6d

R= 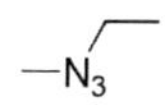

6e

R= 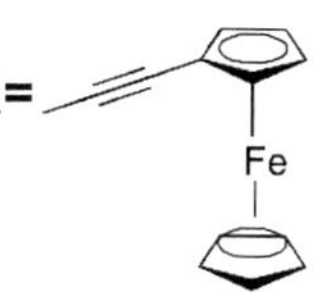

6f

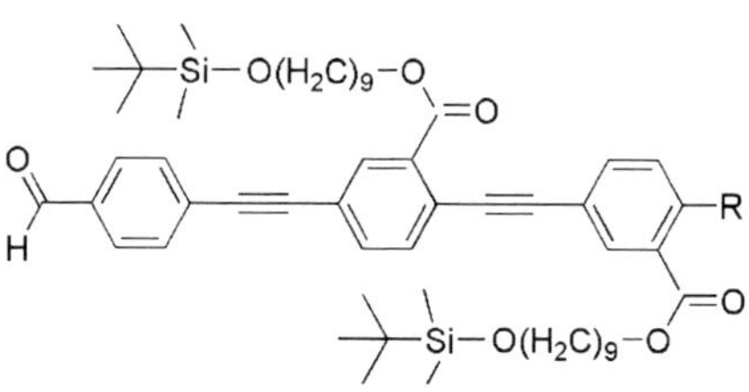

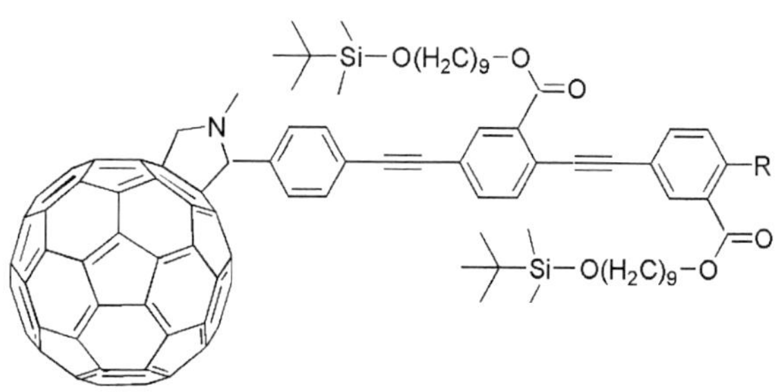

R= 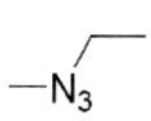

6g

R= 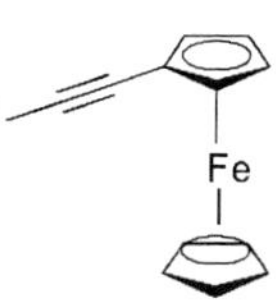

6h

R= 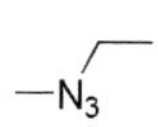

6i

R= 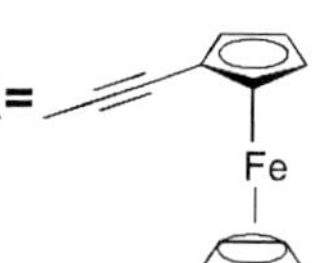

6l

Family 7. Mannose substituted PE oligomers and dendrimer.

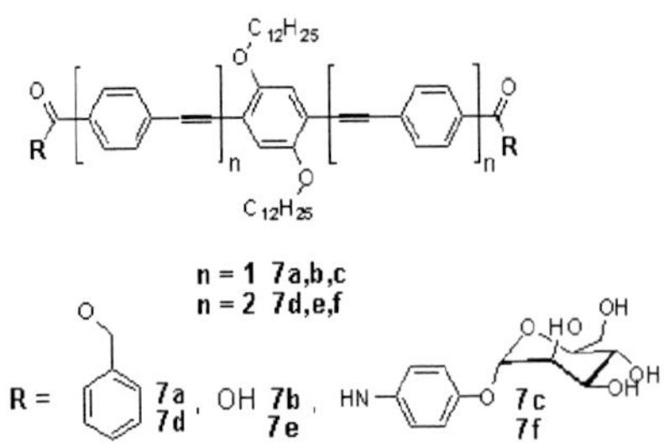

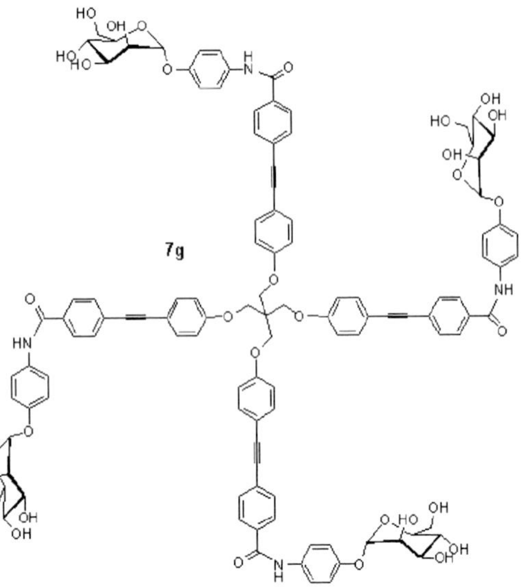

Family 8. (11-undecanol)benzoate ethynylenes.

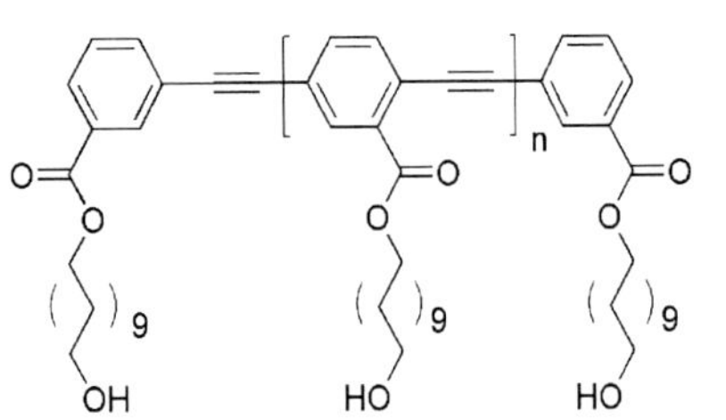

n=1,3,5 8a,b,c

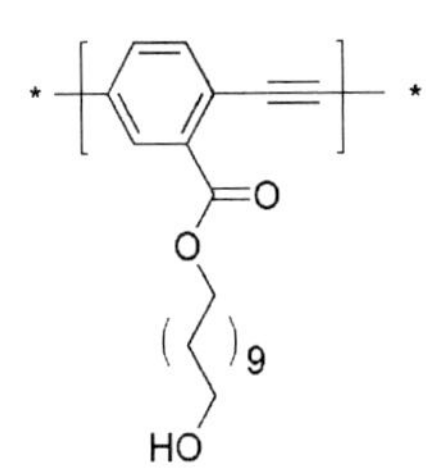

8d

Family 9. Pyridine PE copolymers.

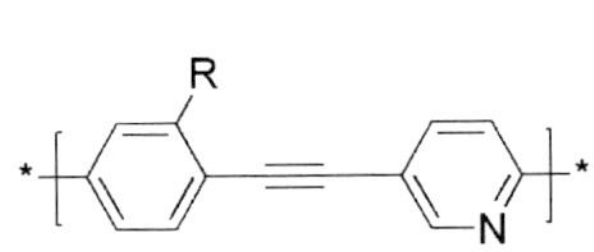

R= $-OC_4$ 9a
R= $-C(O)OC_{11}OH$ 9b

Figure 2. Some representative phenylene ethynylenes studied in our group.

PEs are usually considered as rigid rods due to the stiffness of the backbone. As a consequence, unsubstituted PEs are insoluble and can not be processed. As for other conjugated polymers, lateral substituents can be inserted that increase the conformational entropy of the system enhancing the solubility. These chains are usually long aliphatic groups, either directly attached (real aliphatic chains) or through –O- or –COO- moieties (ether, ester) [13a,b]. When a heteroatom is linked before the aliphatic chain, it is possible also to modulate the electronic properties as the ethers/esters are well known electron-donor/withdrawing groups. The length of the aliphatic moieties is, on the contrary, not relevant in what concerns the intrinsic optical properties but a marked effect in the solid state organization and optoelectronic properties is observed as it will be discussed in the following section. Even when the lateral substituents are present, the polymers of this class of PEs (rigid rod- like) are poorly soluble. They exhibit a high tendency to agglomerate even in solution giving rise to inhomogeneous films. During the synthesis, the longer polymer chains precipitate and low or modest molecular weights are usually obtained. A strategy to obtain more soluble materials and with optical properties competitive with those of the corresponding polymers, is to synthesize oligomers. As it was reported for several PE series [14], the optical properties of a heptamer/octamer compete with those of their homologue polymer due to the saturation of the effective conjugation length. The UV-Vis spectra of a family of rigid rod-like phenyleneethynylene oligomers bearing cholesterol ester lateral substituents are reported in Figure 3 as an example.

The maximum absorption wavelength is centered at 294, 348, 366, 373 and 375 nm for the monomer, trimmer, pentamer, heptamer and polymer, respectively. The wavelength increase in passing from the monomer to the polymer is of 81 nm but of only 2 nm between the heptamer and the polymer as it is better described in the insert.

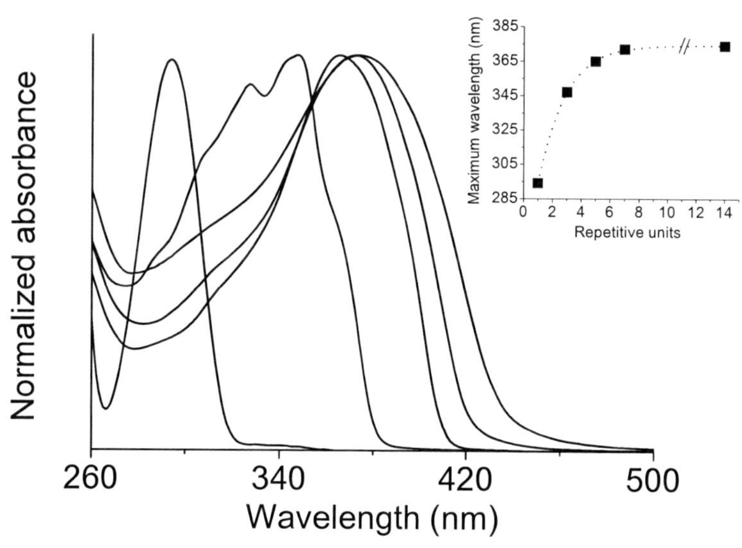

Figure 3. UV-Vis spectra of 1 (a,b,c,d,e) in $CHCl_3$ solution. Insert: graphic of the maximum wavelength vs. number of repetitive unit.

Oligomers thus represent modeling molecules for the understanding of the optical and optoelectronic properties of the corresponding polymer. Moreover, due to their monodispersity, they have been used as standards for the proper determination of the molecular weight of PPEs as poly(styrene) standards give overestimated MWs as a consequence of the incomparable hydrodynamic volume in solution of untangled vs. rod like structures [13, 14]. The optical properties of oligomers are however more markedly affected by termini groups than the corresponding polymer. Figure 4 shows the absorption spectra in $CHCl_3$ solution of trimmers 2a, 2b and 2c. The wavelength maximum shifts from 380 nm for 2a to 384/388 nm for Br/I terminated trimmers. Moreover, for 2b a shoulder at around 420 nm can be seen and likely associated to molecular π–π stacking. Along with the growth of the oligomer length, this effect decreases as the distance between the termini groups and the conjugated backbone increases. Passing from the H pentamer 2g to the Br homologue 2h, only a 3 nm red shift (contra the 8 nm shift of the corresponding trimmers) is observed. However, the termini groups can strongly affect the solubility and molecular organization in solid films, regardless of the conjugated backbone length, as discussed later on.

An alternative approach to obtain soluble phenyleneethynylene polymers is to insert flexible sequences in the main backbone i.e. synthesizing "semirigid" or rod-coil copolymers, see family n. 3 [15, 16]. In this case, the presence of not conjugated sequences breaks the electronic delocalization and the maximum absorption wavelength in chloroform is 384 (3a) and 394 (3b) nm, similar to the maximum of trimmer 2a. As a matter of fact, for these polymers, the conjugated segment is composed of triphenyleneethynylene units. Despite to the presence of quenching atoms (O and S), high quantum yields (0.71 for 3a and 0.59 for 3b) were determined in the chloroform solution.

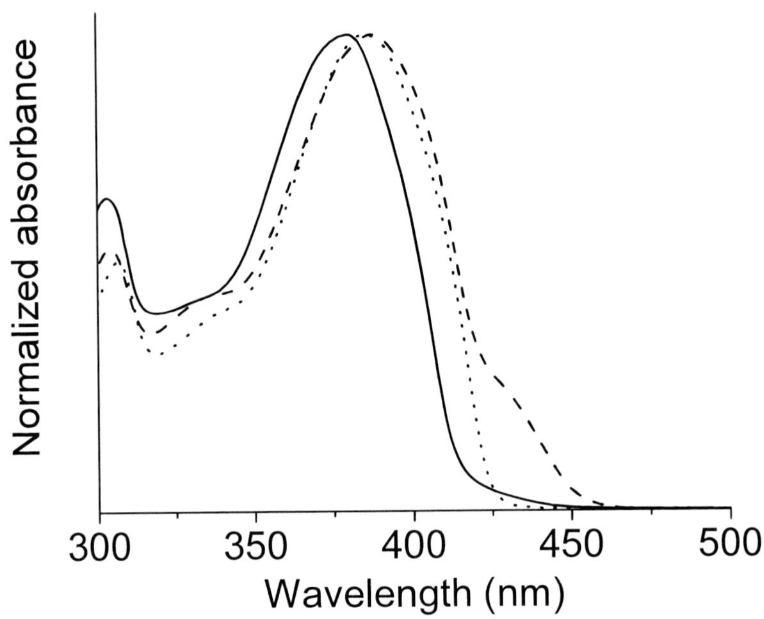

Figure 4. UV-Vis spectra of trimmers 2a (solid line), 2b (dotted line) and 2c (dashed line) in $CHCl_3$.

This emission property combined with the possibility to make layer-by-layer films with poly(aniline) and enzymes through H or S-S bonding as well as dipole-dipole interactions permitted to use this kind of PPEs in oLEDs or sensors [15, 16].

The conjugation of PE polymers can be modulated by adding in the conjugated backbone aryls with an electron-donor/withdrawing atom (family n. 4 and 9). Figure 5 reports the optical properties of family n. 4 in chloroform solution. 4a, 4b, 4c and 4d absorb at around the same wavelength (380 nm), while for 4e a marked red shift (of around 30 nm) is observed because thiophene is involved in the conjugated system increasing the electronic delocalization due to its electron donor character. Concerning the emission properties, the fluorescence spectra of 4a and 4b are excitonic, suggesting that a high molecular order is achieved for these two molecules in the excited state. In agreement with the absorption spectra, their emission maximum is almost the same (431 and 434 nm). 4c, 4d and 4e exhibit on the contrary a broad band at 522, 463 and 514 nm, respectively, giving large Stokes shift (141, 86 and 105 nm). The tendency of the emission maxima for these polymers with respect to their chemical structure is not in concordance with the behavior found for the absorption properties. The discrepancy observed for 4c and 4d can be likely associated with a large dipole moment that is generated in their chloroform solution and that affects markedly the excited state. For the copolymers of family n. 9, a blue shift is observed with respect to the analogous PE homopolymers in agreement with the electron-widrawing character of pyridine [17].

Another interesting family of copolymers is reported in family n. 5 (5e,f,g) where symmetric phenyleneethynylene moieties bear diacetylenic central units. Diacetylene structures are usually considered as by products of the Sonogashira Heck cross-coupling reaction to obtain PPEs, as homocoupling can not be avoided even if the reaction is carried under inert atmosphere [13b].

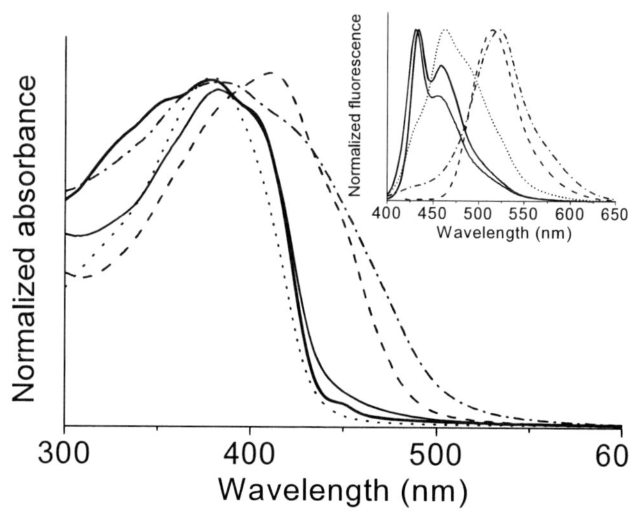

Figure 5. UV-Vis and fluorescence (insert) spectra in chloroform solution of: 4a (solid line), 4b (thick solid line), 4c (dash-dotted line), 4d (dotted line) and 4e (dashed line).

However, in our group, we have isolated and characterized these oligomers. The optical properties in solution are assigned mainly to the phenyleneethynylene units thus for instance comparing the maximum absorption wavelength of the dimmer 5a (350 nm), it is very close to that found for the dodecanoxy PE dimmer [356 nm, 18].

So far, we have discussed the chemical modulation of the optical properties of PE oligomers and polymers in terms of their functionalization aimed to increase the solubility and/or the conjugation of the system. These are two key factors for application in optoelectronics as already mentioned. Specific groups can be inserted in the main backbone or in the lateral substituents also for favoring a defined functional property.

For solar cells application, it is necessary to synthesize "push-pull" molecules. Fullerene presents a high electron affinity that makes it a good candidate as electron-withdrawing group of novel structures with potential photoconducting properties while ferrocene and triazene are electron-donor groups. The fulleropyridines family n. **6** is obtained by dipolar cicloaddition of the corresponding carboxyaldehyde terminating oligomers with C60. The absorption maximum wavelength is red shifted by increasing the conjugated backbone length, passing from 322 nm for 6a to 347 nm and 355 nm for 6d and 6h, respectively. However, the electron-withdrawing effect of fullerene reduces the conjugation, obtaining the maximums for the corresponding fullerene derivatives, at 315 nm (6b, shoulder) 326 nm (6f) and 349 nm (6l). By other side, it expands slightly the absorption range as it can be seen in Figure 6, where the UV-Vis spectrum of 6l is compared with that of its precursor, 6h. The fullerene derivative, in fact, presents, with respect to its precursor, an additional absorption in the UV and a slightly more intense absorption in the visible (see insert) due to the C60 substituent.

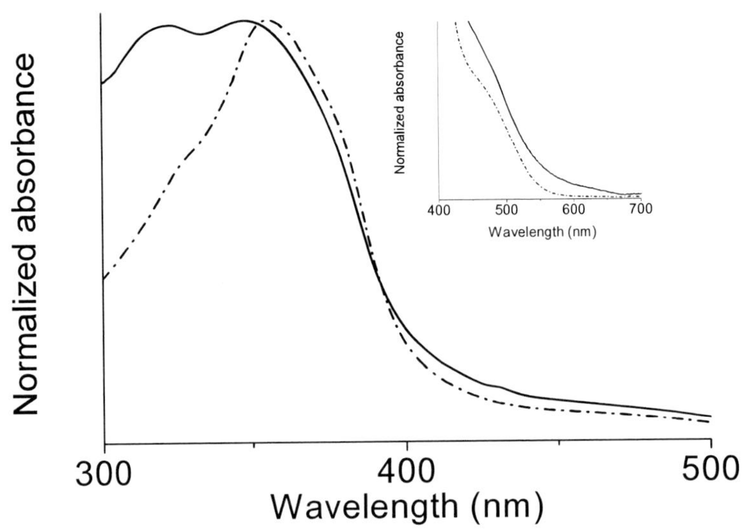

Figure 6. UV-Vis spectra of the fulleropyridine 6l (solid line) and its aldehyde precursor 6h (dash-dotted line) in chloroform solution. Insert: magnification in the visible range.

For fluorescent sensoring or biosensoring applications, the PEs are to be substituted with "receptor" groups that are able to bind or receive (in the host-guest model) an analyte (either chemical or biochemical molecule). As previously mentioned, 3b forms layer-by-layer films where enzymes can be deposited on the top in order to develop optical biosensors [15]. A biosensor is an integrated device constituted of a detection element (transducer) and a receptor unit. In this case, the receptor is the enzyme and the conjugated system is the transducer. The two parts of the device are in contact through not covalent bonds and in different layers. An alternative approach is to synthesize molecules that present in the same structure the two units as the mannose PE derivates of family n. **7**. Mannose is an oligosaccharide with well known capacity to bind specifically to the type I *fimbriae* of *Escherichia coli* bacteria [19]. When the detection of the bacteria by fluorescence is searched, two points must be considered. 1) Mannose is a highly quencher molecule as it contains several OH groups in the structure. 2) If the detection is desired in solid state films (real biosensors), the molecules must be insoluble in water and vice versa for detection in solution (where at least it must be soluble in a solvent miscible with water such as THF, DMF, methanol in order to be added to the bacteria culture). On the basis of these requirements, the family 7c and 7f was synthesized starting from their benzoate protected molecules (7a and 7d), that were later benzyl hydrolyzed to obtain the carboxylic acids (7b and 7e) free for their further condensation reaction with 4-aminophenyl-α-D-mannopyranoside. The dendrimer 7g is obtained starting by a Williamson alkylation of 4-iodophenol with pentaerythritol tetrabromide to obtain a first generation that later undergoes a Sonogashira-Heck cross-coupling with benzyl 4-ethynylbenzoate giving a benzyl protected benzoate dendrimer. After cleavage of the benzyl group, the carboxylic terminated dendrimer is subjected to an amidation reaction with 4-aminophenyl-α-D-mannopyranoside to obtain **7g** in four steps. In general, the quantum yield ϕ decreases when the mannose is substituted in agreement with what expected. Nonetheless, even the shorter oligomer **7c** and despite its low fluorescence properties (ϕ= 0.023 in DMF) permits the detection of the *E.coli* by fluorescence and by Laser Scanning Confocal Microscopy (LSCM, in fluorescence mode) (Figure 7).

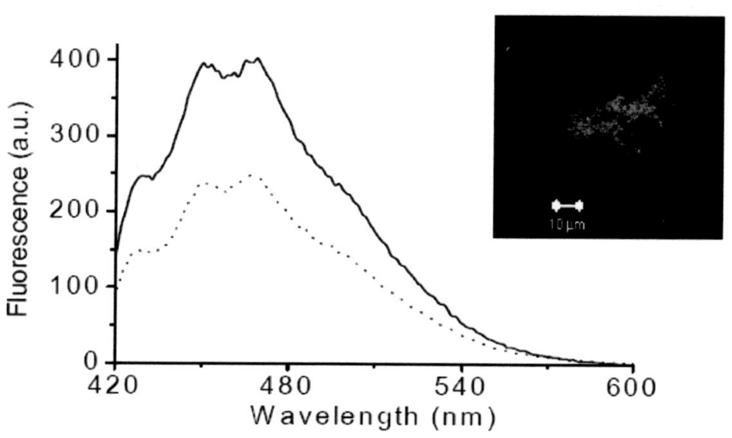

Figure 7. Fluorescence spectra of trimmer 7c in a buffer solution with (bacteria sample, solid line) and without (blank sample, dotted line) *E. coli*. Insert: laser scanning confocal micrograph (in fluorescence mode) of *E.coli* agglomerates assembled with 7c.

The trimmer **7c** emits in the buffer water solution with two peaks at 452 and 469 nm. The fluorescence intensity however increases when it is assembled with the bacteria (that does not emit when it is alone). This result could be explained considering that several washing and centrifugation cycles have been realized in both samples (blank and bacteria sample) in order to be sure to eliminate any oligomer excess. The fluorescence of the blank sample is thus due to the quantity of the oligomer that remains dissolved in the buffer solution. The solubility increases when the bacteria are present as they link the oligomer through the assembly (because of the interactions between the mannose and the *fimbriae*) in fluorescent agglomerates that can be easily detected by laser scanning confocal microscopy (insert).

Molecular and Supramolecular Organization in Solid State

The molecular and supramolecular organization of PEs in solid state is driven by the chemical structure of the molecules and (for the films) by the deposition parameters. For sake of simplicity, the results on bulk and on thin films will be separately presented in the following.

Bulk

According to literature, classic Schlieren textures have not been observed in phenyleneethynylene (PE) polymers, until Weder [20] and Bunz [21] reported for the first time a "nematic" liquid crystal (LC) phase of lyotropic type in the case of alcoxy substituted PPEs (Weder) and a nematic coexisting with a smectic phase in the case of Bunz for ultrahigh molecular weight alkyl substituted PPEs. Bunz [13b] proposes for these mesophases the term "smatic" as they correspond to a nematic phase according to the texture observed by polarized optical microscopy (POM), while their X-ray diffractograms are consistent with a smectic mesophase. However, he remarks that electron diffraction microscopy is required to clearly identify the exact nature of these mesophases. With the exception of these works, classical LC textures are rarely reported for PPEs. The lack of texture in PPEs can be understood on the basis of the high π-π intra-molecular interactions that exist between the polymer chains; inclusive, longer the chains are and aggregation rather occurs. Although, the birefringence observed by POM in the fluids of some substituted PE, their temperature behavior exhibited by differential scanning calorimetry (DSC) combined with the X-ray diffraction patterns (SAXS), indicate that even lineal rigid rod-like PPEs can present LC properties. All these features, however, must be taken in account as a whole for the correct description of the material solid state transitions. In fact, the X-ray patterns in function of temperature as an individual analysis usually do not give relevant information on mesophases because the enthalpies associated with the transitions are insignificant and the PPEs tend to degrade before melting, thus the isotropic state is almost inexistent supporting the theory that π-π interactions are really strong for these materials. Moreover, the interpretation of an X-ray diffractogram can be ambiguous. For instance, the appearance of a broad halo at wide angles with very sharp reflections at low angles in an X-ray pattern can be considered evidence of a LC phase, however it can also be due to a poorly developed microcrystalline phase (in small blocks), embedded in a liquid-like amorphous matrix. This kind of materials is named "sanidic" [22] or board-like according to an interesting work on rod-like polyesters and polyamides (non PPEs type) substituted with long alkyl chains reported by Voigt-Martin [23]. The author finally concludes that electron diffraction combined with high resolution imaging microscopy is a better tool for the understanding of the material transitions. The term sanidic has been adopted to classify also some PPEs as those reported by Bunz, mainly because of their similitude with bricks.

The introduction of side chains in the conjugated backbone allows: 1) to increase the solubility of the polymers, 2) to obtain a specific mesophase or 3) to reduce the transition temperature that could induce molecular orientation both spontaneously or by mechanical techniques such as rubbing, with the aim to obtaining polarized emission light. It is to be mentioned, however, that until now few PPEs are reported to present both light emission and a specific mesophase. At this respect, in our group, we have designed and synthesized polymers and oligomers (as molecular models) that could satisfy these requirements (family n. 8).

The amphiphilic character of the (11-undecanol) benzoateethynylene oligomers and polymer of family n. 8 permitted to obtain Langmuir-Blodgett multilayer films on hydrophilic substrates. According to the X-ray reflection patterns (Figure 8a), the molecules are assembled in the film with Y-type supramolecular arrangement; in other words, molecules adopt a two comb-like organization, where chains are oriented in opposite direction, giving rise to an hydrophilic surface after each transfer [14b, 17, 24]. These well structured films showed Kiessig Fringes and a Bragg peak consistent with a calculated periodic spacing of around 3.7-4.0 nm (Figure 8). The molecular dimension in a totally extended conformation is 2.0 nm long, revealing that the self assembly occurs in double layers with a slight tilting of chains within the film. In powder, the X-ray diffraction patterns of 8d suggest a disordered smectic C phase with a lamellae spacing of 3.5 nm coexisting with a crystal phase. We believe that these crystals are rather a block of molecular aggregates that do not disappear with temperature. Interestingly, the lamellae spacing is of 3.7 nm; value that corresponds to the model already described in films (of two inverted combs). As what concerns the oligomers, the trimmer of the same series has crystal behavior with lamellae spacing of 4.0 nm. The pentamer exhibits a disordered smectic phase with a partial order of the alkyl chains and it is regarded as a hexatic type with a lamellae spacing of 4.2 nm. The heptamer presents rather a disordered smectic C phase with lamellae spacing of 3.9 nm. From this series of molecules, we conclude that the 11-undecanol esters side chains are amphiphilic enough to elaborate very stable Langmuir-Blodgett films through hydrogen bridges formed between the OH at each layer, and this latter origins the particular supramolecular auto assembling in bilayers.

As it will be reported in the following, we also succeeded in depositing 2,5-bis(alcoxy)phenyleneethynylenes 2f, 2i and 2l by layer-by-layer self assembly technique from dilute solutions. By monitoring the film growth by UV-Vis spectroscopy, it was possible to demonstrate a regular transfer of matter at each deposition according to the Lambert Beer law.

(a)

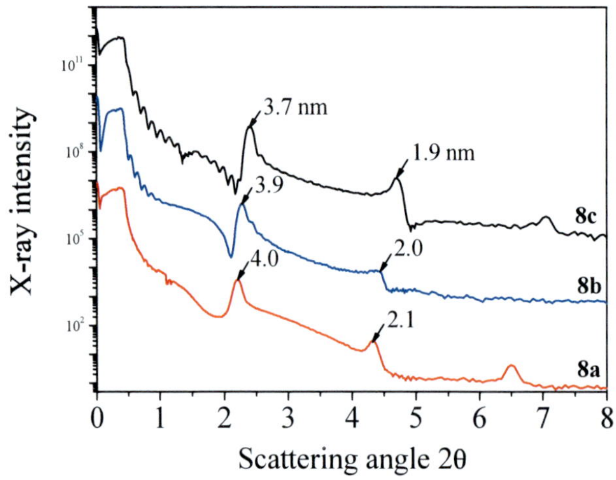

(b)

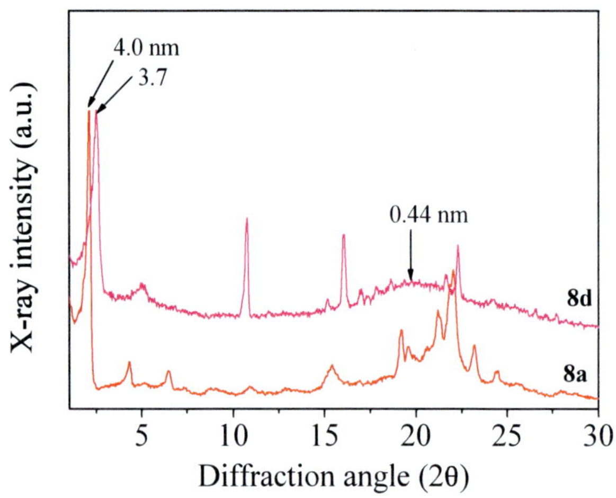

(c)

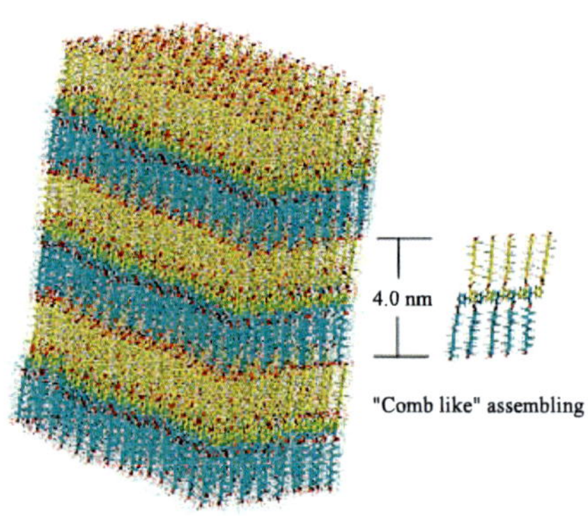

Figure 8. (a) X-ray reflectivity plots of Langmuir-Blodgett multilayer films of 8a, 8b and 8c showing the Bragg peaks corresponding to the periodicity within the films. (b) X-ray diffraction patterns of 8a and 8d powder at 50 °C. (c) Molecular arrangement in “comb-like” bilayers as deduced from both, the X-ray diffraction in powder and X-ray reflectivity in Langmuir-Blodgett films. Note that surface is always highly hydrophilic.

The films presented by AFM a flat and homogeneous topography. However, their X-ray reflection patterns never showed the typical Kiessig Fringes usually obtained in well structured films and that could confirm the molecular order within the layers. Therefore, we realized their X-ray diffraction from powders and we obtained relevant information of the molecular organization. All molecules exhibit at room temperature a periodic structure at small angles indicative of lamellar systems. While the trimmer 2a presents a single first ordered peak, the pentamer 2f, heptamer 2i and the polymer 2l exhibit rather two first ordered peaks, Figure 9a. The lamellae spacing periodicity is equal to 3.54 nm for the peak of 2f, and 3.39 ± 0.12 nm and 2.60 ± 0.88 nm as average value deduced from the two peaks for 2f, 2i and 2l. At higher angles a series of half broad bands covering from $2\theta = 16.0$ to 23.5° with spacing of 5.2, 4.47, 4.5, 4.0 and 3.87 nm, being these reflection bands associated with the order of the dodecanoxy side chains; the average distance between two chains is of 4.0 nm. On the basis of the molecular extended conformation from aliphatic to aliphatic chain, Figure 9c, the series has a length (l) of 3.68 nm, suggesting that the alkyl chains of the trimmer 2a are 16° tilted with respect to the normal, while for 2f, 2i and 2l the two peaks correspond to average angles of 16° and 44 ± 2.0°. By DSC; the trimmer 2a is the only molecule to show well defined melting-crystallization peaks (Total ΔH_{melt} = ±75 J/g) suggesting a crystalline behavior, while 2f (±40 J/g), 2i (±23 J/g) and 2l (±58 J/g) give a series of broad endothermic-exothermic transitions. The relative low melting enthalpy of 2f, 2i and 2l with respect to 2a is an indicative of the high polymorphic character of these materials; hypothesis that was corroborated later by POM. From Figure 9b, it can be seen how very small crystals are forming from the melting and are randomly distributed. After analyzing this series of 2,5-bis(dodecanoxy)phenyleneethynylene molecules, we conclude thus that the term "*sanidic*" proposed by Bunz for PE polymers can also be applied to these oligomers that match better with brick shape.

Polymers can also assemble in double layers with interdigitations of their side chains when a benzoate or di-alcoxy benzene are copolymerized with a non substituted aryl, such as benzene or pyridine. Furthermore, if the substituted lateral chains have an OH and the other aryl is pyridine, the assembling can be carried out by hydrogen bridges with the nitrogen group as is the case of polymer 9b [17], which shows, by X-ray analysis, very particular features such as a single LC phase in function of the temperature and strong π-π interactions between the aryls with spacing of 0.36 nm. The butoxy copolymer 9a presents interdigitation with lamellae spacing of 7.8 nm suggesting an assembling in double layer. However, at high angles, the diffractogram shows similar features to those already discussed for the dodecanoxy series and in other related structures [25] such as bands covering from $2\theta = 13$ to 26° with the difference that in this case, the bands are sharper being the most intense that corresponding to a spacing of 3.98 nm. In general, this copolymer exhibits likely a more crystalline phase than the polymer 2l.

(a)

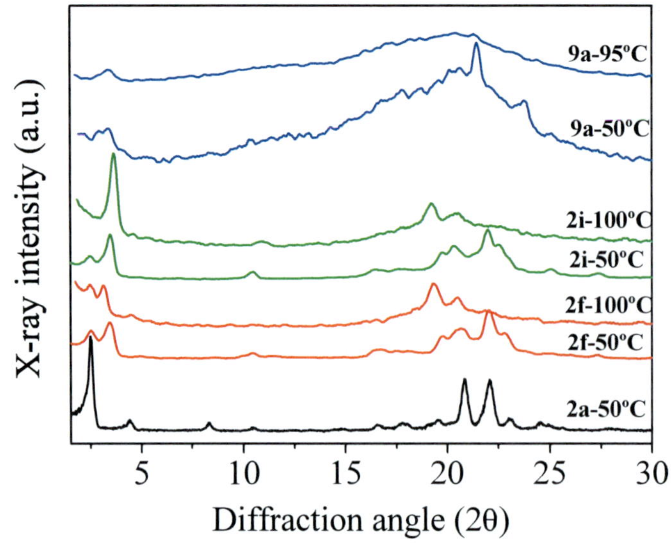

(b)

112 °C

105 °C

85°C

Figure 9. (Continued on next page.)

45°C

(c)

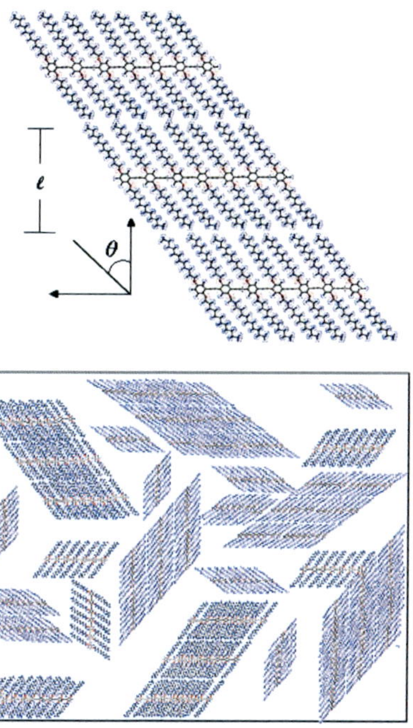

Figure 9. (a) X-ray diffraction patterns of 2,5-bis(dodecanoxy)phenyleneethynylene 2a, 2f, 2i oligomers and the polymer 2l in function of the temperature. (b) Polarized light microscopy photos taken at the temperature indicated during the cooling cycle of the heptamer 2i. (c) Molecular arrangement deduced from the X-ray patterns in powder and picture that schematically represents the formation of small crystals during melting.

Thin Films

As what concerns the solid state films, as previously mentioned, two main factors are important: 1) the chemical structure of the molecules as for the bulk (intrinsic factor) and 2) the experimental conditions for the film preparation (extrinsic factors), including solvent, concentration, deposition technique, surface substrate treatment, annealing, and orientation.

The combination of these two parameters can yield different molecular and supramolecular arrangement that reflects in specific optical, morphological and optoelectronic properties.

Effect of the Chemical Structure

In order to consider the intrinsic effect of the chemical structure of PEs on their molecular and supramolecular assembly, in this section, samples prepared by spin coating from chloroform solutions of the same concentration and under identical spin rate, will be considered for the discussion unless otherwise stated.

The rigid rod-like PEs such as those of families n. **2** give crystalline films even if the density, shape and size of the crystals vary depending on the molecule. As a consequence, the homogeneity changes from sample to sample. Figure 10 presents a representative AFM image.

No effect of the termini group was observed for the trimmers where the morphology and roughness of the films are identical. Nonetheless, as previously mentioned, the presence of the bromide as terminal groups is very important for the butoxy pentamers. 2g, in fact, do not form a continuous deposition while 2h gives very homogeneous films. In order to improve the processability of 2g and rather profiting its intrinsic optical properties, blends of 2g and 2f were prepared [26]. The morphology of the blend changes along with the relative percentage of 2g/2f as shown by the images of Figure 11 obtained by LSCM. The 100/0 film presents a very irregular topography, characterized by holes. Films prepared with 80/20 relation show structures similar to dendrites while a morphology with "crowns" of 1250 microns of diameter is observed when the blend has a 65/35 composition.

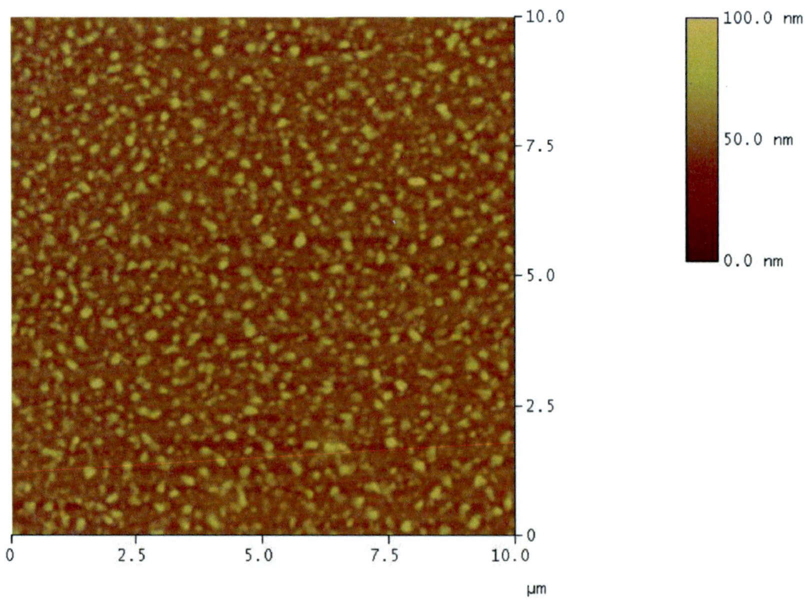

Figure 10. Tapping mode AFM image, of a film of 2m prepared by spin coating on glass substrate.

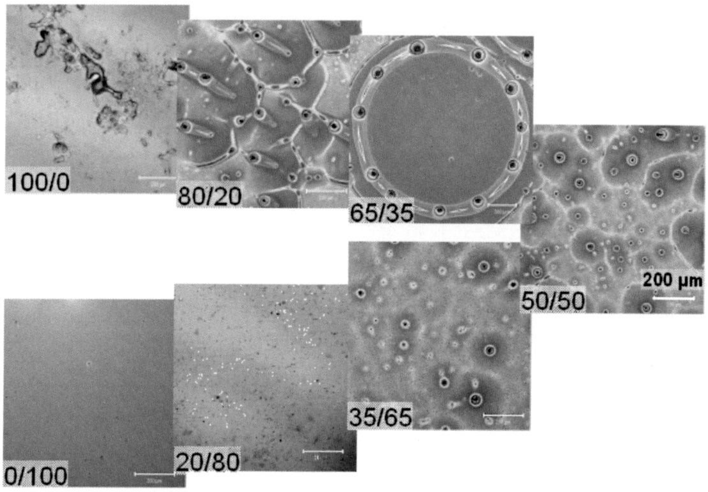

Figure 11. LSCM contrast images of 2g/2f blend films on glass substrate.

The presence of these structures can be ascribed to microphase segregation in the blend. Taking as reference the LSCM image of Figure 12, it is possible to distinguish three zones: the crown zone, the internal zone limited by the crown and the external zone of the crown. The optical textures by POM (Figure 12, right) are consistent with polycrystalline morphology in the three regions where the size, shape and density of crystals are different depending on the zones. This likely provides differences in the macroscopic properties of the material from area to area [27].

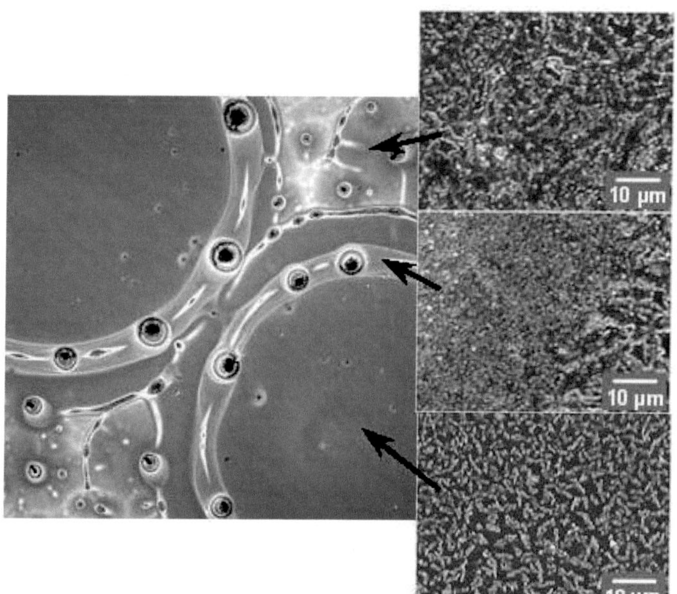

Figure 12. LSCM contrast image (left) and POM optical textures for three zones (right) of 65/35 film on glass substrate.

By increasing the proportion of 2f in the blend, the circular structures vanish improving gradually the homogeneity of the topography up to give a flat surface for the 0/100 film, suggesting that the quality of the films depends on the length of the alkoxy chains as could be expected, considering that the origin of the aggregation is the inter-chain attraction forces between phenyleneethynylene chains [13a]. The length of the alcoxi chains affects the molecular and supramolecular arrangement of PPEs that reflects not only on the morphological but also on the optical properties of the films. Figure 13 shows for instance the UV-Vis spectra of spun films of the dodecanoxy (2a) and butoxy (2d) trimmers.

The film of the dodecanoxy trimmer shows a typical excitonic spectrum indicating a high molecular order in the solid state. On the contrary, the absorption spectrum of the 2d film is very broad with no excitonic features. The baseline is flatter for the dodecanoxy trimmer sample suggesting that the scattering from crystals is very much lower. The presence of long chains seems to favor a regular packing of the molecules that is more evident when two dodecanoxy units are bonded by a diacetylenic group as in the family **5**. In fact, for PPEs of family n. **2**, only the trimmer and pentamer exhibit excitonic spectra. For higher conjugated systems (heptamer and polymer), the spectra are broad and unresolved similarly to the case of **2d**. On the contrary, all the molecules of family n. **5**, despite to the length of the conjugated part, show excitonic spectra. It is well known that diacetylenes present excitonic spectra due to the high molecular order likely associated with the symmetrical structure of these molecules [28]. For the molecules of family n. 5, it is thus not surprising to obtain these spectral features. For 2a (2b,2c) and 2f, it can be argued that they can be assembled as the symmetric oligomers.

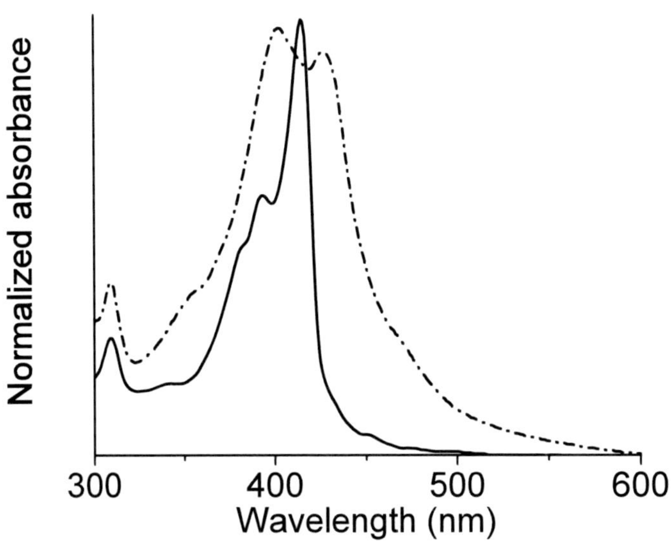

Figure 13. UV-Vis spectra of films of 2a (solid line) and 2d (dash-dotted line) prepared by spin-coating on glass slides.

Despite the fact that poly(diacetylene)s are usually not fluorescent, the presence of the central diacetylenic moiety in the oligomers of family n.5 gives rise to rather high quantum yields in thin films; for instance spun films of 5b exhibit quantum yield of 0.35 *contra* 0.18 of the dodecanoxy pentamer 2f (that is the most comparable corresponding PE oligomer).

The alternation of rigid rod-like PE segments with flexible sequences in the **3** family, gives a granular type morphology (Figure 14). A similar behavior was reported for other conjugated di-block copolymers and due to micelles of rod-coil molecular aggregates [29].

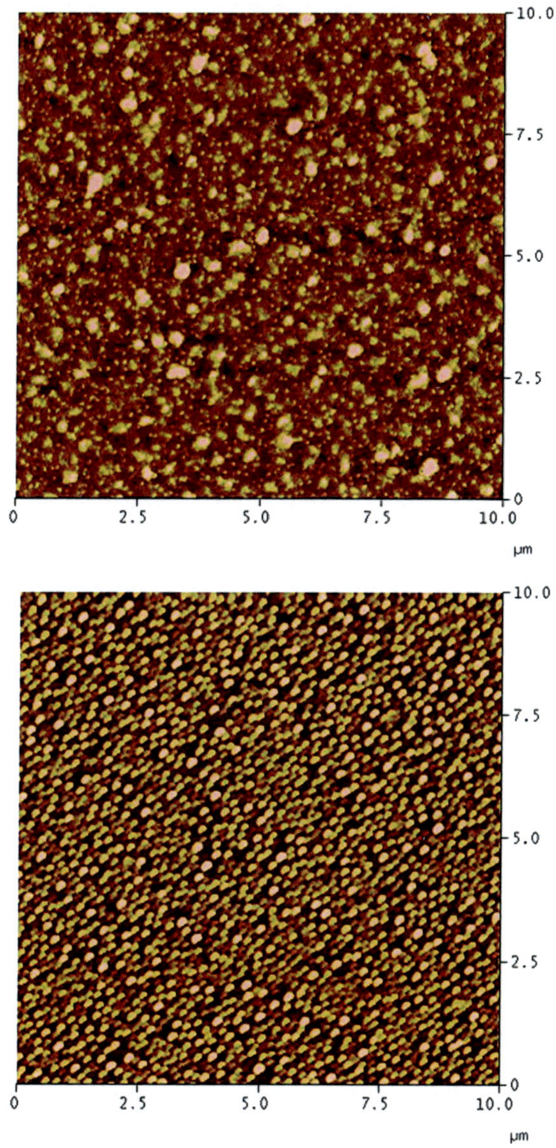

Figure 14. AFM Tapping mode bidimensional image of films of 3a (top) and 3b (bottom) prepared under the same conditions (glass substrates, spin-coating, 2000 rpm, 1 g/l $CHCl_3$ solution).

Finally, molecules of family n. **8** permit the formation of Langmuir-Blodgett films with worm-like morphology with folding zones [14b]. For the pyridine copolymers, the interdigitation of different molecular chains as previously discussed, gives rise to very homogenous films with extremely flat surface [17].

Effect of Film Preparation Conditions

A large variety of extrinsic parameters can affect the molecular and supramolecular assembly of PEs. Their effect is very complex to be resumed in a brief paragraph as they are strictly associated with the intrinsic chemical structure of the molecule. Moreover the change of a parameter (for instance the solvent) can have different results depending on other extrinsic factors, for instance it can be observed an effect of the solvent variation when spin coating is used but not for self-assembly. In this section, only the most representative examples will be presented in view of their effect on the final efficiency of the optoelectronic device.

The surface substrate treatments can be usually divided in hydrophilic and hydrophobic ones as the main goal is to obtain or increase the affinity of the molecule to be deposited on the specific substrate. Glass and quartz slides are commonly hydrophilic substrates but the handling and storing usually change its surface character. Treatments with sulphochromic mixture or piraña solution are the most reported methods to both clean and recover the hydrophilic character of the substrate surface. For obtaining hydrophobic substrates, a thin layer of either aliphatic or aromatic thiols is deposited by self-assembly according to the methodology first described by Netzer and Sagiv [30]. PEs present in general a higher content of hydrophobic groups due to the conjugated backbone. The hydrophilic character can be imparted by the lateral substituents or the copolymerization with polar sequences. For the molecules of Figure 2, the families n. 2, 5 (alcoxi rigid rod-like PEs), 4, 1, (ester rigid rod-like PEs) and 6 (fulleropyridines) are practically hydrophobic while 3, 7, 8 and 9b are amphiphilic. Despite to this different character, all the molecules can be deposited on hydrophilic substrates (glass, quartz and ITO) probably because of electrostatic interactions.

For the rigid rod-like PEs, among all the possible extrinsic factors, only the deposition technique has an effect on the optical and morphological properties of the films. For these molecules, we thus report here on films prepared by spin-coating and self-assembly from chloroform solutions on hydrophilic substrates. The concentration of the solution and other parameters related to each technique (such as spinning rate, immersion time etc.) will be reported only when necessary and generally fixed in order to give a comparable thickness with the two techniques. The optical and morphological properties of the films prepared by spin coating have been described in the previous paragraph. All the molecules can be also deposited by self-assembly for simple dipping of the substrate in the PE solution. By repeating the dipping process several times, it is possible to construct multilayers [31]. The regular built up of the multilayer is driven by the same interactions that occur in the solution (and that eventually are responsible for agglomeration), i.e. π–π interactions between the conjugated backbones. A major factor in the formation of a self-assembled layer, is the solubility of the PE; lower it is, easier is the formation of the film. However, the deposition can be favored by optimizing some experimental parameters as the immersion time of the film in the solution, the concentration, immersion/emersion rate and solvent or post-deposition treatments (as, in the case of the multilayer, the drying of each layer before to be immersed in the solution for the subsequent deposition). As an example, the absorbance (and consequently the thickness) of a layer of 5d decreases with the time of immersion of the substrate in the solution (Figure 15), indicating a high solubility for this system. This limits strongly the process of multilayer, where a maximum of only 4 depositions can be achieved.

For family n. 2, analogously, the trimmers can be deposited with a regular increase of thickness only up to three layers. The polymer 21, that is less soluble, permits on the contrary to repeat the dipping cycles up to 9 times.

When self assembly is used, a higher homogeneity and regular thickness with respect to spin-coating are usually observed for the films. As a term of comparison, in Figure 16, the AFM Tapping mode images of films of 21 prepared by spin-coating and self-assembly are reported.

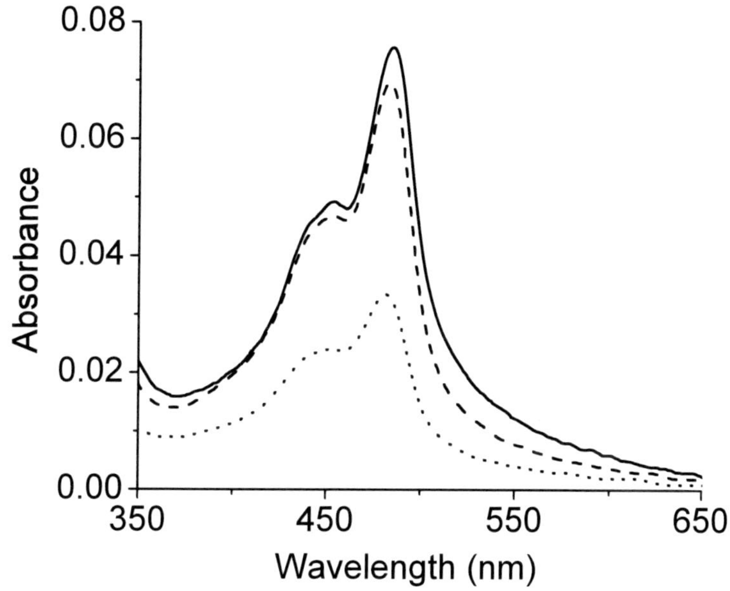

Figure 15. UV-Vis spectra of 5d self assembled films on quartz slides. From top to bottom: 5, 10 and 20 minutes of immersion of the substrate in 2 g/l $CHCl_3$ solution.

The spun film shows a columnar topography as it can be observed in the corresponding height profile, the roughness (Rq) in this scanning area is 20.0 nm. For the self-assembled film, the columns are very sharp appearing rather as needles. Rq is, in this case, equal to 11.0 nm. As a consequence of the higher homogeneity of the self-assembled film, its UV-Vis spectrum presents a very lower baseline and more resolved peaks with respect to the sample obtained by spin-coating.

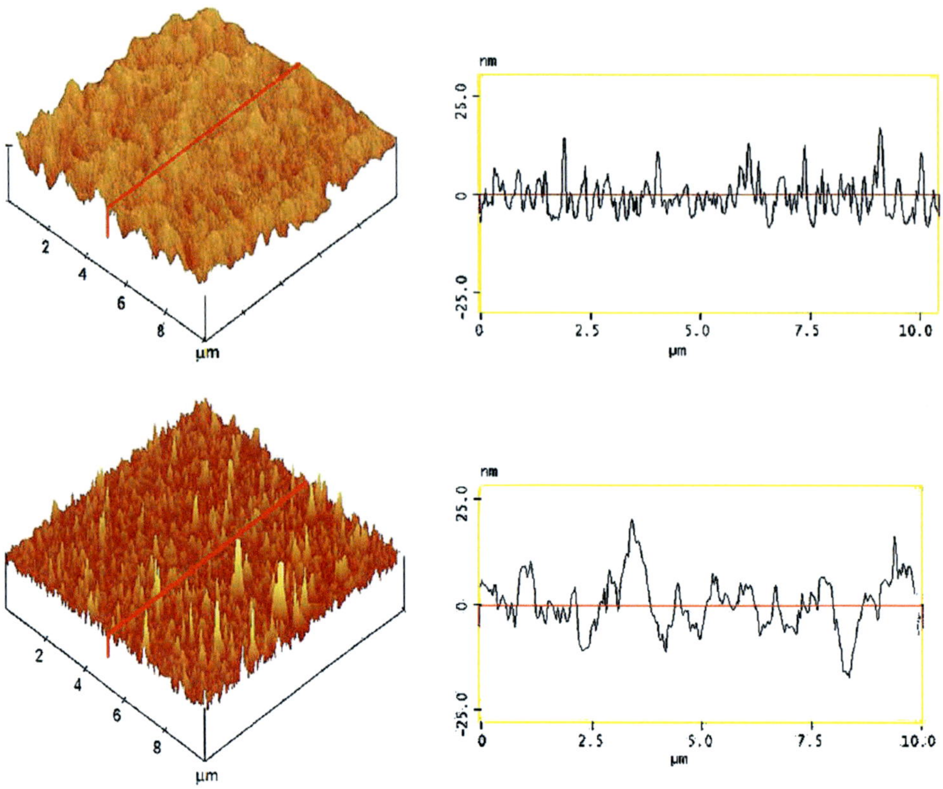

Figure 16. AFM Tapping mode images of films of 2l prepared by spin-coating (top) and self-assembly (9 layers, bottom) from 2 g/l $CHCl_3$ solutions on glass substrate. Left side: tridimensional image; right side: section analysis showing the height profile corresponding to scanning in the line marked as red.

Contrary to what observed for the rigid rod-like PEs, an interesting behavior has been found for the semirigid PE 3b films on changing the experimental conditions of self-assembly. For this polymer, a granular type morphology was observed for films either prepared by spin-coating (see Figure 14) or self-assembly [15]. However, the grain size changes between the two techniques and in generally, starting from chloroform solutions of the same concentration, spin coating gives a more homogeneous distribution in the grain size as it can be deduced by the comparison of the AFM image of Figure 14 with that reported in Figure 17, right side [32], probably because the higher rate for the solvent evaporation that is achieved by spin-coating reduces the agglomeration process.

By repeating the dipping cycles, a multilayer can be formed and the morphology is retained layer-by-layer. A marked effect is observed, however, when the solution concentration (and consequently the thickness) is increased. In Figure 17, the bidimensional AFM image in a 10X10 µm area for films prepared by self-assembly (1 layer) with different thickness is reported. In the right image, a granular morphology is observed as previously discussed. For the thick sample (left side image), the morphology is rather worm-like. Inclusive it seems to be formed of domains where the material “coils”.

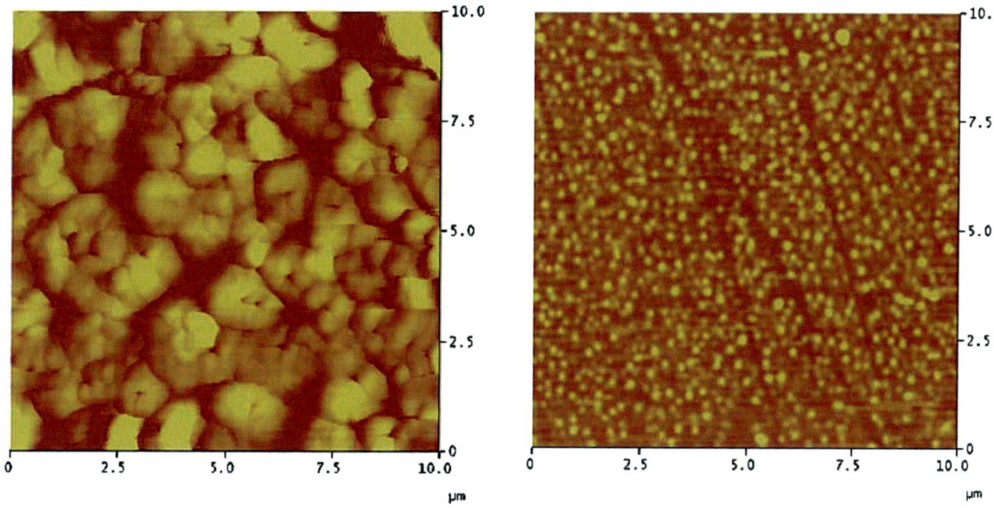

Figure 17. Tapping mode AFM image of films of 3b prepared by self-assembly (1 dipping) from $CHCl_3$ solution of 100 nm (left) and 0.5 nm (right).

The effect of the solvent [33] is observed only on the quantity of deposited material being higher for more volatile solvents (such as chloroform with respect to toluene) and the homogeneity and size of the grains. Toluene gives bigger grains and a broader distribution of sizes. The same result is obtained when a low dipping rate is used and could be ascribed to the fact that both conditions reflect in a higher possibility that the molecules agglomerate.

Effect of the Supramolecular Arrangement on the Optoelectronic Properties of Light Emitting Diodes

In this section we restrict the discussion of the effect of the supramolecular organization of our PEs on the intrinsic optoelectronic properties of light emitting diodes [34] just to give a general panorama.

In Figure 18, the current density-voltage and the bright-voltage characteristics of 2f are reported as example of a typical ITO/PE/Al oLED device behavior. Table 1 reports the threshold voltages for diodes of some representative molecules and related thickness.

The bright is proportional to the current indicating that the quantum efficiency is constant over a wide range of currents and in agreement with the results reported by Zou et al. where this phenomenon was attributed to the orientation of internal dipoles [35].

The threshold voltage corresponds to the minimum bias necessary in order that this orientation occurs allowing the flux of charges in the emitting layer. As the capacity to orient internal dipoles in presence of an electric field (i.e. the polarizability) of a molecule increases directly with its conjugation, the V_{th} values decreases within the family n. 2 passing from the trimmers to the polymers. The effect of the aliphatic chains is not very marked as it can be deduced from the comparison of the V_{th} value for 2b (dodecanoxy) and 2d (butoxy) trimers.

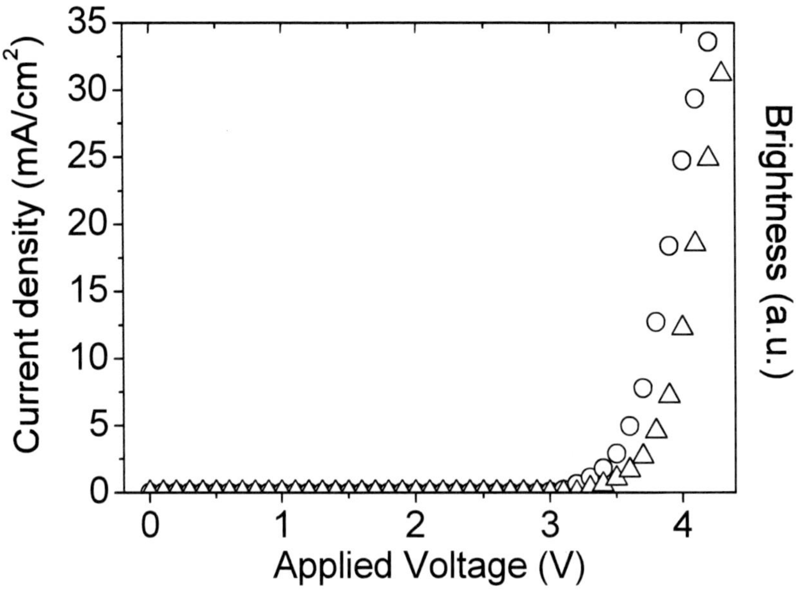

Figure 18. Current density (squares) and bright (triangles) versus applied voltage for a ITO/2f/Al oLED.

The diode of the butoxy polymer 2m presents a very lower value with respect to the dodecanoxy analogue. Nevertheless, the comparison is not reliable in this case due to the different thickness. In fact, it is well known that the threshold voltage depends on the thickness of the emitting layer [36]. So far, the effect of the film preparation technique is not evident. In fact, the V_{th} of oLED of 2l decreases slightly passing from spin-coating to self-assembly. The thickness of the self-assembled layer is lower than that of the spun film and as the tendency of the threshold voltage with thickness goes in the opposite way, this slight difference in the V_{th} values can not be ascribed to thickness effects. The same behavior is observed in the surface conductivity where in general, self-assembled films, give higher values with respect to spin coating samples. As this property is independent on the thickness of the layer, this result confirms the relevance of the difference observed in the threshold voltages. Finally, both experiments suggest a higher conjugation in the self-assembled films that could be ascribed to a more ordered packing of the molecules of these samples in agreement with their better optical and morphological properties. The deposition technique seems to not affect the electroluminiscent properties; for instance for the polymer 2l, oLEDs prepared by spin-coating or self-assembly (with both 100 nm of thickness), give the same EL spectrum with a maximum at 508 nm and a quantum efficiency in the range of 10^{-3} % at rather low threshold voltage (4V). It is to be remarked that so far, self-assembly has not been previously reported as preparation technique for oLEDs. The closest reference being the work by Hong et al. where electrostatic assembly was used for the elaboration of diodes based on poly(4-vinylpyridine-co-vinylpyridinium chloride) [37].

Table 1. Threshold voltages of light emitting diodes of representative PEs of Figure 2.

Molecule	Thickness (nm)	V_{th} (V)
2a	96	4.0
2b	98	3.9
2c	105	4.0
2d	100	3.8
2f	115	3.2
2i	114	2.8
2l	80	2.0
2l*	76	1.8
2m	125	1.8

* OLEDs prepared by self-assembly (9 layers). All the others were obtained by spin-coating.

Recent Advances. Phenyleneethynylene Nano-Organization

Nanotechnology is spreading in the scientific word as the major research area due to the peculiar properties that can be achieved when a material is manufactured at nanoscale. The applications of nanomaterials are very broad including nanomedicine, catalysis, biotechnology, optics and optoelectronics [38]. Pure nanomaterials can be mono- (nanowires), bi- (nanofilms) or three-dimensional (quantum dots and nanoparticles). Hybrid nanomaterials or nanocomposites can also be found in literature as combination of a nanostructure with a macro-material. In the previous sections we presented the chemical modulation of the optical and optoelectronic properties through a proper molecular design of PEs. In this section, we report on the possibility to reach the same goal by reducing the material size in nanostructures.

Gold and silver nanoparticles garnered a lot of attention due to the interesting electrical and optical properties associated with the surface phenomenon called "plasmon wave". The plasmon band can shift in energy and intensity according with the particle size and shape [39]. When the metallic nanocore is passivated with a chromophore, its optical and optoelectronic properties are markedly affected. In particular, the emitting electronic states of fluorophores capped on metallic particles can follow different decaying pathways, reflecting in quenching or enhancement of the fluorescence quantum yield depending on the characteristics of the nanostructure (size and shape of the metallic core, distance between the fluorophore and core, etc.) [39]. PEs have been used in previous works as rigid linkers of gold nanoclusters [40]. In a recent interesting work, organocarbodithioate-functionalized gold nanoparticles were prepared from amminofunctionalized gold nanoparticles and carbodithioate-terminated bis(phenylene)ethynylenes and oligo(phenyleneethynylene)s [41]. The sulfur atoms in the flexible sequences of 3b permit to functionalize gold nanoparticles with this PE polymer by a two phase method adapted from ref. [42]. It is to be mentioned, however, that the purification was a challenge because the particles are soluble in the same solvents of the polymer, thus washing was not possible. Both presented ratio front of zero and classical column chromatography could not be used either. Precipitation/solubilization cycles succeeded in eliminating the excess of polymer but the residual particles became poorly soluble due to the high π–π interactions of the polymer capping agent. Size Exclusion Chromatography (SEC) resulted to be effective not only to separate the excess of polymer (that goes out of the column as first fraction, Figure 19) but also to fractionate the nanoparticles in function of their hydrodynamic volume.

Three fractions (here called, f1, f2 and f3) were collected with an elution time increasing along with a decreasing size. All the solutions are yellow [43]. The optical properties in toluene solutions are resumed in Table 2. In figure 20, their absorption spectra in toluene are reported and compared with that of 3b.

In the high energy region, two peaks can be found and attributed to the polymer absorptions in agreement with the results already published for 3b solutions in chloroform [15] and to the spectrum of the polymer in toluene included for sake of comparison. The maximums however are blue shift with respect to the polymer with an increasing hypsochromism as the size of the nanoparticles decreases. No plasmon band is detected indicating a very small size of the particles (<5 nm) in agreement with what reported for thiol passivated gold nanoparticles [44]. The emission peak (insert) due to the polymer passivant exhibits a blue shift tendency in agreement with the absorption properties. The presence of the gold core reduces the quantum yield relative to 3b (ϕ_{rel}). The quenching effect of gold is well known and ascribed usually to energy or electron transfer [39]. Interestingly, in our case, it is more marked for the first fraction while for f2 and f3 the reduction of the quantum yield is of only 10%.

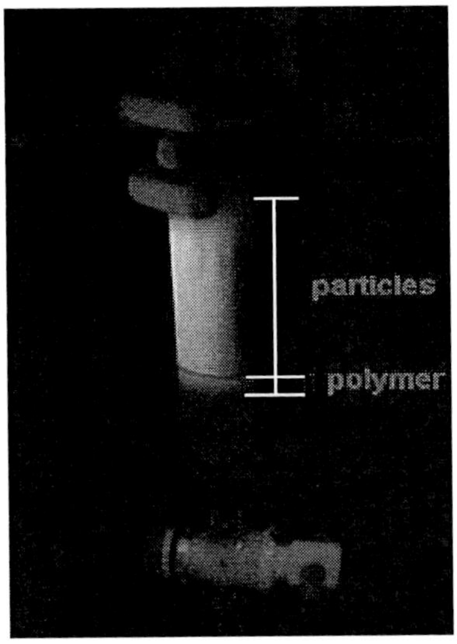

Figure 19. Purification by SEC of gold nanoparticles functionalized with 3b. Photo of the column irradiated at 365 nm. The dark line corresponds to the limit at which the particles are visible under normal light. The polymer fraction is visible only under UV light.

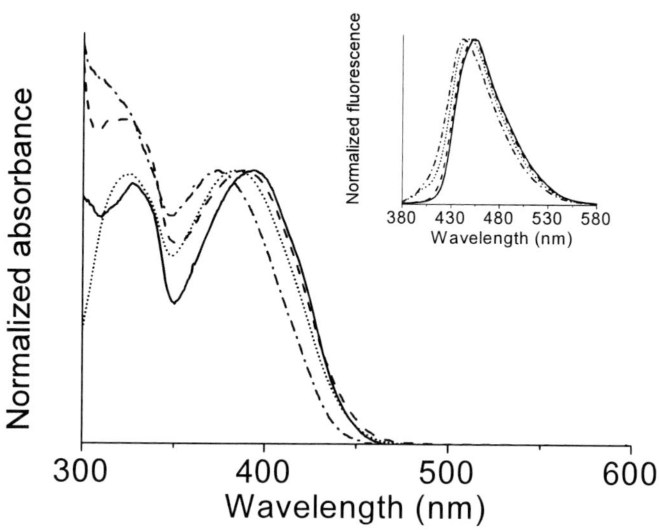

Figure 20. Absorption spectra of toluene solutions of the fractions obtained by SEC: f1, dashed line; f2, dotted line and f3, dash-dotted line and of the polymer (solid line). Inserted figure: Corresponding fluorescence spectra.

Table 2. Optical properties in toluene of gold nanoparticles passivated with 3b and of the polymer

Material	$\lambda_{abs\ max}$ (nm)	$\lambda_{emis\ max}$ (nm)	ϕ_{rel}
f1	324, 388	453	0.76
f2	325, 384	446	0.89
f3	318 (shoulder), 373	440	0.88
3b	326, 393	454	1

The NMR analysis suggests that the functionalization involves the sulfur atoms of both the –C(O)S- and –CH_2-S-CH_2- of the flexible sequences. Accordingly to the macromolecular property of our fluorophore, we hypothesize that the different fractions can be formed as a consequence of the passivation of the gold with different molecular weight fractions of the polymer. However, in order to understand the chemistry of this system, we are actually synthesizing analogous oligomer models that, being monodisperse molecules, could be assembled on gold in a precise (and more easily predictable) manner.

Another interesting approach to modulate the optical (and optoelectronic properties) of PEs at nano-level is to "construct" a phenyleneethynylene film by realizing the synthesis on a substrate by following the idea of "polymer brush" [45]. Although, at least to the best of our knowledge, no reports are so far reported on conjugated polymer brushes. Following the results obtained with the gold nanoparticles, the synthesis is carried out on gold coated glass slides. First the substrate is immersed in a solution of 4-bromobenzenthiol. The thiol group anchors to the substrate surface and the bromide atoms are left outside and free to react, for instance, with the (11-undecanol)2,5-diethynylbenzoate through a Sonogashira-Heck cross-coupling. The second layer terminates with an acetylenic group that undergoes a subsequent Sonogashira-Heck reaction with the (11-undecanol)2,5-dibromobenzoate. This gives a bromide terminated layer, which permits making another coupling with the acetylene monomer growing the polymer as schematically depicted in Figure 21.

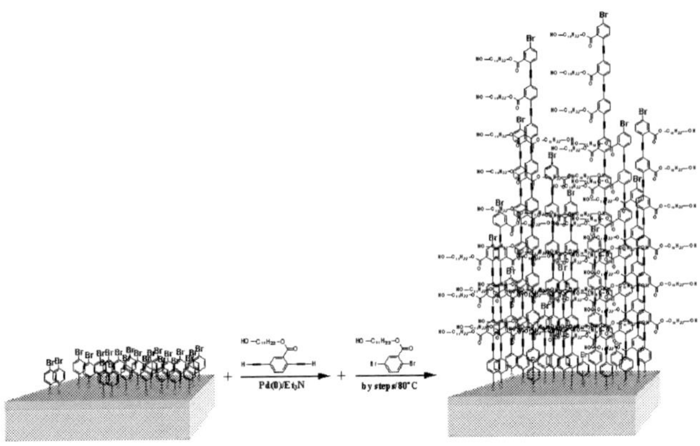

Figure 21. PE polymer brush on gold surface.

Figure 22 reports the excitation and fluorescence spectra of the polymer brush of Figure 19. In the excitation spectrum, a main peak at 395 nm can be observed that gives rise to the emission at 520 nm. These wavelengths are 20 nm lower than those of the (11-undecanol)benzoateethynylene 8d deposited by Langmuir-Blodgett on glass [14b], indicating a slight decrease in the conjugation length. However, the bands, especially the fluorescence peak, are narrower suggesting a higher order.

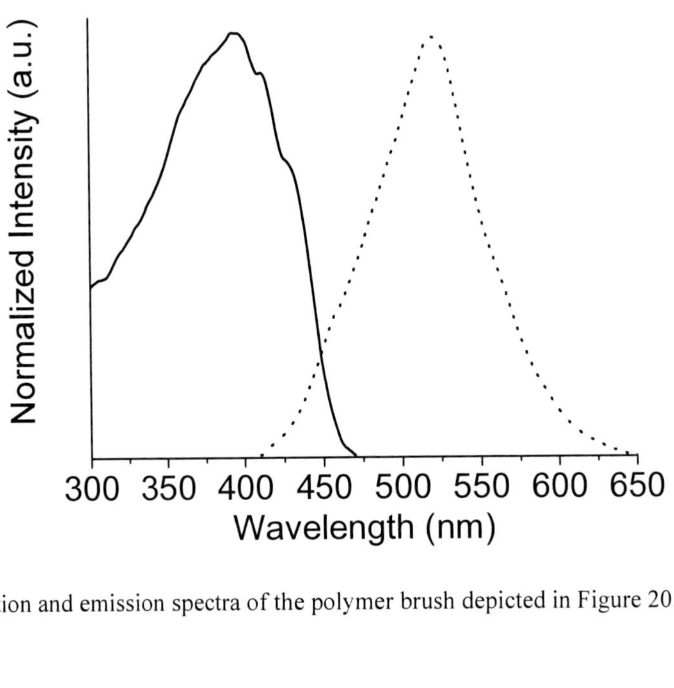

Figure 22. Excitation and emission spectra of the polymer brush depicted in Figure 20.

The morphological analysis by AFM is consistent with the polymerization process. As shown in Figure 23, the topographic image of the surface film reveals a columnar morphology as previously reported, however it is noticed that the columns heights are very irregular. A polymerization always gives a broad distribution of weight. In this case, as expected, we observe a broad distribution of height.

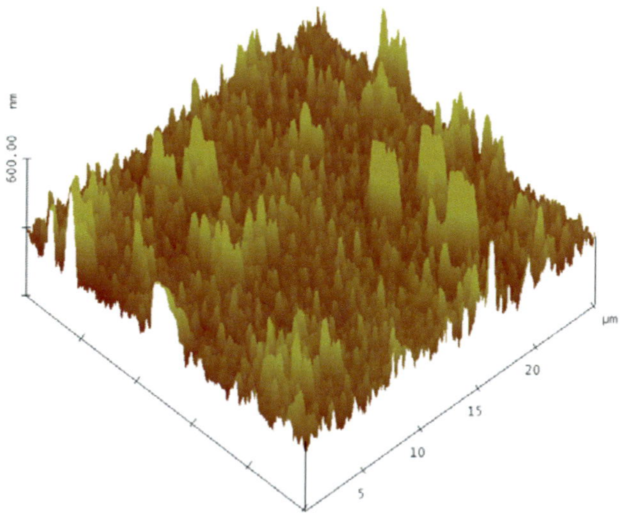

Figure 23. Tridimensional Tapping mode AFM image of the polymer brush depicted in Figure 20.

As the termini groups are bromide,another aryleneethynylene could be additionally grown on this polymer brush. This synthetic strategy opens the way to constructing well-defined conjugated structures, vertically organized with respect to the substrate.

References

[1] See for instance as a general review on oLEDs, *Organic Light Emitting Devices. Synthesis, Properties and Applications.* Müllen, K.; Scherf, U. Ed. Wiley-VCH Verlag GmbH &Co KGaA, Weinheim (Germany) 2006.

[2] Dimitrakopolous, C.D.; Malefant, P.R.L. *Adv. Mater.* 2002, *14*, 99.

[3] Shirakawa, H.; Louis, E. J.; MacDiarmid, A. G.; Chiang, C. K.; Heeger, A.J. *J. Chem. Soc. Chem. Comm.* 1977, 579.

[4] Friend, R.H.; Gymer, R.W.; Holmes, A.B.; Bourroughes, J.H.; Marks, R.N.; Taliani, C.; Bradley, D.D.C.; Dos Santos, D.A.; Brédas, J.L.; Lögdlund, M.; Salaneck, W.R. *Nature* 1999, *397*, 123.

[5] Sariciftci, N. S. *Prog. Quantum Electron.* 1995, *19*, 131.

[6] Pei, Q.; Chi, G.; Zhang, C. ; Yang, Y.; Heeger, A.J. *Science* 1995, *269*, 1086.

[7] Michaelis,A.; Berneth, H.; Haarer, D.; Kostromine, S. ; Neigl, R.; Schmidt, R. *Adv. Mater.* 2001, *13*, 1825.

[8] Swager, T.M. *Acc. Chem. Res.* 1998, *31*, 201.

[9] Gerard, M. ; Chaubey, A. ; Malhotra, B.D. *Biosensors & Bioelectronics* 2002, *17*, 345.

[10] Moore, J.S. *Current Opinion in Solid State & Materials Science* 1996, *1*, 777.

[11] As a very exhaustive referente on supramolecular assembly of π-conjugated systems, see Hoeben, F.J.M.; Jonkheijm, P.; Meijer, E.W.; Schenning, A.P.H.J. *Chemical Reviews* 2005, *105*, 1491.

[12] Scout, J.C.; Kaufman, J.; Brock, P. ; DiPietro, R.; Salem, J.; Goitia, J. *J. Appl. Phys.* 1996, *79*, 2745.

[13] See for instance the following reviews and references therein. A) Giesa, R., *JMS-Rev. Macromol. Chem. Phys. C* 1996, *36*, 631. b) Bunz, U.H.F. *Chem. Rev.* 2000, *600*, 1605. c) Swager, T.M. *Acc. Chem. Res.* 1998, *31*, 201. d) Tour, J.M. *Chem. Rev.* 1996, *96*, 537. e) Moore, J.S. *Acc. Chem. Res.* 1997, *30*, 402.

[14] A) Schumm, J.S.; Pearson, D.L.; Tour, J.M. *Angew. Chem. Int. Ed. Engl.* 1994, *33*, 1360. b) Arias-Marín, E.; Arnault, J.C.; Guillon, D.; Maillou, T.; Le Moigne, J.; Geffroy, B.; Nunzi, J.M. *Langmuir* 2000, *16*, 4309.

[15] Vazquez, E.; Esquivel Aguilar, A.; Moggio, I.; Arias, E.; Romero, J.; Barrientos, H.; Torres, J.R.; Reyes Vega, M. *Mat. Sci. Eng. C* 2007, *2*, 787.

[16] Espinosa-González, C.; Moggio, I.; Arias-Marín, E.; Romero-García, J.; Cruz-Silva, R.; Le Moigne, J.; Ortiz-Cisneros, J. *Synth. Met.* 2003, *139*, 155.

[17] Arias-Marin, E. ; Le Moigne, J. ; Guillon, D. ; Moggio, I. ; Geffroy, B. *Macromolecules* 2003, *36*, 3570.

[18] Gonzalez-Rojano, Norma; Arias-Marin, Eduardo; Navarro-Rodriguez, Damaso; Weidner, Steffen. *Synlett.* 2005, *8*, 1259.

[19] (a) Firon, N.; Ofek, I.; Sharon, N. *Infect. Immun.* 1984, 43, 1088. (b) Li, J.; Zacharek, S.; Chen, X.; Wang, J.; Zhang, W.; Janczuk, A.; Wang, P. G. *Bioorganic & Medicinal Chemistry* 1999, 7, 1549. (c) Qian, X.; Metallo, S. J.; Choi, I. S.; Wu, H.; Liang, M. N.; Whitesides, G. M. *Anal. Chem.* 2002, *74*, 1805. (d) Ofek, I.; Hasty, D. L.; Sharon, N. *FEMS Immunol. Med. Microbiol.* 2003, *38*, 181.

[20] Steiger, D.; Smith, P.; Weder, C. *Macromol. Rapid Commun.* 1997, 18, 643

[21] Kloppenburg, L.; Jones, D.; Claridge, J.B.; Zur Loye, H.C.; Bunz, U.H.F. *Macromolecules* 1999, 32, 4460.

[22] Dierking, I. Wiley-VCH (Germany) 2003.

[23] (a) Voigt-Martin, I.G.; Simon, P.; Bauer, S.; Ringsdorf, H. *Macromolecules* 1995, 28, 236. (b) Voigt-Martin, I.G.; Simon, P.; Yan, D.; Yakimansky, A.; Bauer, S.; Ringsdorf, H. *Macromolecules* 1995, 28, 243.

[24] Arias, E. ; Maillou, T. ; Moggio, I. ; Guillon, D. ; Le Moigne, J. ; Geffroy, B. *Synth. Met.,* 2002,*127*, 229.

[25] (a) Li, H.; Powell, D.R.; Hayashi, R.K.; West, R. *Macromolecules* 1998, 31, 52. (b) Weder, C.; Wrighton, M.S. *Macromolecules* 1996, 29, 5157.

[26] Barrientos, H.; Arias, E.; Moggio, I.; Romero, J.; Rodriguez, O.; Giorgetti, E.; Del Rosso, T. *Synth. Met.* 2004, *147*, 267.

[27] Gránásy, L.; Pusztai, T.; Börzsönyi, T.; Tóth, G.; Tegze, G.; Warren, J.A.; Douglas, J.F. *J. Mater. Res.*, 2006, *21*, 309.

[28] As a general review on no linear optics of polydiacetylenes, see for instance Molecular *Nonlinear Optics. Materials, Physics and Devices*. Zyss, J. Ed. Academic Press Inc. USA 1994, chapter 10.

[29] Lee, M.; Cho, B.-K.; Zin, W.-C. *Chem. Rev.* 2001, *101*, 3869.

[30] Netzer, L; Sagiv, J. *J. Am. Chem. Soc.* 1983, *105*, 674.

[31] When the self-assembly process is repeated to give a multilayer, it is generally established to use the term "layer-by-layer" technique. However, in this chapter, we keep the term "self-assembly" for both mono and multilayer formation for sake of simplicity.

[32] It is to be remarked that AFM image of Figure 17 right, presents a granular type morphology similar to that reported in ref. [15] but with a higher homogeneity in the grain distribution and density. This difference is due to the fact that the film studied in the cited paper was prepared by manually dipping. This procedure gave reproducible results in terms of morphology, roughness and absorbance in different samples prepared by the same operator but differences were observed passing from one operator to another. For this reason, in the continuation of the work with this polymer, the mechanical arm of a Langmuir-Blodgett equipment was used in order to make reproducible the dipping and emersion process, regardless of the operator. The fact that the dipping/emersion rate was fixed and the process was realized always with a stable and continuous way, allowed obtaining a more homogeneous morphology.

[33] It is to be noted that, for the moment, we have studied good solvents for the conjugated part (such as $CHCl_3$, CH_2Cl_2, THF, toluene) of the copolymer. According to literature, differences in the morphology and supramolecular arrangement could arise when a good solvent for the coil polar sequence would be used. This point will be considered for future works.

[34] In order to have a descriptive representation of the general behavior of the molecules in optoelectronic devices, the results here reported are related to oLEDs not optimized with respect to the injection process and lifetimes, i.e. the intrinsic optoelectronic properties are evaluated.

[35] Zou, D.; Yahiro, M.; Tsutsui, T. *Jpn. J. Appl. Phys.* 1998, *37*, L1406.

[36] *Semiconductor Device Physics of Conjugated Polymers*, Greenham, N.C.; Friend, R.H. Solid State Physics Vol. 49 Academic Press Inc. USA, 1995.

[37] Hong, H.; Sfez, R.; Vaganova, E.; Yitzchaik, S.; Davidov, D. *Thin Solid Films* 2000, *366*, 260.

[38] *Metal Nanoparticles: Synthesis Characterization & Applications*. Fedlheim D.L.; Foss C. A. Marcel Dekker, Inc. USA, 2002.

[39] Thomas, G.K.; Kamat, P.V. *Acc. Chem. Res.* 2003, *36*, 888.

[40] a) Osifchin, R:G:; Mahoney, W:J:; Bielefeld, J.D.; Andres, R.P.; Henderson, J.I., Kubiak, C.P. *Superlattices Microstruct.* 1995, *18*, 283. b) Andres, R.P.; Bielefeld, J.D.; Henderson, J.I.; Janes, D.B.; Kolagunta, V.R.; Kubiak, C.P.; Mahoney, W.J.; Osifchin, R.G. *Science* 1996, *273*, 1690. c) Brousseau, L.C.; Novak, J.P.; Marinakos, S.M.; Feldheim, D.L. *Adv. Mater.* 1999, *11*, 447.

[41] Park, T.-H.; Therien, M.J. *Organic Letters,* 2007, *9*, 2779.

[42] van Herrikhuyzen, J.; Janssen, R.A.J.; Meijer, E.W.; Meskers, S.C.; Schenning, A.P.H.J. *J. Am. Chem. Soc.* 2006, *128*, 686.

[43] The reaction mixture is dark purple due to the formation of bigger particles that precipitate during the reaction or when the reaction mixture is concentrated to be charged in the column. The formation of these particles reduces the yield of the reaction but, at least at this moment, can not be avoided.

[44] Busby, M.; Chiorboli C.; Scandola, F. *J. Phys. Chem. B,* 2006, *110*, 6020.

[45] Pyun, J.; Kowalewski, T.; Matyjaszewski, K.; *Macromol. Rapid Commun.* 2003, *24*, 1043 and references therein.

In: Biomimetic and Supramolecular Systems Research
Editor: Arturo H. Lima
ISBN: 978-1-60456-405-1

Chapter IX

Microstructural Influence on Flexure Strength of a Ceromer Reinforced by Two Types of Fibers (Polyethylene and Glass)

Silvana Marques Miranda Spyrides[1] * and Fernando Luiz Bastian**
*Department of Prosthesis and Dental Materials, School of Dentistry,
Federal University of Rio de Janeiro, Rio de Janeiro, RJ, Brazil
**Program of Metallurgical and Materials Engineering,
COPPE, Federal University of Rio de Janeiro, Caixa Postal 68505,
Rio de Janeiro, RJ, Brazil, CEP 21941-972COPPE/UFRJ

Abstract

In this study, the microstructures of a ceromer (Artglass®) reinforced by either glass fibers (GlasSpan®) or polyethylene fibers (Connect®) were characterized and compared and the influence of the fiber reinforcement on the flexural strength of the resulting products evaluated. With this objective, seven bars of each material were produced. One bar of each material was separated for microstructural analysis. The microstructural samples were subjected to metallographic polishing and finishing, and then analyzed using optical microscopy at different magnifications. The images obtained were treated using an image processing computer program (Image Pro Plus) in order to quantify the microstructure by calculating the mean diameter and mean volume fraction of fibers. The flexure tests were made by three-point bending, using six samples of each material. After statistical analysis, the results showed that the mean diameter of the glass fibers (4μm) was smaller than the polyethylene ones (23.6 μm). The mean volume fraction of glass fibers (0.42) was larger than that of the polyethylene fibers (0.28) and the mean center-to-center distance between fibers was smaller in the glass fibers material (33 μm) than in the polyethylene fibers material (61 μm). The flexural strength of both glass and

[1] Mail address: Av. N. S. de Copacabana, 195/310, Copacabana CEP 22020-000 Rio de Janeiro – RJ – Brasil, Email: silmms@globo.com.

polyethylene fiber-reinforced materials was statistically equal, despite the fiber volume fraction being statistically larger in the fiber glass material.

Keywords: *Fiber Reinforced Composites, Ceromer, Microstructure, Mechanical Properties, Metal Free Fixed Prosthesis*

1. Introduction

Fiber-reinforced composites were developed as an alternative to conventional fixed partial prosthodontic, metaloplastic or metaloceramics, to fit small edentulous spaces. This system, comprised of a fiber-reinforced ceromer allows a conservative, less invasive and more esthetic preparation of the supporting teeth [1,2].

These prostheses have a substructure formed by long fibers impregnated with resin matrix which provide strength and rigidity, and a particulate composite cover that improves wear resistance and esthetics [3].

Several types of fibers have been used and studied for dental purpose, and it seems that glass and polyethylene fibers are the most widely used and studied due to their transparency, and ability to keep the original color of the covering composite [4,5,6,7].

Polyethylene fibers have shown high mechanical properties if the load is applied in tension; however, if the load is in compression it shows lower resistance. On the other hand, the glass reinforced material have shown the same properties regardless of the direction of loading so their bending properties are higher than those of polyethylene fibers [6].

Fibers act as reinforcement, increasing the flexural strength, the fracture resistance and the tensile strength of the polymer matrix, but they do not increase the compressive strength [8,9]. Failures of fiber-reinforced ceromers are due to their poor adhesion or cohesion at the fiber/matrix interface.

Mechanical properties of the fiber-reinforced composites are influenced by some factors listed by BEHR [10] as follows: the orientation and the quantity of fibers, the impregnation of fibers within the polymeric matrix, the adhesion of fibers to the polymeric matrix and the properties of fibers versus properties of the polymeric matrix. Another factor that may affect the properties is the position of the fiber reinforcement [11].

The properties of the reinforced fibers differ according to the type of fibers and their architecture [12]. The properties of the covering composites vary according to their composition, not only in the organic matrix but also in the dispersed inorganic phase (type, size and distribution of particles). According to FREILICH et al. [12], ELLAKWA et al., [13] this variation in the composite composition and the type and architecture of fibers affects the properties of fiber-reinforced composites.

ELLAKWA et al., [13] verified that different composite compositions did not produce significant differences in flexure strength of polyethylene fiber-reinforced composites. In a previous work, SPYRIDES and BASTIAN [14] observed that the variation between glass and polyethylene fibers did not produce significant changes in the flexure strength of composites reinforced with these two types of fibers.

The impregnation of the fibers is required to provide close union of the fiber to the polymeric matrix and thus for the strength of the composite [15]. Poor impregnation leads to failure in transferring the load from the matrix to the fiber. On the other hand, spaces between them may occur, which may lead to a considerable decrease of tensile strength and the elasticity modulus of the fiber composites when compared with the theoretical strength and modulus of such composites [16]. These regions increase the water absorption and, thus, decrease the mechanical strength of the composite [17]. Porosity between matrix and fibers diminishes the capacity of load support of fiber-reinforced composites [5,13,15].

According to FREILICH [12], due the necessity of having all fibers fully wetted by the resins, the fiber volumes are generally limited to less than 50%. Pre-impregnated unidirectional dental fiber composites incorporating approximately 45% of glass fibers have flexure strength in the range of 600 and 1,000 MPa. These values are about ten times higher than those of composites without fiber reinforcement, and this represents the primary mechanical benefit of using fiber reinforcement in a prosthetic framework for dental purpose.

VALLITTU and NARVA [5], VALLITTU [15] demonstrated that there is a relationship between the amount of fibers within the polymeric matrix and the increase of flexure strength on the fiber reinforcements.

Larger amounts of fibers result in better properties. However, an excess can lead to an insufficient wetting of the resin, which represents less resistance and poor characteristics of manipulation. To reach an optimum strength, it's necessary to have a high volume fraction of fibers, thorough wetting and a uniform distribution of fibers within the matrix resin [6].

Due to the mechanical demands that fiber-reinforced fixed partial dentures are submitted during the masticators function, it is known that their clinical behavior depends directly on their microstructure and components. As a consequence, the purpose of this study was to evaluate, quantify and compare the microstructure and its influence on flexure strength of a ceromer reinforced by either glass or polyethylene fibers.

2. Materials and Methods

A microstructural and flexure strength chacterization of fiber reinforced ceromer was performed. Manufacturers and compositional information of the tested materials are summarized in table 1.

Fourteen rectangular specimens (30 mm length, 4 mm height and 4 mm width) of Artglass, seven reinforced by polyethylene fibers and seven by glass fiber were produced by inserting the material into a silicone mold (Stern Tek, Sterngold) in three increments. One impregnated fiber layer was placed between the first and second layer of material. Each layer was photocured in the *UniXS* unit for 90 seconds and after the last cure the bar was removed from the silicone mold and photocured for 180 seconds.

Table 1. Investigated materials

Material	Composition	Manufacturer
Artglass Dentine*	Second-generation laboratory composite resin. Multifunctional metacrilic ester matrix (30% by weight) filled with glass particles of silicium, barium and aluminum oxides and silanized (70% by weight with 1 μm medium particle width) and photoiniciators.	Heraeus Kulzer Germany
Connect*	Braided polyethylene fiber (UHMW-PE) 2mm width, coated by plasma gas.	Kerr Corporation USA
GlasSpan*	Silicium, aluminum and boro oxide braided silanized glass fiber with 2mm width.	GlasSpan Inc. USA
Connect Resin*	Fluid resin composed of non cured metacrilic ester monomer, photoiniciators, inorganic fillers and additives for stabilization.	Kerr Corporation USA

*Brand name.

2.1. Microstructural Characterization

2.1.1. Samples Preparation

Two bars previously prepared, one with glass fiber and the other with polyethylene reinforcement were randomly selected and cut in small blocks each measuring 5 X 4 X 4 X mm. These blocks were embedded in epoxy resin (Ciba Geisy 311 with 10% of 24 catalizer) with the transversal, longitudinal and longitudinal base surface exposed as shows Figure 1.

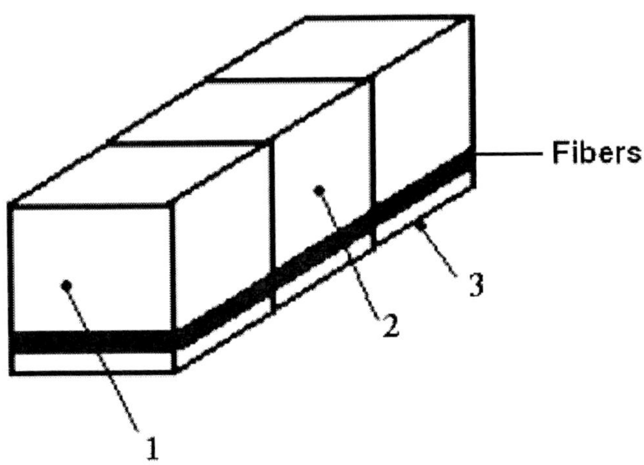

Figure 1. Schematic representation of produced material bar showing the position of the fibers layer and the cuts: transversal (1), longitudinal (2) and longitudinal base (3).

After the resin cure, the samples was finished in an industrial polishing machine (DP - NAP - Struers model Knut Rotor) with a series of emery papers (600, 800 and 1,200 grit - 3M Company), under current tap water refrigeration.

After that, fine metallographic polishing was carried through by pan cloth with 1μm and 0,5μm diamond paste (AP paste F - Struers Metallographic Equipment) in industrial polishing machine (Prazis APL – 4). Final cleanness after the polishing process was performed by an ultrasonic device.

2.1.2. Acquisition, Treatment of the Images and Quantitative Characterization

After polishing and finishing, samples of Artglass/Connect and Artglass/GlasSpan were observed at the fiber region using optical microscopy (Olympus BX60M) in three sections with magnifications of 100X, 200X and 500X. Digital photographs in JPEG format were acquired with a photographic camera (CCD COMU/Snappy), with the intention of obtaining the mean volume fraction, diameter and centre-to-centre spacing of the fiber reinforcement region of those materials.

The acquired pictures were then treated with Adobe Photoshop 5.0 program (Image/Adjustments/Equalize) in order to compare shape, size and distribution of fibers, and also detect if there was any void or porosity next to fibers or in the fiber/matrix interface.

Six acquired pictures (500X) of the glass fibers region and another six (200X) of the polyethylene fiber region were subjected to a new treatment with Corel Photo Paint program (Magic wand, brush, Eraser, Fill tool and Zoom) with the objective of leaving the fibers with white and the matrix with black color. After that these pictures were processed with an Image Pro Plus program with the purpose of obtaining statistically the average diameter and volume fraction of each type of fiber.

Using the obtained values of fibers mean diameter (μm) and the fibers mean volume fraction (%) the mean centre to centre distance between fibers (λ) were calculate with the following equation [18]:

$$\lambda = L \text{ x } \frac{1\text{-}Vv}{Vv}$$

where λ is the mean centre to centre distance between fibers, L the fiber mean diameter (μm) and Vv the fibers mean volume fraction (%).

2.2. Flexure Strength

Flexure testing was performed according to ANSI/ADA No. 27 – 1993 specification 7.8.1.1. Six bars of each material were used. The bars were subjected to manual grinding with 800 grade emery paper (3M Co.). The tests were performed in an Instron testing machine, model 4204, at a crosshead speed of 0.5 mm/min. The specimens were placed on a three-point bending test device with 20mm between the supports as illustrated in figure 2.

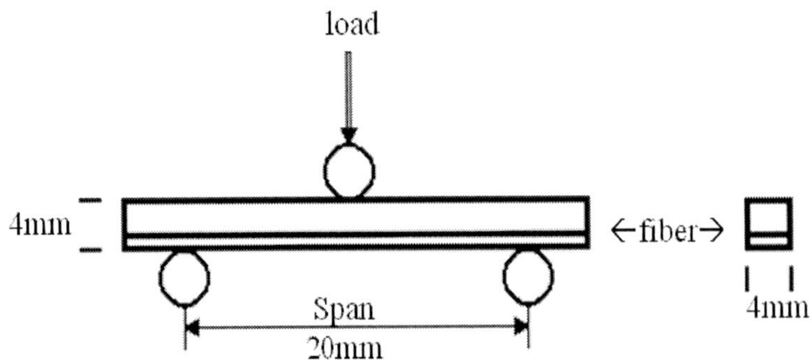

Figure 2. Diagrammatic representation of three point flexure testing device.

From the results of the flexure tests it was possible to calculate the flexural strength, deflexion and elasticity modulus of the test specimens using the following equation:

$$\sigma = \frac{3Pl}{2bh^2}$$

where: σ is the flexure strength, l the span between the supports, b the width, h the height of the sample and P the fracture load.

Means, standard deviations and coefficients of variation were computed (Excel 2000 Microsoft Office). All data were statistically analyzed with t student test for averages comparison between the groups. The level of significance was set at $\alpha = 0.05$.

3. Results and Discussion

3.1. Microstructural Characterization

The observations of the acquired images in transversal, longitudinal and longitudinal base cuts (figures 3 and 4) shows that the shape, size and distribution of fibers were different between the two studied materials and that the fibers had a reasonable impregnation by the polymeric matrix in both glass and polyethylene reinforced composites.

In transversal sectional views, figures 3a, 4a 5a, 5b, 6a and 6b, the glass fibers showed circular transversal shape, a relative uniform distribution and appearance of being totally involved by polymeric matrix, meanwhile the polyethylene fibers showed irregular and elliptical shapes, lager size, less uniform distribution, with some areas without fibers alternating with areas of high concentration of fibers. This irregular distribution of polyethylene and glass fibers can be attributed to the braid architecture of both. In places where fibers cross each other, greater proximity between fibers occurs.

In present study, the shape and distribution of fibers in transverse section were similar to the observed in the study of GOLDBERG and BURSTONE [19], BEHR et al [10] and BAE et al [20]. It also appeared that wetting of glass fibers to the polymeric matrix was similar to that of study of BAE et al [20] which used GlasSpan fibers impregnated with Aelitefil.

In longitudinal section views, figures 3b and 4b, the glass fibers showed a range of ovalized shapes. This happened due to the position of cutting. In this same section, the polyethylene fibers presented varied shapes with tendency to elongated forms.

In longitudinal base section, figure 3c and 4c, it was possible to observe the braid architecture of glass fibers. The polyethylene fibers didn't show the braid architecture probably due to the larger diameter of its fibers in relation to the glass fibers.

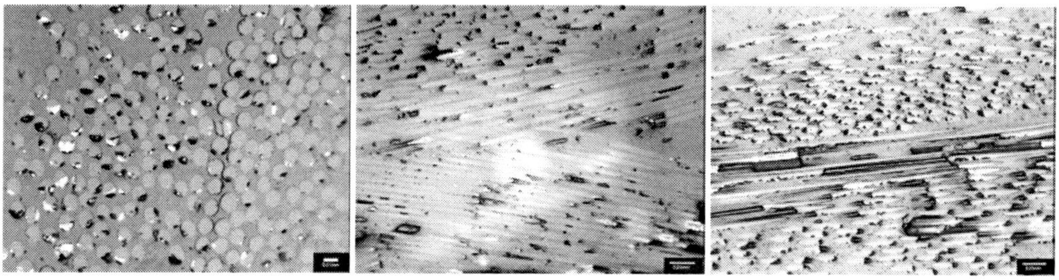

(a) (b) (c)

Figure 3. Optical microscopy views of Artglass/GlasSpan material (a) transversal section X 500, (b) longitudinal section X 200 and (c) longitudinal base section X 200.

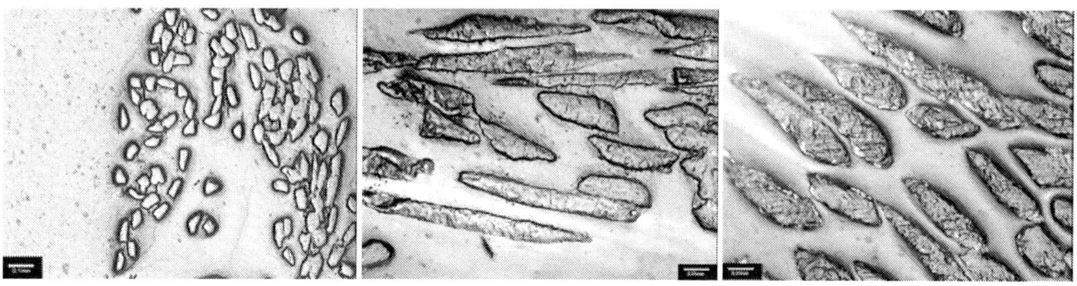

(a) (b) (c)

Figure 4. Optical microscopy views of Artglass/Connect material (a) transversal section X 100, (b) longitudinal section X 200 and (c) longitudinal base section X 200.

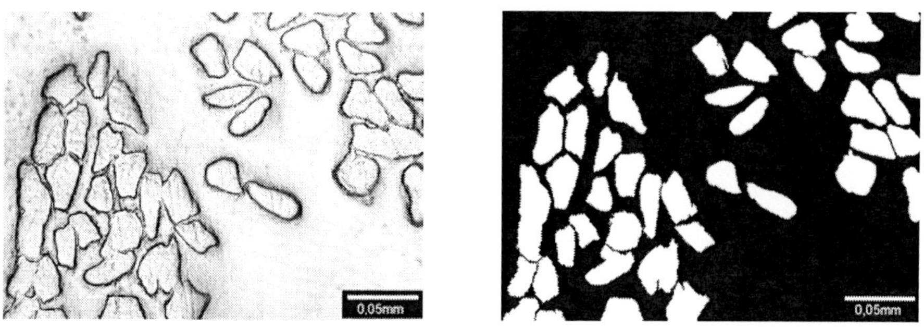

(a) (b)

Figure 5. Optical microscopy views of Artglass/Connect material (fibers region) with magnification X 200 in transversal section, (a) treated image using Photoshop 5.0 program and (b) treated image using the Corel Photo Paint program.

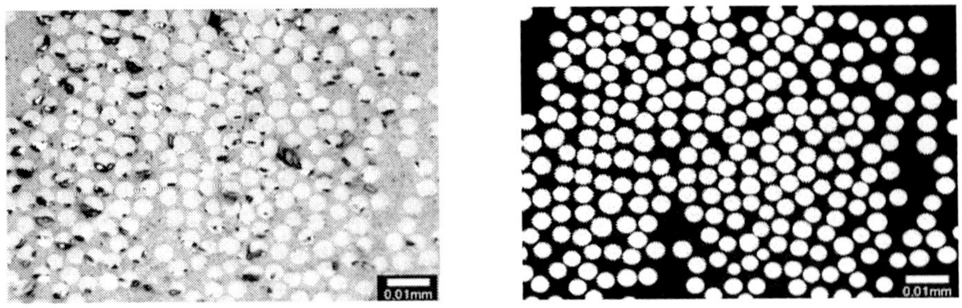

(a) (b)

Figure 6. Optical microscopy views of Artglass/GlasSpan material (fibers region) with magnification X 500 in transversal section (a) treated image using 5.0 Photoshop program and (b) treated image using the Corel Photo Paint program.

The values of fiber diameter, fiber number (per square mm) and fiber volume fraction are shown in table 2.

The Student t-test exhibited significant differences ($p < 0.05$) for all means values.

Table 2. Values of fiber diameter (μm), fiber number (per mm^2) and fiber volume fraction (%). Means values, standard deviations (sd), p-values and coefficients of variation (CV)

	Material	Mean	p-value	s.d.	C.V.
Diameter (μm)	*GlasSpan*	4.0	0.000004*	0.5	12.5
	Connect	23.6		6.0	25.4
Number in 1 mm^2	*GlasSpan*	31385.2	0.00000028*	2051.2	6.53
	Connect	640.8		141.9	22.1
Volume fraction (%)	*GlasSpan*	42	0.0002*	3.44	8.2
	Connect	28		4.84	17.3

* significant difference ($\alpha = 0.05$).

According to CHAWLA [21], smaller diameter of the fibers leads to lower probability of defects. In present study (Table 2) the mean diameter of the glass fibers (4μm) was smaller than the polyethylene ones (23.6 μm).On the other hand, the mean centre to centre distance between fibers was smaller in the glass fibers material (33 μm) than in the polyethylene fiber one (61 μm). Also, the mean volume fraction of glass fibers 0.42 (42%) was larger than that of the polyethylene fibers 0.28 (28%)

GOLDBERG and BURSTONE [19], VALLITU [15], BEHR et al. [10] performed microstructural evaluation by the incinerating or dissolution methods. The weight was calculated before and after the incinerating or dissolution process in order to get the total weight of fibers and thus the total fibers weight fraction. These methods seem to be imprecise, due to the fact that by incinerating or dissolving the materials there is a possibility of finding particles and resin rests together with the fibers themselves. Besides, these two methods do not allow observing inhomogeneous regions in the composite materials. Another problem is that these two methods consider that all fibers have an uniform distribution in the composite matrix which is not always true.

In our investigation, the fiber glass content (42%) was very close to the reported by GOLDBERG and BURSTONE [19] (43% and 45%) which used the dissolution method. However, the present results (42%) were completely different from the results of BERH et al. [10] (28% and 12%) which used the incineration method.

The results of BAE et al. [20] of mean volume fraction for both glass and polyethylene fibers was far smaller than the results found in present study. This probably happened due to the method used in the BAE et al [20] study. In that work, the authors weighed the material as a whole, the fiber reinforcement plus the composite. As a result, the mean volume fraction of fibers was very small. However, this method is not used in engineering because the mean volume fraction must represent the structural region of the reinforcement and not the whole material itself.

One of the main reasons for different fiber content in the literature is the method of measurement. However, different methods of sample preparation, other types of fibers or fiber wetting also can cause differences in fiber content [10].

A problem with the obtained values of volume fraction of fibers, as already pointed by BEHR et al [10], is that the fibers are not cut completely axially or transversally during the preparation of the samples. As a result, "the sectional view of the fiber is not circular but instead elliptical, which results in incorrect volume content of fibers". The same authors (BEHR et al) also pointed that "the distribution of fibers in the matrix is not always homogeneous. A sectional view with a representative low or high fiber content could falsify the fiber volume calculation". The first point could not be avoided in present work, so there is some error in the obtained values of volume fraction, mainly in the polymeric fiber material. The second point was avoided through measurements on different regions of the samples and subsequent averaging of the values.

3.2. Mechanical Behavior Characterization - Flexure Strength

The values of flexure strength obtained are listed in table 3.

The t student test exhibited no statistical differences for flexure strength ($p = 0.20$) between groups.

Artglass/Connect presented the largest coefficient of variability in all tests indicating a higher dispersion.

During the flexure tests Artglass/GlasSpan presented an elastic behavior with only one maximum load peak before breaking, while Artglass/Connect presented an elastoplastic behavior and the occurrence of two load peaks before breaking as can be seen in Figure 7. The load used to calculate the flexure strength of this material was the first load peak from which the material was already in process of cracking by delamination process.

During the tests, the bars showed different behavior: The glass fiber bars (Figure 8a) developed cracks in the tensile part of the central region of the bar with propagation into de compression side; while the polyethylene fiber bars (Figure 8b) showed not only these cracks in the central portion but also fracture in other parts of the bar, mainly next to one of the supports.

Table 3. Values of flexure strength (σ) of Artglass/GlasSpan and Artglass/Connect. Mean values, standard deviations (sd), p-values and coefficients of variation (CV)

	Material	Mean	p-value	s.d.	C.V.
Flexure Strength (σ) (MPa)	Artglass/GlasSpan	131.02	0.20	9.95	7.59
	Artglass/Connect	116.81		23.51	20.12

* significant difference (α = 0.05).

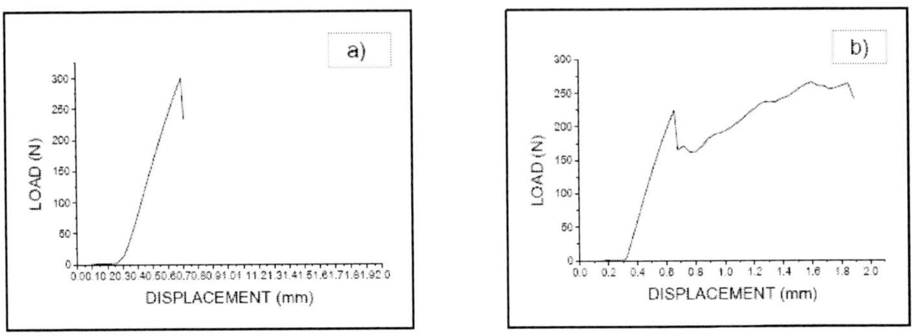

Figure 7. Typical load vs. displacement records in the flexure test a) Artglass/GlasSpan and b) Artglass/Connect.

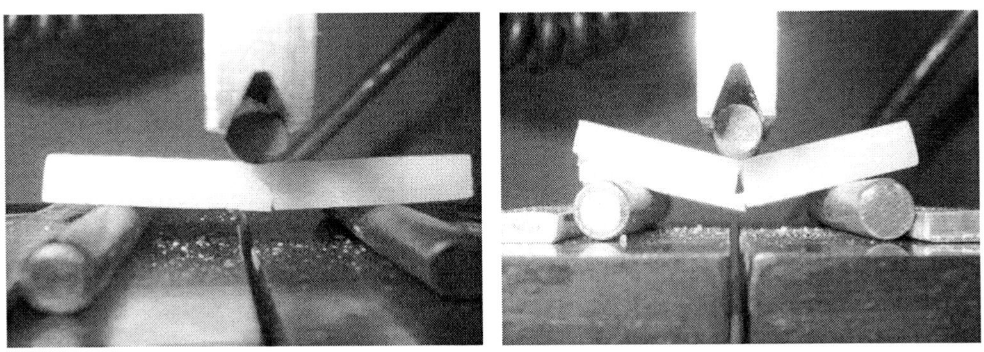

(a) (b)

Figure 8. Flexure testing (a) fiber glass reinforced bar (b) polyethylene fiber reinforced bar.

After the tests, it was observed that Artglass/GlasSpan did not show delamination between the particulate composite and fibers, which means that this material probably reached its maximum and possible flexure strength. On the other hand, Artglass/Connect showed cracks at the interface with delamination that could had resulted from porosities, impurities, air captured or even lack of adhesion at the interface. This can indicate that this material had reached its limit of shear bond strength at this interface, with consequent cracking, as shown in Figure 9

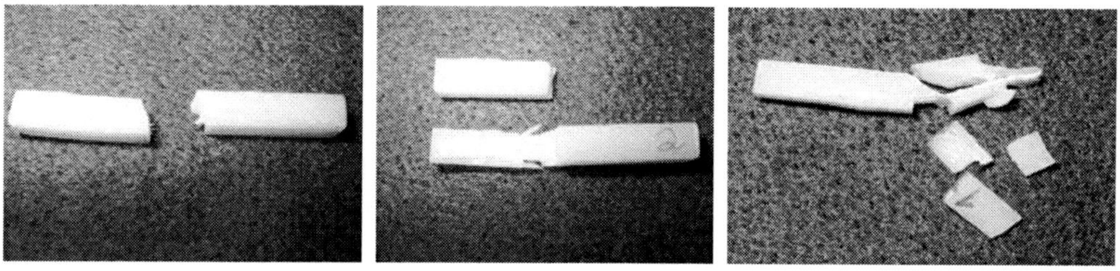

(a) (b) (c)

Figure 9. Cracked bars after flexure testing (a) glass fiber reinforced, (b) and (c) polyethylene fiber reinforced.

The flexure strength of the two materials was considered statistically equal despite of Artglass/Connect had presented delamination and supported a slightly larger load. In terms of material development, it can be presumed that this material still presents potential for improvement through further studies on its fiber/matrix interface.

The mean values of flexure strength of ceromer reinforced with polyethylene fiber (116.81 MPa) found in present work was smaller than the mean value of ELLAKWA et. al. [22] work for the same material (261.1 MPa). This probably happened due to their fibers position which was also in the tensile side of the bars but closer to the base.

In BEHR et. al. [10] work the flexure strength values found for composites reinforced with previously impregnated glass fibers Vectris, 618.0 MPa, and manually impregnated Fibrekor, 585.0 MPa, are very high not only in relation to the values found in present work but also to the values found by ELLAKWA et al.[22] and BAE et al [20]. This great difference may be due to the fact that in BEHR et al.[10] study the authors tested bars with fibers impregnated without coverage of particulate composite. This fact can be evidenced comparing the values found by BAE et al. [20] and CHONG and CHAI [23] for the fiber glass Vectris with coverage of composite Targis, respectively 296.0 MPa and 84,0 MPa, and for fiber glass Fibrekor with coverage of composite Sculpture, 203.0 MPa and 165.0 MPa.

VALLITU [5,12] demonstrated that there is a relationship between fiber content and the flexure strength. The same fact had been reported by GOLDBERG and FREILICH [6] which found that higher fiber content caused higher strength. Despite this, FREILICH [12] limited the fibers volume fraction to 50% in order of having all fiber wetted by the resin.

GOLDBERG and BURSTONE [19] also found that the highest fibers content lead to an increase of flexure strength contradicts the results of BEHR et al. [10] and BAE et al. [20] which did not obtain any increase in flexure strength when the amount of fibers had increased.

In the present study the mean volume fraction of glass fibers 0.42 (42%) was larger than that of the polyethylene fibers 0.28 (28%). Moreover, when the flexural strength results of these two types of fiber reinforced materials are compared, we can see that despite the larger mean volume fraction of the glass fiber material and the higher tensile strength of the glass fibers according to FREILICH et al. [12], the flexural strength of both glass and polyethylene fiber reinforced materials was statistically equal.

6. Conclusion

The flexural strength of both glass and polyethylene fiber-reinforced materials was statistically equal, despite the fiber volume fraction being statistically larger in the fiber glass material. The elastic behavior and the type of cracking, without any delamination observed in the glass fiber reinforced ceromer, indicates that the fibers had a good distribution and adhesion within the ceromer, probably reaching the maximum possible strength. On the other hand, the elastoplastic behavior and cracking by delamination of the polyethylene reinforced ceromer indicates that the development of this material needs improvement, by increasing the fiber adhesion within the ceromer, better distribution of fibers and a higher mean volume fraction of fibers.

Acknowledgements

The authors thank Mr. Renan Hufnagel Bella for the session of his dental prostheses laboratory. Also, to CNPq for continous support and IMA-Macromolecule Institute for the use of equipments.

7. References

[1] Shannon, A. (1997). Fiber-Reinforced Composite Bridge: Inlay-to-Inlay Technique. *Dentistry Today*, 48-53.

[2] Winters, K.L. (1999). Using a fiber-reinforced ceromer for fixed restorations. *Dent. Today*, 70-73.

[3] Freilich, M.A; Duncan, J.P; Meiers, J.C; Goldberg, A.J. (1998). Preimpregnated, fiber-reinforced prostheses. Part I Basic rational and complete-coverage and intracoronal ficed partial denture designs. *Quint. Int. v.29, n.11*, 689-696.

[4] ramos, v; runyan, d.a; christensen, L.C. (1996). The effect of plasma-treated polyethylene fiber on the fracture strength of polymethyl methacrylate. *J. Prosthet. Dent., v..76, n.1*, 94-96.

[5] Vallitu, P.K and Narva, K. (1997). Impact strength of a modified continuous glass fiber-polymethyl methacrylate. *Int. J. Prosthod., v.10, n.2*, 142-148.

[6] Goldberg, A.J., Freilich M.A., (1998, march).Fiber-reinforced Composites. *Dental* Advisor, v.15, n.2, 2-5.

[7] Honda, M.I; Vieira, D; Guedes, A.P.P. (2001, nov.) Restaurações com reforço de fibras. Dental Special [on-line serial]. 1-8, www.Dentalspecial.com.br

[8] Stipho, H.D. (1998). Effect of glass fiber reinforcement on some mechanical properties of autopolimerizing polymethyl methacrylate. *J. Prosthet. Dent., v.79, n.5,* 580-584.

[9] Belvedere, P.C. (1998). Single-sitting, fiber-reinforced fixed bridges for the missing lateral or central incisors in adolescent patients. *Dent. Clin.of Nor. Am., v.42, n.4*, 665-682.

[10] Behr, M; Rosentritt, M; Lang,R; Handel,G. (2000). Flexural properties of fiber reinforced composite using a vacuum/pressure or a manual adaptation manufturing process. *Journal of Dentistry, v.28, n.7*, 509-514.

[11] Ellakwa, A.E; Shorthall, A.C; Shehata, M.K; Marquis, P.M. (2001). The influence of fibre placement and position on the efficiency of reinforcement of fibre reinforced composite bridgework. *J. Oral Rehabil., v.28, n.8*, 785-91.

[12] Freilich, M.A., Meiers, J.C., Duncan, J.P.,Goldberg,A.J.(2000), *Fiber-Reinforced Composites,* Hong Kong, ed. Quintessence Publishing Co, Inc.

[13] Ellakwa, A.E., Shorthall, A.C., Shehata, M.K., Marquis, P.M. (2001), Influence of veneering composite composition on the efficacy of fiber-reinforced restorations (FRR). *Oper. Dent, v. 26, n.5,* 467-75.

[14] Spyrides, S.M.M. And Bastian F.L. (2004). In vitro comparative study of the mechanical behavior of a composite matrix reinforced by two types of fibers (polyethylene and glass). *Materials Science and Engineering C, v. 24*, 671-677.

[15] Vallitu, P.K. (1999).Flexural properties of acrylic resin polymers reinforced with unidirectional and woven glass fibers. *J. Prosthet. Dent., v.81, n.3*, 318-326.

[16] Vallitu, P.K. (1998). Some aspects of the tensile strength of unidirectional glass fiber polymethyl methacrylate composite used in dentures. *J. Oral Rehabil., v.25*, 100-5.

[17] Vallitu, P.K; Ruyter, I.E; Ekstrand, K. (1998). Effect of water storage on the flexural propertires of E-glass, and silica fibers acrylic resin composite. *Int. J. Prosthodont, v.11*, 340-350.

[18] Underwood, E.E. (1970). *Quantitative Stereology*. Massachusetts, Addison-Wesley Publishing Company.

[19] Goldberg, A.J. and Burstone, C.J. (1992).The use of continuous fiber reinforcement in dentistry. *D. Material, v.8*, 197-202.

[20] Bae, J. M; Kim, K. N; Hattori, M; Hasegawa, K; Yoshinari, M; Kawada,E; Oda, Y. (2001). The flexural properties of fiber-reinforced composite with light-polimerized polymer matrix. *Int. J. Prosthodont, v.14, n.1*, 33-39.

[21] CHAWLA, K. K. (1987). *Composite Materials Science and Engineering*. Germany. ed. Springer-Verlag New York Inc.

[22] Ellakwa, A.E; Shorthall, A.C; Shehata, M.K; Marquis, P.M. (2001). Influence of veneering composite composition on the efficacy of fiber-reinforced restorations (FRR). *Oper. Dent., v. 26, n.5*, 467-75.

[23] Chong, K.H. and CHAI, J. (2003). Strength and mode failure of unidirectional and bidirectional glass fiber-reinforced composite material. *The International Journal of Prosthodontics, v.16, n.2*, 161-166.

In: Biomimetic and Supramolecular Systems Research
Editor: Arturo H. Lima
ISBN: 978-1-60456-405-1

Chapter X

Skeleton of *Tetraclita Rufotincta*: A Novel Biomaterial for Tissue Engineering Applications

Gross-Aviv Talia[1], Astachov Liliana[1], Kantarovich Keren[2], Bar Ilana[2], and Vago Razi[1 7]

1. Department of Biotechnology Engineering, Ben-Gurion University of the Negev, Beer Sheva, 84105 Israel
2. Department of Physics, Ben-Gurion University of the Negev, Beer Sheva, 84105 Israel

Abstract

Fabrication of an adequate scaffolding material having apt architecture, mechanical properties and biocompatibility for optimal cell-material recognition and biosynthesis is one of the great challenges of tissue engineers. In this study, we introduced a novel porous biomaterial of marine origin, the skeleton of *Tetraclita rufotincta* (Crustacea, Cirripedia) as a potent scaffolding material for bioengineering applications. We chose this biomaterial due to its porous architecture, chemical composition and since it is an abundant species along the tidal zone of the Red Sea and other oceans. We showed that the main inorganic constitute of the *Tetraclita rufotincta* (*T. rufotincta*) skeleton is $CaCO_3$ in the form of calcite. Light microscopy, scanning electron microscopy (SEM) and XTT cell proliferation assays of mesenchymal stem cells (MSCs) that were seeded on *T. rufotincta* biomatrices have shown fast adhesion and high proliferation rates. Histochemical, immunoflourescence and quantitative real time polymerase chain reaction (qPCR) surveys revealed that MSCs differentiated into chodrogenic lineage and subsequently secreted chondorgenic extra cellular matrix constitutes without addition of any inducer. We suggest that crystalline skeletal deriviatives of *T. rufotincta* are bioactive, enhance chondrogenesis of MSCs and therefore could be exploited for tissue engineering applications in particularly for *ex-vivo* cartilage bioengineering.

[7] Corresponding author: Razi Vago Prof. Tel: 927 8 6477181 Fax: 927 8 6472983 email: rvago@bgu.ac.il

Introduction

A major task of contemporary medicine is restoration of human tissues and organs lost due to diseases and trauma. Therefore, intense efforts are invested in tissue engineering applications, for developing biological substitutes that will adequately restore, maintain or improve the injured tissue [1, 2, 3, 4].

Recently, adult stem cells became a subject of steadily increasing research, due to their many advantages such as: rapid proliferation and expansion of the cells, differentiation capabilities into different cell lineage and simple harvesting procedure from adult tissues (i.e., bone marrow, fat, muscles and skin) [5, 6]. Mesenchymal stem cells (MSCs) are developed from the embryonic mesoderm and can differentiate into different stromal cell lineages such as osteoblast, adipocytes, myocytes and chondrocytes [7]. The cells were first identified by Friedenstein *et al.*, in the 1970s [8, 9] and isolated from the bone marrow in the 1990s by Caplan [10]. The isolation procedure is based on the adhesion characteristics of MSCs to polystyrene tissue culture and the discarding of nonadherent cells. The detailed procedure, which is widely used, has been reviewed by Alhadlaq and Mao [11]. MSC pluripotent populations obtained from most tissues commonly express a number of surface receptors including CD29, CD44, CD49a-f, CD51, CD73, CD105, CD106, CD166, and Stro1 and lack expression of definitive hematopoietic lineage markers including CD11b, CD14, and CD45 [7, 12, 13, 14].

Differentiation of MSCs into different lineage is usually dictated by immediate environmental signals. Addition of growth promoting factors or cytokines to the medium is the most common way to induce differentiation of MSCs. For example: chondrogenesis is induced by the transforming growth factors (TGF) super family including TGF - β1, TGF - β2 and TGF - β3, while osteogenesis is stimulated by exposure to a combination of dexamethasone, L-ascorbic acid 2-phosphate and β-glycerophosphate [6]. Each differentiation lineage has typical markers: Chondrogenic differentiation up regulates the expression of type II collagen and glycosaminoglycan (GAG), where both of them are activated by the SRY (sex determining region Y)-box 9 (*SOX9*) transcription factor [6, 15, 16]. Osteogenesis of MSCs is marked by the *runt homeodomain protein 2 (Runx2)* transcription factor, precipitation of hydroxyapatite [$Ca_{10}(PO_4)_6(OH)_2$], up-regulation of alkaline phosphatase activity and expression of type I collagen. Late osteogenesis markers are sialoprotein, osteocalcin, osteopontin and osteonectin [6, 17, 18]. The main adipogenic differentiation markers are the transcription factor peroxisome proliferator-activated receptor-γ2 and lipid vacuoles synthesized [6].

Stem cell-based bioengineering strategies, however, present new challenges as it is inextricably tied with the provision of a three-dimensional (3D) local microenvironment that supports the maintenance and differentiation of the cells. Functional bioengineered tissues, therefore, must be created in the presence of special scaffolding materials as similar as possible to the tissue's natural milieu [4].

There are two main groups of scaffolding materials: synthetic and natural. The most common synthetic polymers are the poly(glycolic acid) (PGA), poly(L-lactic acid) (PLLA) and their co polymer poly(lactic-co-glycolic acid) (PLGA). The synthetic polymers can be easily produced and their properties can be tailored for specific applications. They have been found to be biocompatible and have a certain degradable characteristic. However, cells have a limit adherence capability to these matrices, do not easily create tissue structures with a specific predefined shape for implantation and the scaffolds are mechanically unstable [17, 19, 20].

The natural type includes materials such as collagen [21, 22], alginate [23, 24], chitosan [25] and silk [26]. These materials are biodegradable, have low toxicity and low inflammatory response and have proper surface molecular cues to stimulate stem cells differentiation. However they have poor mechanical properties and do not readily contribute to the creation of tissue structures with a specific predefined shape for implantation. These characteristics became disadvantages in cartilage and bone tissue engineering due to the heavy load bears in those areas and poor integration between the new and the native matrix [27,17,19]. The natural group of materials also includes the natural inorganic skeletons produced by marine invertebrates. These skeletal materials are usually made of $CaCO_3$ in the crystalline form of calcite, aragonite and silicated SiO_2 materials combined with $Mg(OH)_2$ [28]. These crystalline materials have superior mechanical properties and structural elements and contribute significantly to specific biological functioning by providing organizational and spatial cues towards morphogenesis. In addition, the geometry of spaces within the biomineralized framework is specific, uniform and inter connected, so a wide range of structural designs are available [29].

Several researches are studying the natural coral exoskeleton as bone graft substitutes mostly of the reef building species that include *Porites*, *Acropora*, *Lobophyllia*, *Goniopora* and *Pocillopora*, [30, 31]. These coralline biomaterials have pores measuring between 100 and 250 μm [32, 33], allowing fast cellular invasion, adherence and proliferation [34, 35]. Coral skeleton derivatives are known to be biocompatible materials *in vivo* and *in vitro* [30, 32, 36]. This kind of bioceramic exhibits osteogenic differentiation of MSCs into osteoblasts, which make bone matrix on the ceramic surface [36, 37, 38]. However, most of the coralline skeletal material is harvested from limited marine sources [30].

In the Gulf of Eilat (Red Sea) there are two species of *Tetraclita* occupying the intertidal zones of rocky shores: *Tetraclita rufotincta (T. rufotincta)* and *Tetraclita achituvi (T. achituvi)*. The former inhabit the entire *Tetraclita* belt, including the upper, middle and to a lesser extent the lower parts, while the latter is restricted to the lower part of the belt [39]. *T. rufotincta* is a sessile barnacle, a type of arthropod belonging to subphylum Crustacea, Cirripedia. It has an external skeleton of calcareous plates that form a truncate form, in which the animal is enclosed. The side plates, called parietes, rest on a basal disk firmly cemented to the substratum [40]. The plates include the calcite mineral, as shown by X-ray diffraction [41] and electron diffraction [42], and they increase in area by mineral deposition at their sides and lower edges [40].

The objectives of this study were to investigate the *T. rufotincta* exoskeleton as a novel biomaterial for *in vitro* tissue engineering. To this end, we studied the material characteristics and then examined the interaction of MSCs with this crystalline porous material. We followed a number of parameters that indicate the differentiation of MSCs into chodrocytes and the *ex-vivo* formation of carltilage-like tissue under the control of this 3D microenvironment.

Materials and Methods

Scaffold Preparation

T. rufotincta were collected from the upper belt of the shallow zones, adjacent to the interuniversity institute of marine science (IUI), Eilat, Israel. The animal was removed and the shells were bleached with commercial hypochloric solution and sliced. Scaffolds were prepared by polishing to a thickness of approximately 1 mm and an area of 0.5 cm^2, using an 8" grinder machine (model SPT 900). The organic and inorganic residues were removed by immersing the scaffolds in analytical H_2O_2 solution (Gerdrogen 30 % by weight; Riedel-de Haen, Germany) for 10 min and subsequently in dilute commercial hypochloric solution over night. The residues of hypochloric solution were washed away with distilled water (DW). Samples were then immersed in dilute Penicillin-Streptomycin-Neomycin (PSN) antibiotic mixture (Invitrogen Corporation, USA) with DW (1:3) for at least 5 days at 37 °C in a humidified 5 % CO_2 incubator. The scaffolds were autoclaved (121 °C, 30 min) and dried overnight at 60 °C before seeding.

Density, dry weight, wet weight, and buoyancy were calculated using Satorius LA2305 digital balance fitted with density tool kit. The porosity of five scaffolds was estimated according to the ASTM C373 [43].

Raman Spectroscopy

Raman spectra of the biomaterial were monitored using a confocal Raman microscope (Jobin Yvon, Labram UV HR) equipped with Olympus BX41 optical microscope, driven by Labspec 4.14 software. Raman scattering was excited by a diode laser at 784.79 nm with laser power of about 10 mW. The dispersive spectrometer was equipped with a 600 lines/mm grating and combined with an air cooled charged coupled device (CCD) for Raman signal detection. A TV camera provided the possibility of recognizing the biomatrix surface, focusing the laser beam on the point to be measured. The focusing was done by a x50/0.75 microscope objective to spots of ~1.5 μm diameter. A confocal pinhole of 100 μm diameter, before the entrance slit to the spectrograph, rejected fluorescence and Raman signal from out of focus planes.

Cell Culture

MSCs (cell line ATCC/CRL 12424), passage 3 were seeded statically at a concentration of 10,000 cells per scaffold on the *T. rufotincta* biomatrices. Cells were cultured in Dulbecco modified Eagle's medium (DMEM) (Biological Industries, Israel) supplemented with 4.5 g/l D-glucose, 1.5 g/l sodium bicarbonate (Sigma), 1 mM sodium pyruvate (Biological Industries, Israel), 10 % (v/v) fetal calf serum (Biological Industries, Israel), 1 % L-Glutamine (Biological Industries, Israel), and 1 % Biomyc-3 antibiotic solution (Biological Industries, Israel). Cells were cultured at 37 °C in a humidified 5 % CO_2 incubator. The medium was replaced every two days. Analyses were performed at the following time intervals: 4, 7, 10, 14, 21 and 28 days.

Scanning Electron Microscopy (SEM)

Stock solution of fixative was prepared using 2 % paraformaldehyde in DW and 2.5 % glutaraldehyde in 0.1 M buffer phosphate. Biological samples were rinsed with phosphate buffered saline (PBS) (Biological Industries, Israel) and immersed in the fixative at 37 °C for 30 min. After fixation the samples were washed and maintained in PBS at 4 oC. 24 hours prior to the scanning, samples were washed twice with PBS for 10 min and soaked in a serial gradient of ethanol: 50, 75, 90 and 95 %, for 10 min each and three times in 100 % analytic ethanol for 10 min. Then the samples were immersed in hexamethyldisilazane (HMDS) (Electron microscopy sciences, PA) and ethanol solutions in different volume ratios (1:2, 1:1 and 2:1, respectively). Solution residues were evaporated over night. Morphology of the samples was visualized using scanning electron microscopy (SEM) (JEOL, JSM-5610-LV). The SEM micrographs of unseeded scaffolds and energy dispersive spectroscopy (EDS) were monitored using field emission SEM (JEOL, JSM-7400F).

Histochemical Analyses

At each time point, samples were stained with safranin O for GAG content, von Kossa for bone nodules formation and oil red O for fat or lipids, according to procedures described below. The samples were photographed using a stereoscopic microscopy imaging system (Nikon, SMZ 1500) fitted with a digital camera (Nikon, DXM1200).

Safranin O Staining

A stock solution of 1 % safranin O (Sigma-Aldrich; 0.5 g in 100 ml of isopropanol) was prepared. Samples were rinsed three times with PBS, fixed with 12 % formaldehyde for 5 min and then stained with safranin O stock solution for 15 min. Serial ethanol gradients (50, 75, 90 and 95 %, and twice 100 %) were used to rinse staining residues from the sample.

Von Kossa Staining

At the experiment day a fresh stock solution of 5 % silver nitrate (Sigma-Aldrich) and 5% sodium thiosulfate (Sigma-Aldrich) were prepared in DW. Samples were rinsed with 150 mM NaCl three times, fixed in 95 % ice cold ethanol, rinsed with DW and incubated under a UV light in the presence of silver nitrate stock solution for 30 min. Following the incubation the samples were rinsed three times with DW and incubated in the presence of sodium thiosulfate stock solution for 2 min.

Oil Red O Staining

A stock solution of oil red O (Sigma-Aldrich; 0.5 g in 100 ml of isopropanol) was prepared and passed through a filtering paper. 6 ml of the solution were mixed with 4 ml of DW, left for 1 hour at room temperature, and filtered through a 0.2 μm prior to use. Samples were washed with PBS, fixed with 12 % formaldehyde in PBS for 15 min, stained with 300 μl of oil red O stock solution for 1 hour at room temperature and rinsed three times with DW.

Immunofluorescence Analyses

Samples were rinsed gently with PBS and fixed with cold 4% glutaraldehyde for 10 min at 4 °C. Fixed samples were rinsed 3 times with PBS (5 min per wash) and nonspecific antibody binding sites were blocked by incubating samples in blocking solution containing 3 % Carnation instant nonfat dry milk (Nestle, Switzerland) in PBS for 15 min. The samples were placed in a humid chamber and stained with goat polyclonal Immunoglobulin G (IgG) against type I collagen (Santa Cruz Biotechnology, USA) or goat polyclonal IgG against type II collagen (Santa Cruz Biotechnology, USA), diluted in a 1:50 ratio with PBS at 4 °C, over night. Three washes (5 min per wash) with PBS were performed and then the samples were covered with a 1:50 diluted solution of donkey anti-goat IgG fluorescein isothiocyanate (FITC) conjugated affinity purified antibody (Santa Cruz Biotechnology, USA) or rhodamine red-X-conjugated affiniPure donkey anti goat IgG (H+L) (Jackson Immunoresearch Laboratories, INC., USA) in PBS for 60 min at room temperature. After addition of the secondary antibodies, samples were washed again 3 times with PBS and photographed using an inverted fluorescence microscope (Nikon, Eclipse TE300) with FITC or rhodamine filters.

Prolifration Analysis

Cell proliferation analysis was carried out using a cell proliferation kit, XTT based colorimetric assay (Biological Industries, Israel) according to the manufacturer protocol. Calibration curve was plotted using counted MSCs. Absorbance was measured at 450 nm (EKx808; Biotek Inc.)

Expression of Genes Encoding Molecular Markers by Quantitative Real Time Polymerase Chain Reaction (qPCR)

RNA extraction: Total RNA of tissue culture on biomatrices was extracted using Tri-Reagent solution (MRC, USA), according to the single step acid-phenol guanidinium extraction method described by Chomczynski and Sacchi, 1987 [44]. Intact RNA integrity was tested on 1.7 % agarose gel containing 0.25 μg/ml ethidium bromide. The concentration was quantified spectrometrically at 260 and 280 nm wavelengths (NanoDrop ND-1000, USA).

Complementary DNA (cDNA): 2 μg of total RNA was converted into cDNA using EZ-First strand cDNA synthesis Kit (Biological Industries, Israel) according to the manufacturer protocol. At the end of the procedure the cDNA was diluted to a concentration of 20ng/μl.

qPCR: Gene specific primers were designed using the primer express™ software (V2.0, Applied Biosystems, USA), thus one of the paired primers was placed at the junction between two exons. The oligonucleotides were checked using PubMed Nucleotide BLAST and purchased from MWG (Germany).

Measuring of specific mRNA expression were done using SYBER green chemistry and monitored by an ABI Prism 7000 Sequence Detection System (Applied Biosystems, USA). For each primer an optimal concentration (Table 1) was determined, which was sufficient for gene amplification without getting primer dimmers. The quantity of cDNA for each gene was determined according to standard curve at the accepted value after 25 cycles, which were required to reach threshold fluorescence signal level ($Ct \approx 25$) and similar efficiency (appear as resembled slope in the standard curve). A 20 μl mixture of samples containing 10 μl platinum® Syber® green qPCR super mix-UDG with ROX (Invitrogen, USA), 5 μl of cDNA and complete amount of diethyl pyrocarbonate (DEPC) treated with double DW were prepared. For polymerase activation the samples were heated to 95 °C for 15min. DNA was then amplified for 40 cycles of 95 °C for 15 min followd by 60 °C for 30 s. Glyceraldehyde 3-phosphate dehydrogenase (GAPDH) was selected as a reference gene. The expression levels were exhibited relative to the expression of the gene in the fourth day of the experiment. All analyses were preformed at least in duplicate and two independent experiments were performed.

Statistical Analysis

Statistical analysis was preformed for XTT and qPCR using student's *t-test*. Values $p<0.05$ were considered to indicate statistically significant differences.

Results

Skeletal Properties

The structure of *T. rufotincta*'s skeleton builds from three main layers: the internal parietes, the center which is highly porous and the external parietes, as shown in the light microscopy photograph of Figure 1. The measured pores diameter varies between 0.5-2 mm (Figure 1).

The SEM images of the center layer (Figure 2A) and external layer (Figure 2B) of *T. rufotincta*, measured at magnifications of x10,000 and x11,000, respectively, revealed a high porosity structure made of crystalline material. The porosity was found to be 35 % ± 1 % and the average bulk density 2.71 ± 0.01 g/cm^3. EDS analysis of the compositional element concentration of the two crystals, marked as pt1 and pt2 in Figure 2B reveals the presence of Ca, C and O (Figs. 2C and 2D) and small amount of other elements. The proximate ratio between the dominant three elements which build the skeletal of *T. rufotincta* is 1:1:3 respectively (added table in Figure 2). The composition of the *T. rufotincta* was also analyzed by Raman spectroscopy.

Table 1. Primers for quantitative real time polymerase chain reaction (qPCR) analysis of gene expression

Gene	Accession Number	Forward and reverse primers	Primer concentration (nM)	Quantitiy of cDNA (ng)
SOX9	NM_011448	F: 5'-ACCAGTACCCGCATCTGCAC-3' R: 5'-CTCGTTCAGCAGCCTCCAG-3'	300	25
Runx2	NM_0092820	F: 5'-TGGCGTCAAACAGCCTCTTC-3' R: 5'-TGGTGCTCGGATCCCAAA-3'	700	5
Type II collagen	NM_031163	F: 5'-TGAAGACCCAGACTGCCTCAA-3' R: 5'-GAGCCGCGAAGTTCTTTTCTC-3'	300	25
GAPDH	NM_008084	F: 5'-CCTGGAGAAACCTGCCAAGTAT-3' R: 5'-GATGCCTGCTTCACCACCTT-3'	500	2.5

The representative Raman spectrum exhibits Raman shifts of 155, 280, 713 and 1087 cm^{-1} (Figure 3). The first two shifts correspond to lattice modes, while the latter two to CO_3^{-2} ion modes, i.e., the ν_1 symmetric CO stretch and the ν_4 in-plane bending, respectively. Comparison of these Raman shifts with the standard Raman shifts of the polymorphs of $CaCO_3$ revealed that the spectrum corresponds to that of calcite [45, 46].

Tissue Morphology and Proliferation

In order to study the distribution and density of the cells on the 3D biomatrices, SEM analyses were carried out. From the SEM micrographs of Figs. 4A and 4D, it is evident that even one day post seeding, MSCs already adhere to the *T. rufotincta* skeleton, stretch on the biomatrix surface and form a few cell-cell contacts. After 4 days of culture (Figs. 4B and 4E), the cells proliferate, invade the pores and create more cell-cell contacts. Furthermore, after 7 days (Figs. 4C and 4F) tissue had been formed across the pores of the biomatrix. Most of the cells were stretched, covering the surface of the biomatrix while forming a uniform layer, resembling fibroblast like tissue [10]. Also, a few rounded cells are observed within the tissue (Figure 4F).

Cell proliferation is demonstrated in Figure 5. Although 10,000 cells/matrix were seeded, the 4 days profile assay revealed only 5000 cells/matrix. Since the cells were seeded in a random suspension it is likely that some were lost into the void area. However, there was a significant ($p<0.05$) higher cell number after 7 days of culture. The number of cells at the seventh day was 5 times higher than the number of cells at the fourth day, revealing the high proliferation capacity. The ability of MSCs to proliferate did not decline even 14 and 21 days post seeding. After 21 days of culture another significant peak in the profile was observed. Following 28 days, the acute proliferation halted and a non significant decrease in cell number was detected.

The set of photographs taken by the stereomicroscope show tissue across the *T. rufotincta* pores (Figure 6). At the first days of culture, bridges of tissue created from one side of the pore to the second are observed. The beginning of bridges formation can be seen in Figs. 6A and 6B, where morphology of half circle created by the growing tissue is encountered. After approximately 10 days of culture some of the pores were totally filled with tissue [6I, white arrow].

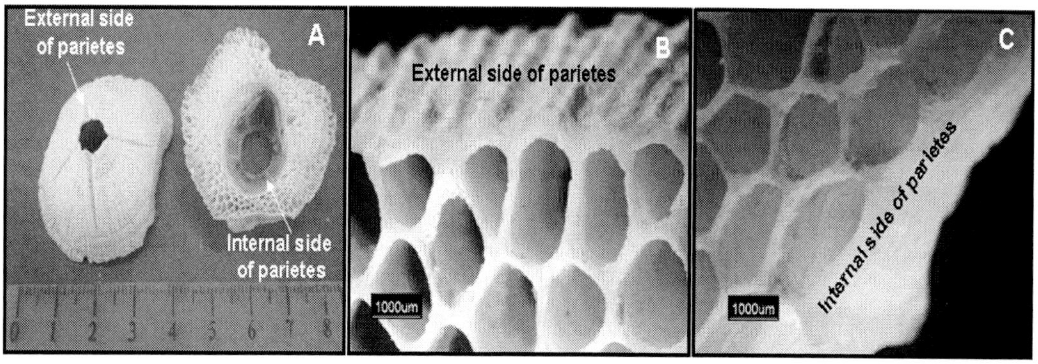

Figure 1. Light microscopy photographs of the skeleton of *T. rufotincta.* A: photograph of *T. rufotincta* parietes. B and C: Stereoscope micrographs of the external and internal sides of *T. rufotincta* parietes.

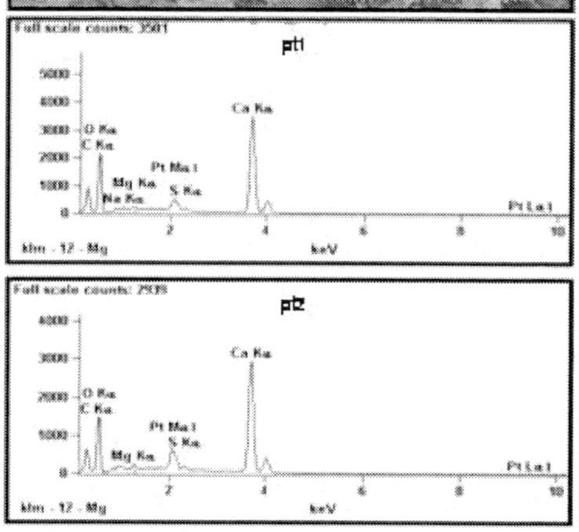

%ATOM	C	O	Ca
pt1	17.01	66.59	16.4
pt2	15.73	65.24	19.03

Figure 2. Scanning electron micrograph obtained from *T. rufotincta* biomatrices. A: The surface of the center layer at low (x10,000) and high (higher right corner, x45,000) magnifications showing the morphology. B. The surface of the external layer at low (x11,000) magnification. C and D: Diagrams illustrating the energy dispersive spectroscopy of the two crystals shown in B and indicated as pt1 and pt2, respectively. The atoms ratio is shown in the added table.

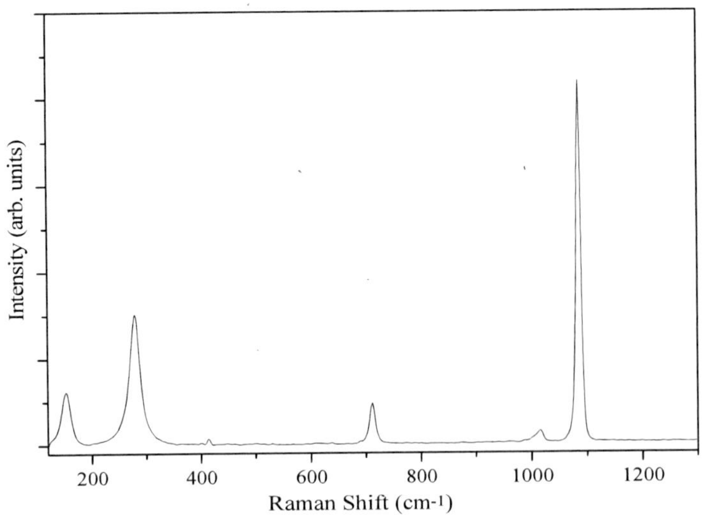

Figure 3. Representative Raman spectrum obtained from *T. rufotincta* biomatrices.

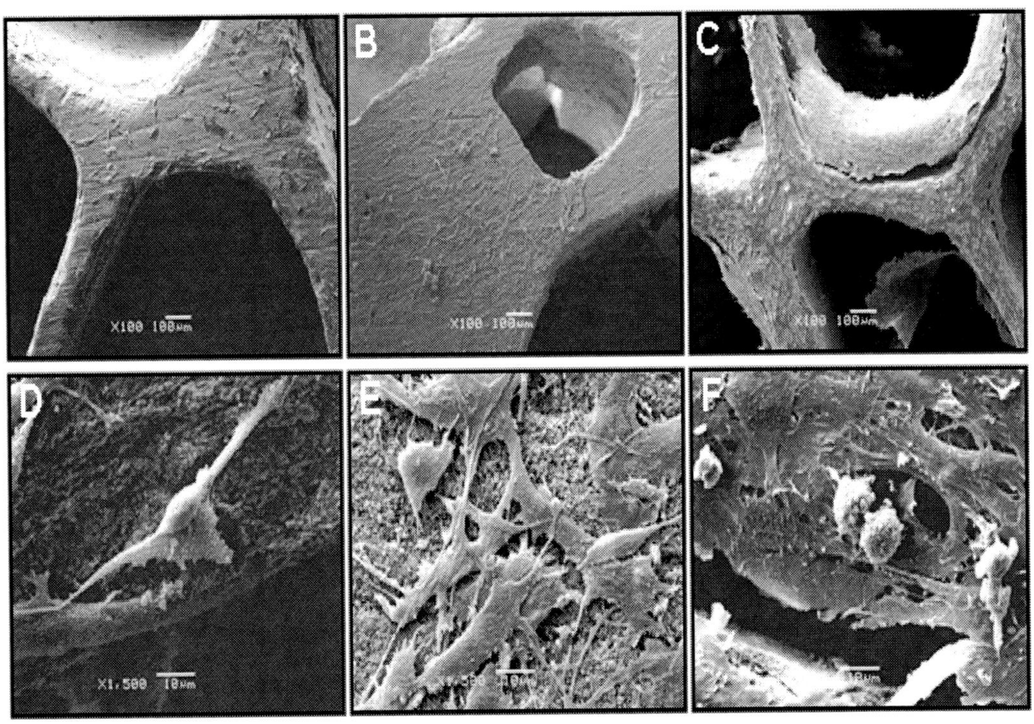

Figure 4. SEM micrographs of *T. rufotincta* biomatrices seeded with MSCs at a magnification of x100 (upper row) and x1,500 (lower row). A and D: 1 day post seeding. B and E: 4 days post seeding. C and F: 7 days post seeding.

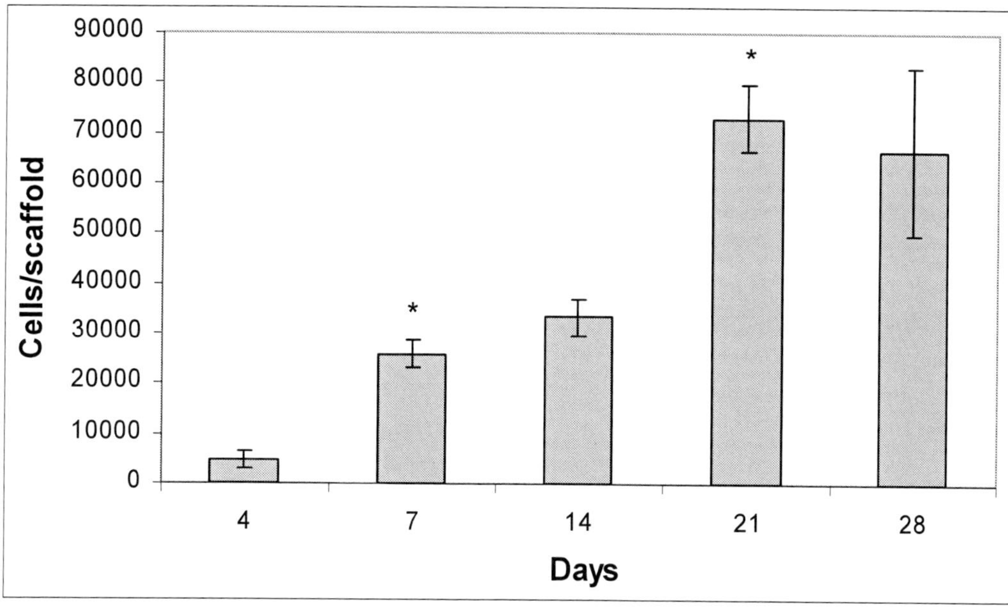

Figure 5. Cell proliferation analyses of MSCs seeded on *T. rufotincta* biomatrices. Columns represent means ± S.D. The stars represents significant ($p<0.05$) differences between the signed column and the prior time point column.

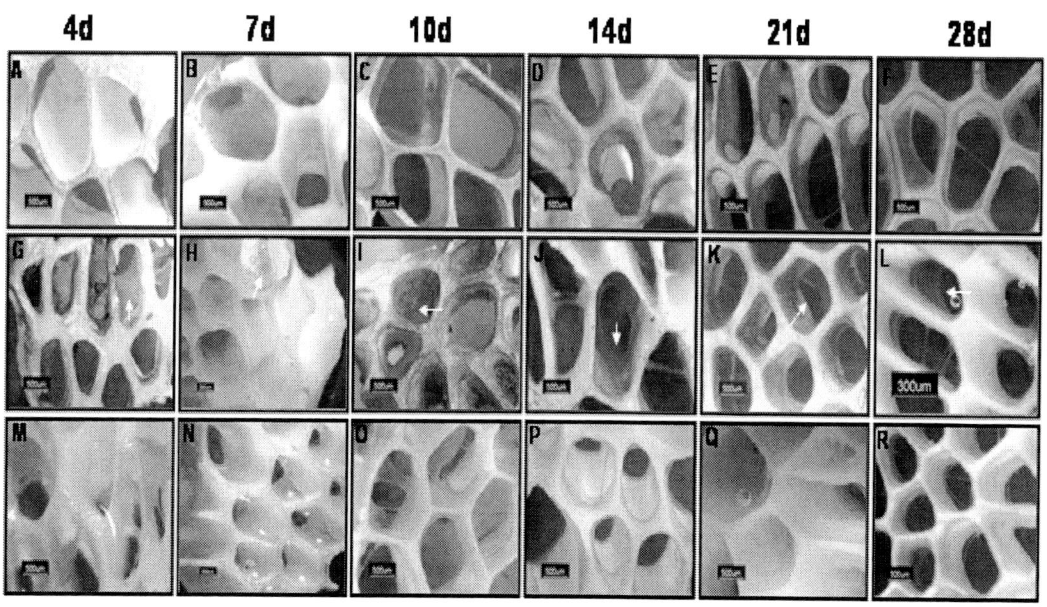

Figure 6. Streoscope photographs at different days post seeding (4, 7, 10, 14, 21 and 28 days) showing: (A-F) GAG expression in the formed tissue; (G-L) phosphate incorporation into newly formed tissue on *T. rufotincta;* The white arrows indicate the presence of formed tissue; (M-R) fat aggregate staining in the formed tissue. Calibration line represents 500 μm.

Differentiation

Chondrogenesis of MSCs is indicated, mainly, by the appearance of GAG and type II collagen in the extracellular matrix (ECM) [15]. GAG staining with safranin O, exhibited by red color, can be seen after 4 and 7 days of culture (Figure 6A and 6B). From day 10 onward the staining became stronger and at day 14 GAG aggregates are observed (Figs. 6C and 6D). In addition, the formed tissue across the pores, at days 21 and 28 indicates existence of GAG (Figs.6E and 6F).

Also, a minor expression of type II collagen was observed after 7 days (Figure 7, bottom row). However, 10 days post seeding there is a noticeable detection of type II collagen and the staining became glowing at days 14 and 21 of the culture, indicating the existence of type II collagen in the ECM.

Relative mRNA expression of type II collagen was detected through the experiment (Figure 8). A moderate increase in expression levels was observed between days 4 and 14 followed by a significantly increase at day 21 to approximately 265 % ($p<0.05$, *t*-test) and then a slight decrease on day 28. These results agree with the immunoflouresence type II collagen observations.

Oil red O staining was used to determine if the biomatrices may influence MSCs adipogenesis (Figs. 6M-6R). No dark red aggregates of fat formation were observed in any time point and the intensity of the staining was not seen to differ during the course of the experiment.

Phosphate incorporation into newly formed tissue, which indicates osteogenesis of MSCs [47], was examined by von Kossa staining (Figs. 6G-6L). Brown staining of von Kossa was hardly seen on the formed tissue, except to a small extent of stain near the side wall of the pore. This trend observed during all the time points of the experiment. The white arrows in each photograph indicate the appearance of an unstained formed tissue.

In each time point, the formed tissue on *T. rufotincta* was detected with a specific antibody for type I collagen. The same tissues were also photographed with light microscopy to ensure correct positioning of the immunofluorescent surveys (shown in the small photographs of Figure 7, upper row). During the experiment, immunofluoresce staining showed that MSCs did not produce type I collagen in the newly formed tissue.

In addition, relative mRNA expression of two transcription factors was examined to assess MSCs differentiation. The first was *SOX9* for chondrogenesis and the second *Runx2* for osteogenesis (Figure 9). The expression of *SOX9* was low through the first seven days of culture. Afterwards, the expression peaked till 14 days post seeding and then decreased. The expression of *Runx2* remained relatively constant with insignificant increase in days 10 and 28. The expression of *SOX9* was higher from those of *Runx2* from day 10 till 21. At day 28 the expressions of both transcription factors were similar.

Discussion

Scaffolding materials for tissue engineering should provide the necessary environment for cells to anchor adhere, proliferate and differentiate in a given area [17].

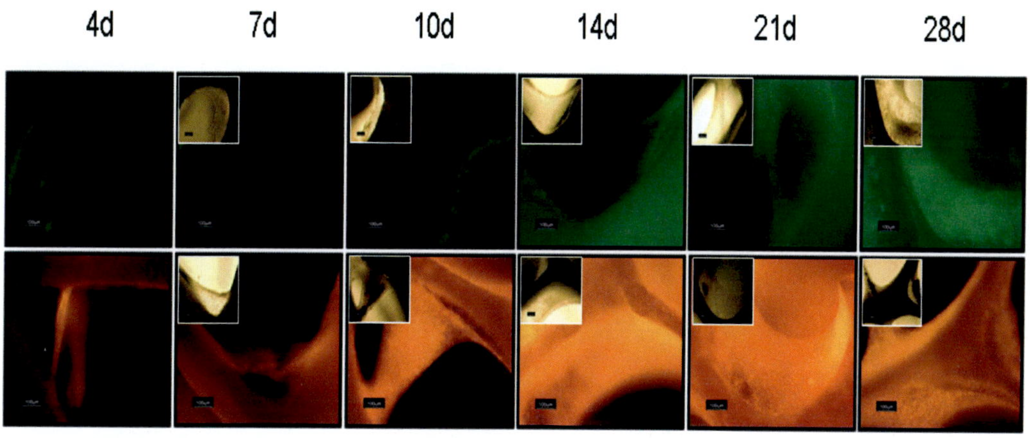

Figure 7. Light microscopy photographs at different days post seeding (4, 7, 10, 14, 21 and 28) showing: type I collagen (upper row) and type II collagen (bottom row) in MSCs tissue culture on *T. rufotincta*. The small images show the tissue without staining.

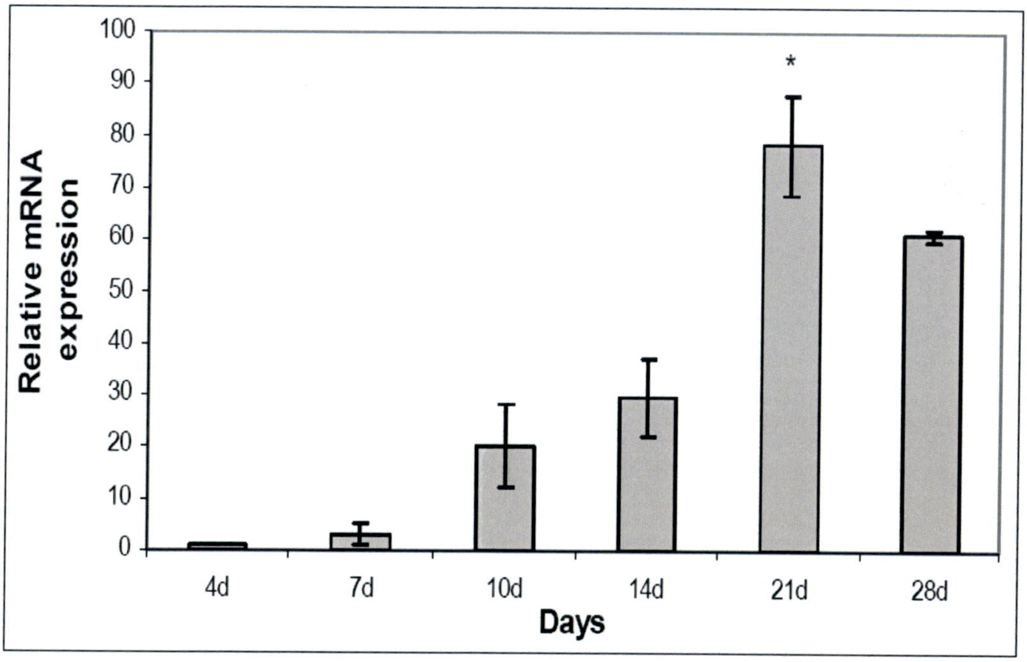

Figure 8. mRNA expression of type II collagen in MSCs seeded on *T. rufotincta* biomatrices on different days post seeding. Expression levels are shown as fold changes ±SD, relative to the expression level on day 4. Columns represent means ± S.D. The stars represent significant level.

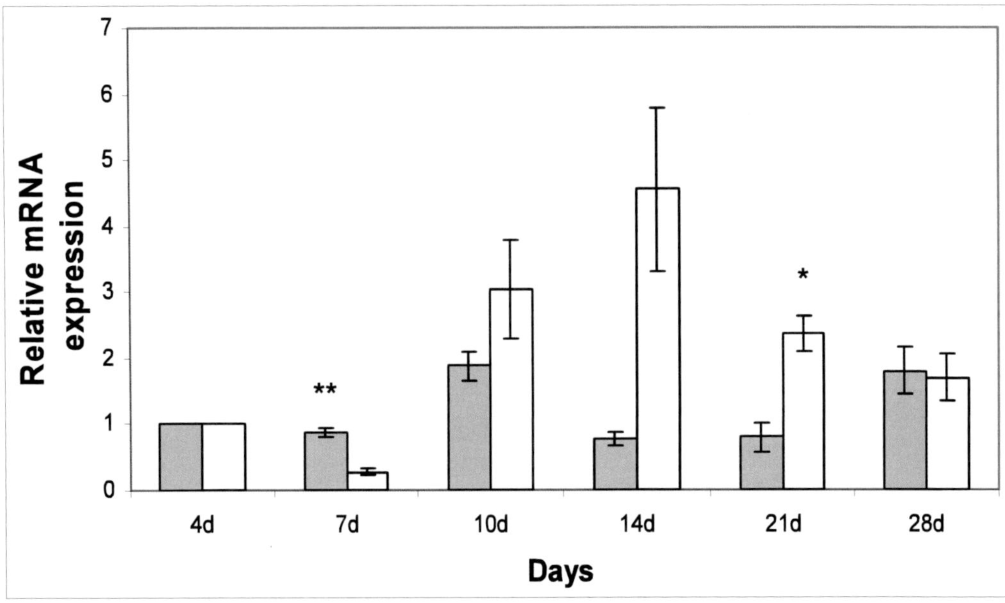

Figure 9. mRNA expression of *SOX9* (white) and *Runx2* (Grey) in MSCs seeded on *T. rufotincta* biomatrices on different days post seeding. Expression levels are shown as fold changes ±SD, relative to the expression level on day 4. The asterisks represent significant differences according to *t*-test. Columns represent means ± S.D. The stars represent significant level.

The present study introduces for the first time, skeletal material of *T. rufotincta* for tissue engineering applications. *T. rufotincta* is an abundant species in the Gulf of Eilat and other oceans, therefore, large quantities can be easily harvested with low environmental impact. The material has various advantages, it is made of $CaCO_3$, a crystalline form that is known to be biocompatible *in vivo* and *in vitro* [30, 32, 36]. Its high porosity, allows tissue ingrowth, and permit mass and heat nutrient exchange. This kind of a highly porous microenvironment may also permit invasion of vasculature, neural and lymphatic systems to encourage fast recovery, if implanted *in-vivo*. Although it is a natural product, the geometry of the pores within the bioscaffold is specific, modular and uniform. The modular calcite crystalline arrangement resembles other natural inorganic skeletons produced by marine invertebrates hence possesses high compressive strength and low tensile strength [31]. These mechanical properties are important for specific biological functioning by providing organizational and spatial cues towards morphogenesis, especially in hard tissue engineering [29, 30, 31]. In addition, handling and treatment of this class of scaffolds such as shaping to size, machining and sterilization could be easily executed due to their mechanical and physical properties. This is in contrast to other known natural scaffolding material, such as chitosan, alginate and collagen [17, 19].

In this study MSCs were seeded on the *T. rufotincta* skeleton. Our results imply that *T. rufotincta* promotes cell adhesion and proliferation and exhibits cytocompatibility. The rapid proliferation accelerates the tissue development within the *T. rufotincta* pores.

In this case cell-biomaterial contacts were carried in a static manner in contrast to cases were dynamic seeding or surface modifications with sticky Arg-Gly-Asp (RGD) motif are necessary to facilitate adhesion to materials such as collagen, PGA and PLLA [48, 49].

As a result, the seeding procedure on *T. rufotincta* does not require high cell densities, which is advantageous when the source of the cells is autologous or allogeneic and there is a need to sacrifice healthy tissue for the treatment of a disfunctioning tissue.

In addition, we assume that the *T. rufotincta* biomaterial provides a comfortable environment which encourages cell-cell, cell-ECM interactions and tissue development. The fact that the cell could adhere and proliferate within this novel 3D biomaterial means that the rapid interaction between cells and material is followed by triggering of stringent molecular signaling, prior the development of the tissue.

It is widely accepted that MSCs differentiation occurs with the presence of inducers, like growth factors and cytokines. However, several studies suggested that differentiation of MSCs on coralline calcium carbonate biomatrices could be achieved without the addition of any inducers [50, 37]. It has been suggested that the unique 3D crystalline morphology, together with its chemical composition of $CaCO_3$ may play an important role in cell adhesion, proliferation, and differentiation.

It is important to note that phosphate was not detected in the formed tissue, except of small amounts of precipitation along the pores perimeter (Figure 6G-6L). Also, Type I collagen, which is a specific bone ECM protein [47], was not detected in *T. rufotincta* during all experimental period (Figure 7, upper row). Moreover, the expression levels of *Runx2*, osteoblasts transcription factor, remained lower than the levels of *SOX9* during the first 21 days of the experiment. These results indicate that most of the MSCs do not differentiate to the osteogenenic lineage. This finding disagrees a previous study, which showed that MSCs seeded on calcium carbonate in the form of calcite (from natural limestone) exhibited osteogenesis after 20 days of culture [50]. The calcite that Fricain et al., [50] used in their study, was obtained from aragonite (coral) modified to calcite by heating. In this way, the architecture of the scaffold was maintained, but the modified coral had a lower pore diameter ($\approx$ 150 μm) than *T. rufotincta* (> 500 μm). Therefore, we assume that the geological morphology and the phase shift of the scaffold could influence the lineage of the developed tissue.

Previous studies suggested that MSCs and neurons seeded on aragonite crystalline scaffold of marine origin may have the capacity to directly exploit calcium from the biomaterial [37, 51]. The authors postulate that this intimate calcium dependent interaction is a causative factor leading to the fast adhesion, differentiation and tissue development here as well.

However, in case of *T. rufotincta* due to the big pore size the created tissue stretched across the pores and most of the tissue is not in direct contact with the calcite crystals. We assume that this kind of organization reduces the direct cell-material interactions and calcium uptake resulting in chondrogenic differentiation rather than the aragonite derived osteogenic differentiation. The evidence of GAG and type II collagen after seventh days of culture and their accumulation till 21 days post seeding strengthen this assumption. In addition, *SOX9* was expressed as early as seven days after onset of the experiment. Its expression is crucial because *SOX9* has a pivotal rule in the transition from mesenchyme-to-chondrocyte morphology and in converting condense MSCs into chondrocytes [52, 16]. In contrast, *Runx2* is expressed in the late condensation stage of chondrogenesis and then has substantially decreased expression in proliferating chondrocytes followed by increased expression again in prehypertrophic chondrocytes and hypertrophic chondrocytes, while the expression of *SOX9* decreased [16]. The motions in the expression of *SOX9* and *Runx2* on *T. rufotincta* also suggest that chondrogenesis of MSCs occurred. The condensation stage continues approximately till the seventh day, followed by attendance of proliferating chondrocytes. A maturation of chondrocytes into prehypertrophic chondrocytes could occur just after 21 days of culture. This process happened without addition of any additives to the medium or any mechanical stimuli to encourage MSCs differentiation. Hence, we postulate that *T. rufotincta* biomatrices have favorable potency for converting MSCs to chondrocytes. We believe that addition of growth factors or mechanical stimuli may prevent the maturation process of chondrocytes and it will provide a favorable system for cartilage tissue engineering. However, further research should be carried out in this direction.

In summary, we conclude that the skeletal material of *T. rufotincta* could be used as scaffolding material for tissue engineering, which provides a porous 3D microenvironment for cell adhesion and proliferation, and formation of tissue culture. Even more important is the finding that *T. rufotincta* biostimulated the distribution, proliferation and the specific differentiation of MSCs to chondrocytes. We believe that this scaffolding material could be a step forward towards creation of 3D *ex-vivo* cartilage construct.

References

[1] Makarand, V.R. and Sittinger, M. (2002) Tissue engineering: advances in *in vitro* cartilage generation. *Trends biotechnol.* 20 (8), 351-356.

[2] Gao, J. Yao, J.Q. and Caplan, A.I. (2007). Stem cells for tissue engineering of articular cartilage. *Proc. Inst. Mech. Eng. [H].* 221 (5), 441-50.

[3] Vunjak-Novakovic, G. and Kaplan, D.L. (2006) Tissue engineering: the next generation. *Tissue Eng.* 12 (12), 3261-3263.

[4] Freed, L.E. Guilak, F. Guo, X.E. Gray. M.L. Tranquillo, R. Holmes, J.W. Radisic, M. Sefton, M.V. Kaplan, D. and Vunjak-Novakovic, G. (2006). Advanced tools for tissue engineering: scaffolds, bioreactors, and signaling. *Tissue Eng.* 12 (12), 3285-305.

[5] Eberli, D. and Atala, A. (2006). Tissue engineering using adult stem cells. *Methods Enzymol.* 420, 287-302.

[6] Marion, N.W. and Mao, J.J. (2006). Mesenchymal stem cells and tissue engineering. *Methods Enzymol.* 420, 339-361.

[7] Phinney, D.G. Prockop, D.J. (2007). Concise review: mesenchymal stem/multipotent stromal cells: the state of transdifferentiation and modes of tissue repair--current views. *Stem Cells.* 25 (11), 2896-902.

[8] Friedenstein, A.J. Chailakhjan, R.K. and Lalykina, K.S. (1970). The development of fibroblast colonies in monolayer cultures of guinea-pig bone marrow and spleen cells. *Cell Tissue. Kinet.* 3, 393-403.

[9] Friedenstein, A.J. Gorskaja, J.F. Kulagina, N.N. (1976). Fibroblast precursors in normal and irradiated mouse hematopoietic organs. *Exp. Haematol.* 4, 267-274.

[10] Caplan, A.I. (1991). Mesenchymal stem-cells. *J. Orthop. Res.* 9 (5), 641-650.

[11] Alhadlaq, A. and Mao, J.J. (2004). Mesenchymal stem cells: Isolation and therapeutics. *Stem Cells Dev*. 13 (4), 436-448.

[12] Dennis, J.E. and Caplan, A.I. (2004) Bone marrow mesenchymal stem cells. In: S. Sell (Eds.), *Stem cells handbook* (pp. 104-117). New Jersey, USA: Humana Press Inc.

[13] Pittenger, M.F. Mackay, A.M. Beck, S.C. Jaiswal, R.K. Douglas, R. Mosca, J.D. Moorman, M.A. Simonetti, D.W. Craig, S. and Marshak, D.R. (1999). Multilineage potential of adult human mesenchymal stem cells. *Science* 284, 143-147.

[14] Payushina, O.V. Domaratskaya, E.L. and Starostin, V.I. (2006). Mesenchymal stem cells: Sources, phenotype, and differentiation potential. *Biol. Bull.* 33 (1), 2-18.

[15] Hu, J.C.Y. and Athanasiou, K.A. (2003). Structure and function of articular cartilage. In Y.H. An, K.L. Martin (Eds.), *Handbook of histology methods for bone and cartilage* (pp. 73-95). New Jersey, USA: Humana Press Inc.

[16] Kronenberg, H.M. (2003). Developmental regulation of the growth plate. *Nature*. 423, 332-336.

[17] Vats, A. Tolley, N.S. Polak, J.M. and Gough, J.E. (2003). Scaffolds and biomaterials for tissue engineering: a review of clinical applications. *Clin. Otolaryngol*. 28, 165-172.

[18] Zelzer, E. and Olsen, B.R. (2003). The genetic basis for skeletal diseases . *Nature*. 423 (6937), 343-348.

[19] Temenoff, J.S. and Mikos, A.G. (2000). Review: tissue engineering for regeneration of articular cartilage. *Biomaterials*. 21, 431-440.

[20] Kinner, B. and Spector, M. (2002). Mesenchymal cell culture: cartilage. In: A. Atala, and R.P. Lanza (Eds.), *Methods of tissue engineering* (pp. 317-331). California, USA: Academic Press.

[21] Freyria, A.M. Cortial, D. Ronziere, M.C. Guerret, S. and Herbage, D. (2004). Influence of medium composition, static and stirred conditions on the proliferation of and matrix protein expression of bovine articular chondrocytes cultured in a 3-D collagen scaffold. *Biomaterials*. 25, 687-697.

[22] Pieper, J.S. van der Kraan, P.M. Hafmans, T. Kamp, J. Buma, P. van Susante, J.L.C. van den Berg, W.B. Veerkamp, J.H. and van Kuppevelt, T.H. (2002). Crosslinked type II collagen matrices: preparation, characterization and potential for cartilage engineering. *Biomaterials*. 23, 3183-3192.

[23] Stevens, M.M. Qanadilo, H.D. Langer, R. Prasad and Shastri, V. (2004). A rapid-curing alginate gel system: utility in periosteum-derived cartilage tissue engineering. *Biomaterials*. 25, 887-894.

[24] Ma, H.L. Hung, S.C. Lin, S.Y. Chen, Y.L. and Lo, W.H. (2003). chondrogenesis of human mesenchymal stem cells encapsulated in alginate beads, *Inc. J. Biomed. Mater. Res.* 64A, 273-281.

[25] Secheriest, V.F. Miao, Y.J. Niyibizi, C. Westerhausen-Larson, A. Matthew, H.W. Evans, C. Fu, F.H. and Suh, J.K. (2000). GAG-augmented polysaccharide hydrogel: Anovel biocompatible and biodegradable material to support chondrogenesis. *J. Biomed. Mater. Res.* 49, 534-541.

[26] Wang, Y. Kim, H.J. Vunjak-Novakovic, G. and Kaplan, D.L. (2006). Stem cell-based tissue engineering with silk biomaterials. *Biomaterials*. 36, 6064-6082.

[27] Mooney, D.J. and Silva, A.E. (2007). Tissue Engineering: A glue for biomaterials. *Nat. Mater.* 6, 327 – 328.

[28] Ruppert, E.E. and Barnes R.D. (1994) *Invertebrates zoology* (6 Ed.) USA: Saunders College Publishing.

[29] Green, D. Walsh, D. Mann, S. and Oreffo, R.O.C. (2002). The potential of biomimesis in bone tissue engineering: lessons from the design and synthesis of invertebrate skeletons. *Bone*. 30 (6), 810-815.

[30] Demers, C. Hamdy, C.R. Corsi, K. Chellat, F. Tabarizian, M. and Yahia, L. (2002). Natural coral exoskeleton as a bone graft substitute: A review. *Bio-med. Mater. Eng.* 12, 15-35.

[31] Shors, E.C. (1999). Coralline bone graft substitutes. *Orthop. Clin. North. Am*. 30 (4), 599-613.

[32] Hu, J. Fraser, R. Russell, J.J. Ben-Nissan, B. and Vago, R. (2000). Australian coral as a biomaterial: Characteristics. *J. Mater. Sci. Technol.* 16, 591-595.

[33] Vuola, J. Bohling, T. Kinnunen, J. Hirvensalo, E. and Asko-Seljavaara, S. (2000). Natural coral as bone-defect-filling material. *J. Biomed. Mater. Res*. 51, 117-122.

[34] Vago, R. Plotquin, D. Bunin, A. Sinelnikov, I. Atar, D. and Itzhak, D. (2002). Hard tissue remodeling using biofabricated coralline biomaterials. *J. Biochem. Bioph. Meth.* 50, 253-259.

[35] Kreklau, B. Sittinger, M. Mensing, M.B. Voigt, C. Berger, G. Burmester, G.R. Rahmanzadeh, R. and Gross, U. (1999). Tissue engineering of biphasic joint cartilage transplants. *Biomaterials*. 20, 1743-1749.

[36] Ohgushi, H. and Caplan, A.I. (1999). Stem cell technology and bioceramics: from cell to gene engineering. *J. Biomed. Mater. Res.* 48, 913-927.

[37] Abramovitch-Gottlib, L. Geresh, S. and Vago, R. (2006). Biofabricated marine hydrozoan: A bioactive crystalline material promoting ossification of mesenchymal stem cells. *Tissue Eng.* 12 (4), 729-739.

[38] Mygind, T. Stiehler, M. Baatrup, A. Li, H. Zou, X. Flyvbjerg, A. Kassem, M. and Bünger, C. (2007). Mesenchymal stem cell ingrowth and differentiation on coralline hydroxyapatite scaffolds. *Biomaterials* 28(6), 1036-1047.

[39] Appelbaum, L. Achituv, Y. and Mokady, O. (2002). Speciation and the establishment of zonation in an intertidal barnacle: specific settlement vs. selection. *Mol. Ecol.* 11 (9), 1731-1737.

[40] Simkiss, K. and Wilbur, K.M. (1989). *Biomineralization - cell biology and mineral deposition*. San Diego, California, USA: Academic press, Inc.

[41] Bourget, E. (1977) Shellstructure in sessile barnacles. *Nat. Can.* 104, 281-323.

[42] Nousek, N.A. (1984). Shell formation and calcium transport in the barnacle *Chthamalus fragilis*. *Tissue Cel.l* 16, 433-442.

[43] ASTM C373-88. Standard Test Method for Water Absorption, Bulk Density, Apparent Porosity, and Apparent Specific Gravity of Fired Whiteware Products. American Society for Testing and Materials: Philadelphia, USA, 1999.

[44] Chomczynski, P. and Sacchi, N. (1987). Single-step method of RNA isolation by acid guaninium thiocynate-phenol-chloroform extraction. *Anal. Biochem.* 162 (1), 156-159.

[45] Perez, F.R. and Martinez-Frias, J. (2003). Identification of calcite grains in the Vaca Muerta mesosiderite by Raman spectroscopy *J. Raman Spectrosc.* 34 (5), 367-370.

[46] Nave, D. Rosenwaks, S. Vago, R. Bar, I. (2004) Pulsed laser deposition of marine origin material: Preparation and characterization of $CaCO_3$ particles and CaO nanocrystals. *J. Appl. Phys.* 95, 8309-8313.

[47] Anselme K., (2000). Osteoblast adhesion on biomaterials. *Biomaterials.* 21 (7), 667-681.

[48] Freed, L.E. and Vunjak-Novakovic, G. (2000). Tissue engineering bioreactors. In R.P. Lanza, R Langer, and j. Vacanti (Eds.), *Principles of tissue engineering* (2^{ed}, pp. 143-156). California, USA: Academic press.

[49] Alvarez-Barreto, J.F. and Sikavitsas V.I. (2007). Improved mesenchymal stem cell seeding on RGD-modified poly(L-lactic acid) scaffolds using flow perfusion. *Macromol. Biosci.* 7 (5), 579-588.

[50] Fricain, J.C. Bareille, R. Ulysse, F. Dupuy, B. and Amedee, J. (1998). Evaluation of proliferation and protein expression of human bone marrow cells cultured on coral crystallized in the aragonite or calcite form. *J. Biomed. Mater. Res.* 42 (1), 96-102.

[51] Baranes, D. Cove, J. Blinder, P. Shany, B. Peretz, H. and Vago, R. (2007). Interconnected network of ganglion-like neural cell spheres formed on hydrozoan skeleton. *Tissue Eng.* 13 (3), 473-82.

[52] Barna, M. and Niswander, L. (2007). Visualization of cartilage formation: Insight into cellular properties of skeletal progenitors and chondrodysplasia syndromes. *Dev. Cell* 12 (6), 931-941.

In: Biomimetic and Supramolecular Systems Research ISBN: 978-1-60456-405-1
Editor: Arturo H. Lima

Chapter XI

Biomimicking Devices Consisting of Proteins: Biomemory and Image Sensor

Jeong-Woo Choi[1,2*], Junhong Min[3] and Byung-Keun Oh[1,2]
[1]Department of Chemical and Biomolecular Engineering and [2]Interdisciplinary Program of Integrated Biotechnology, Sogang University, # 1 Shinsoo-Dong, Mapo-Gu, Seoul 121-742, Republic of Korea, [3]Gachon Bionano Research Institute, Kyungwon University, Bokjung-Dong, Sujung-Gu, Seongnam 461-701, Republic of Korea

Abstract

Biomimetic based bioelectronics have emerged as a breakthrough with great potential capable of generating new concepts and technologies for the development of the next generation electronics. We have developed several biomimicking devices consisting of protein such as biomemory and image sensors, and so on. Bioelectronic device for information storage composed of ferredoxin was based on the electron transport function of plant photosynthesis. The information storage function of the fabricated device was investigated by applying the reduction potential as a write function and measuring the stored reducing charge as a read function. And a biomolecular photoreceptor consisting of bacteriorhodopsin(bR)/flavin complex Lamgmuir-blodgett(LB) films was developed for color recognition. By mimicking the early process of color perception in the human visual system, the light-sensitive biological pigments with photoelectronic conversion function were chosen and used as the main components of the device. bR and flavin were deposited onto patterned ITO glass, and a 9-pixel photoreceptor with a sandwich-type structure of ITO/bR/flavin/electrolyte gel/Pt was constructed. When simple color images were projected onto the photoreceptor, the shapes and the colors of projected images were successfully recognized by the knowledge-based algorithm, and the processed images were visualized onto the monitor connected to the photoreceptor through a data acquisition system and personal computer. And a bioelectronic device consisting of bR as photoreceptor and 6-nitrospiropyran as noise filter was developed for the extraction of desired image. In the constructed bioelectronic device, the increment of the intensity ratio of transmitted lights in comparison with that of input lights was obtained when various input lights of the different light intensity were irradiated onto the films. The proposed devices here can be combined and extended to the development of biorobots with the fundamental researches in the fields of brain/neuroscience, material science, and mechanics, which joint activities will create a new field of biomechatronics. With the whole new world of electronics, the concept of a biocomputer or a biorobot will no longer appeared as a science fiction, but will be realized in this century.

Introduction

A tremendous flurry of exciting progress in electronic devices and systems has been made in the last few decades. The rudimentary integrated circuits containing a few components were replaced with today's version, incorporating millions of active devices on a single chip. In spite of the dramatic changes in the manner that devices are constructed an interconnected, the essential principles and technologies for the device fabrication and operation are consequently as relevant today as it was 30 years ago. [1,2]

Nowadays, the microprocessors in which the highly advanced current electronic technologies are integrated are essentially implanted in almost every part of industrial, consumer products, and appliances such as computers, sensors, control systems, robots, vehicles, TVs, refrigerators, and microwave ovens. Although current electronics can be fabricated using nano technologies in various fields, there are still several inherent problems to be resolved for the near future demands; faster speed of process by parallel processing, bigger memory size by multi-level system, analog information system like brain. [3]

For many years, it has been considered that digital microprocessors are good at handling sequential calculations, but no good at analog information processing, such as visual patterns and auditory sounds because the internal structure and the operating system of the current microprocessors themselves are essentially binary digital, dealing with only '0' and '1' states. Of course, the technology for digital recording of analog data has been remarkably developed, and some of those are already familiar to use in the forms of digitally mastered music, digital cameras, and voice synthesis chips, etc. However, it remains still impossible to sustain the analog information of a drawing that reminds one of a certain photograph.

Comparing with the living organisms, which inherently have dynamic and searchable information processing functions, the current digital microprocessors with comparable capabilities do not exist today and will not be readily built in the near future. Unlike the silicon-based microprocessors, the information processing units in living organisms are multi-dimensional ones with a tremendous number of complex cross references; vertical and horizontal interconnection among a variety of sources for getting analog information, such as visual, auditory, olfactory, gustatory, somatosensory, and even emotional. [4]

The frontiers in the fields of electronics have tried to find out a breakthrough to revolutionize the current world of microprocessors that are designed and operated along the line of the *Von Neumann architecture*. Although the Von Neumann architecture has major strength that the information processing is simple and thereby the control is uncomplicated, the main drawback lies in the bottleneck due to the sequential processing architecture, especially, when a large volume of information must be processed.

Fortunately, it has become relatively easy to try out various architectures owing to the great strides in microprocessors. For this reason, there is growing interest in various *non-Von Neumann style architectures*. However, the parallel processors that have been proposed up to now are often little more than *multiple Von Neumann style*, rather than virtually different ones from the basic ideas of Von Neumann. A truly non-Von Neumann and genuinely revolutionary architectures still remain to be found, and become one of the most challenging subjects to the researchers in the fields of electronics.

From the viewpoint of the electronics, the most noteworthy fact emerging from the life sciences is that the living organisms have computer-like information processing mechanisms, which are totally different from those of the current microprocessors using Von Neumann style architecture. Actually, the living organisms are highly developed biochemical machines regulated by the extremely advanced computer-like central processing units. Even in a case of single cell, the various life activities can be controlled by a well organized recipe consisting of thousands of biochemical reactions. Of course, the higher-level organisms, such as the vertebrates, are also able to perceive a bundle of information with several sensory organs and properly respond to the corresponding stimuli through the reliable analysis performed by the highly organized central processing unit, the brain. That is the main reason why the living organisms have been considered as the most powerful information processors in the world [5]. For this reason alone, it has been the ultimate dream for the electronic engineers to reduce the gap between the today's microprocessors and the nature's living information processors by mimicking some attractive features of the living organisms, such as self assemble capabilities of molecules by specific molecular recognition, self-replication, self-repair, multi-dimensional analog information processing, self-adaptation (learning), parallel information processing, and pattern recognition/matching.

Since the late 1970's the frontiers who eagerly want to find out new concepts for the next generation microprocessors have had much more interests in these amazing functions of the nature's living information processors. Owing to the great progress in the life sciences, the characteristics of various biological structures have been elucidated, and thereby the distance between the nature's living information processors and the conventional silicon-based microprocessors has decreased. The convergence between these two broad disciplines is not a transient fashion, but it is only natural that the fusion between them will continue to deepen and expand with the passage of time.

Bioelectronics has emerged as a breakthrough with great potential capable of generating new concepts and technologies for the development of the next generation electronics [3,6-8]. The main concept was inspired from the fact that an individual biomolecules, especially proteins, could be used as a basic unit of the electronic device, so that the integration scale of the IC (Integrated Circuits) could be increased by several orders of magnitude. It is naturally expected that the memory capacity of the chips would be dramatically improved by something like 10 million or 100 million folds comparing with that of the current silicon based chips.

With a bioelectronics research began in the early 1980s, the research groups related to the topics have emerged in the United States, Japan, and Europe. The experts in the various fields, such as life sciences, physics, chemistry, chemical engineering, electronics, and computer science, began to get together for these fascinating research subjects, and opened the field of new hybrid technology.

The possible application areas of bioelectronics are very diverse. The primitive and fundamental application would be biosensors. Since the development of the enzyme based biosensor for glucose detection first described by Clark in 1962, there has been an impressive proliferation of applications involving a wide variety of target molecules [8-11]. One of the most remarkable advancement in the history of biosensors is a DNA probe for genetic information diagnosis. With the great strides in genetics and the related fields, a dramatic progress has been made in DNA and RNA sensors. Passing through its beginning stage, it is now commercially available in the form of chip. Another importance topic in biosensors would be a biomolecular photoreceptor that could be used as an image capture device [12-15]. Nowadays, biosensors are one of the main streams in bioelectronics, and are successfully used in the various fields, such as bioreactor operations, environmental monitoring personal health care, and medical/clinical diagnosis, etc. Another possible candidate to be realized in the near future would be biodiodes or biotransistors. In the late 1970s, ideas concerning biodevices and biochips emerged, which were essentially related with the construction of electronic devices using biological molecules, such as proteins, as the smallest functional unit [7]. By appropriately aligning such molecules, it was thought that the extremely small devices could be built, in which the constituents of device could be protein complexes, molecular wires, and interface molecules, etc. It was reported that various artificial molecular diodes have been fabricated by mimicking the electron transfer mechanisms in the biological systems, such as photosynthetic reaction center [13,16-22]. Since a biotransistor could be used as a basic unit of a memory device, it would be a key for the development of biochips or biocomputers.

Therefore, the mimetic device using biological materials was simply reviewed with 3 major devices, 1. image information processor with hard ware types, 2. color recognition device using 2 biological materials, 3. photodiode mimicking photo synthetic reaction in biosystem.

1. Bioelectronic Device for the Photosignal Enhancement

The parallel processing without software is one of the amazing functions of biosystems consisting of biological molecules because it is possible to operate mega-information such as an image with high speed using this capability. The hardware-type processors have been investigated to develop a parallel processor for image information processing. The processor mimicking the brain, which is memory-based computer, has been studied. The brain can self-organize the memorized data and program for the functions of learning, processing and read/write. In order to combine the domains of memory and program, the processor should be developed as a hardware-type. The properties of multi-states of biomolecules provide the crucial functions, such as extraction, summation, threshold, and filtering, essentially used in optoelectronic devices. The photo excitation transport and the reversible photochemical reaction in LB films have been studied and the applicability to an optical parallel processor has also been suggested [23].

The recent studies related with the input systems using biomaterials have been focused to the integrated system capable of information processing such as pattern/moving direction recognition. An artificial photoreceptor that consisted of bacteriorhodopsin films was constructed, and the photoswitching function and the differential response to light intensity were reported [24]. It was reported that an image sensor consisting of 256 artificial photoreceptors was capable of the rapid and selective detection of moving objects in image sensing. The optical image captured and processed by this photoreceptor was simultaneously visualized by forwarding the signals to a two-dimensional display [14]. As mentioned above, this bacteriorhodopsin protein can be used to develop the artificial photo receptor due to their stability and photoelectric properties.

The photoelectrical properties of bacteriorhodopsin were applied directly for pattern enhancement device concept, fused with photochromic properties of spiropyran. The objective of usage of spiropyran on bioelectronic device consisting of bacteriorhodopsin was to apply the function of image extraction to photoreceptor by the enhancement of signal to noise ratio, available to parallel-process image information with hardware-type. The signal to noise ratio has been studied with the prospect of signal amplification in the several field [25]. In the proposed image extraction system, noise is cut off, not amplifying a signal, unlikely the conventional image extraction systems in which the signal is amplified while noise is sustained. The proposed bioelectronic device with image extraction function has following characteristics.

1) Detection of visible light signal
2) Enhancement of singal-to-noise ratio by reducing noise; which can reduce the size of the memorized data by eliminating noise.
3) Control of enhancement degree of signal-to-noise ratio by simple method
4) Positive image extraction; the lighter part in the desired image information is extracted

The principle of proposed bioelectronic device was as shown in Figure 1. The absorbance of spiropyran in the region of visible light was changed by the irradiation of UV light by its photochromic property. The photomerocyanine, the coloration state of spiropyran by UV light, has the absorbance peak in the region of visible light which can be changed by the irradiation of visible light. This absorbance peak decreases proportionally to the irradiated visible light intensity and its irradiation time. If two input lights of different light intensity (I_{10}/I_{20}) irradiate on the photomerocyanine films, the absorbance of two parts irradiated by lights with different intensities is obtained as different values. By Lambert-Beer's law, the ratio of the transmitted light becomes,

$$(I_1 - I_2) = (I_{10} - I_{20})e^{(A_1 - A_2)} \tag{1}$$

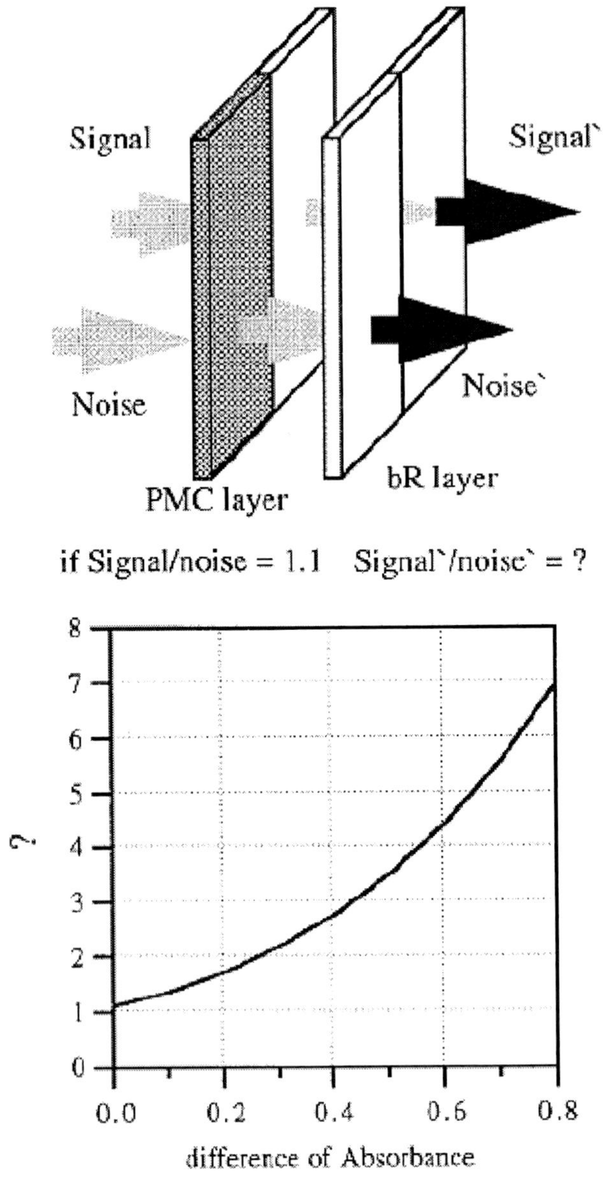

Figure 1. Conceptual principle of signal to noise enhancement of bioelectronics.

If I_{10}/I_{20} is larger than unity, (A_2-A_1) becomes larger than zero due to the photochromic property of photomerocyanine. Due to $e^{(A_1-A_2)}$, the intensity ratio of transmitted light (I_1/I_2) is larger than that of input light (I_{10}/I_{20}). Thus, the increment of intensity ration of transmitted light in comparison with that of input light can be obtained through the photomerocyanine films with irradiation time when two input lights of the different intensities are irradiated onto the films [15]. Spiropyran was deposited onto one side of glass slide by spin coating or LB technique, and then layer of bacteriorhodopsin was formed by LB method or electrophoretic sedimentation technique onto the other side of the glass on which Indium-Tin oxide(ITO) was pre-deposited as shown in Figure 2 [26].

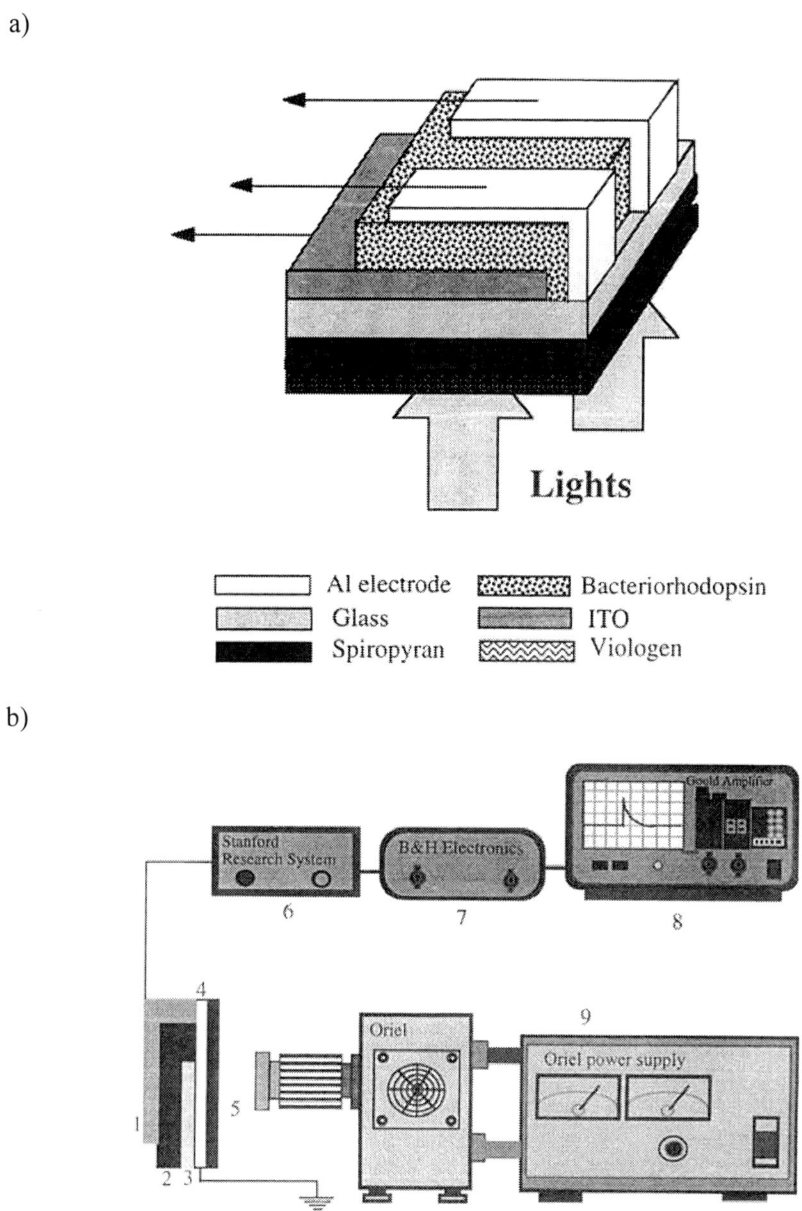

Figure 2. Schematic diagram of (a) bioelectronic device consisting of bacteriorhodopsin and spiropyran for the enhancement of intensity ratio of two input lights.

The absorbance spectra of spiropyran were investigated as UV irradiation time after deposition. Figure. 3 represented the photochromic property of spiropyran films (10 layers) fabricated by the spin coating technique. The absorbance at 550 nm was increased when UV light was irradiated, which is the typical photoreaction process, the transformation from spiropyran to photomerocyanine. Its absorbance at 550 nm then decreased when visible light (near 550nm) was irradiated.

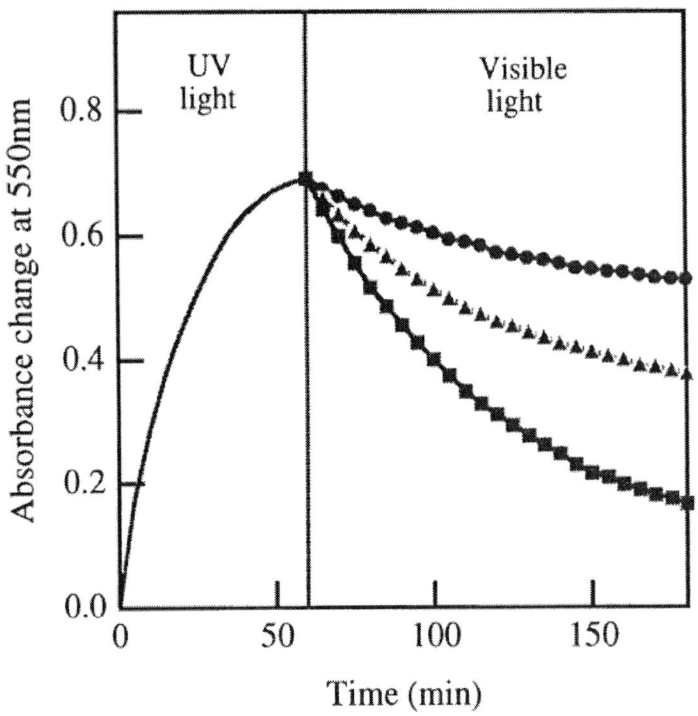

Figure 3. Photochromic effect of spiropyran films (10 layers) with respect of UV light (300-400nm, $1mW/cm^2$) and visible of 550nm and intensity of $1mW/cm^2$ (■), 0.5 mW/cm^2 (▲) and $0.2mW/cm^2$ (●).

This process of the decoloration of the films by conversion of photomerocyanine to spiropyran is proportional to the intensity of the irradiated visible light. Consequently, the intensity of the visible light transmitted via photomerocyanine films was increased as the visible light was irradiated onto the films, due to the decoloration of photomerocyanine proportional to the intensity of irradiated visible light [15]. This implies spiropyran film can be applied to bioelectronic device as noise filter by the enhancement of signal to noise ratio.

The photoelectronic properties of bacteriorhodopsin layer formed on the ITO glass by LB techniques and electrophoretic sedimentation methods was confirmed as shown in Figure 4. The bacteriorhodopsin in purple membrane is grown by the bacterium when the oxygen concentration becomes too low to sustain the generation of ATP via oxidative phospholylation [27,28] The opto-electronic device consisting of bacteriorhodopsin films for information processing have been developed according due to the fact as follows.

1. Long term stability of the protein to thermal and photochemical degradation.
2. Picosecond photochemical reaction times less than 5 ps.
3. High quantum yields of forward and reverse direction that permit the use of low light levels for switching.
4. Wavelength-independent quantum yields.
5. A large shift in absorption spectrum accompanying photochemistry that accurate and reproducible assignment of state.
6. The ability to form thin films of bacteriorhodopsin with excellent optical properties.

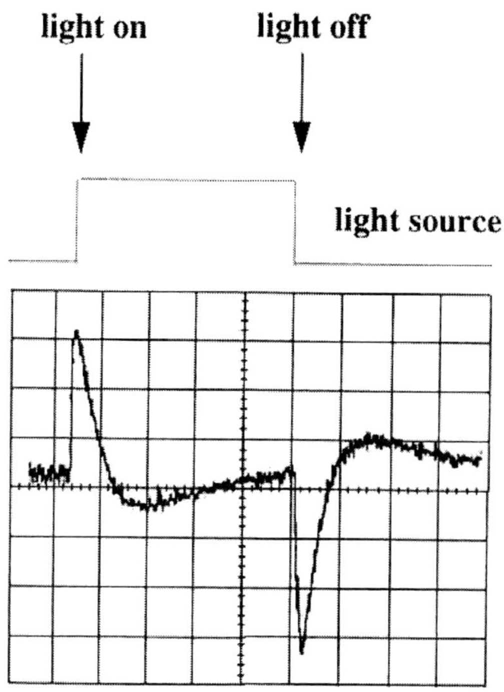

Figure 4. Typical photo induced electrical response of bacteriorhodopsin films.

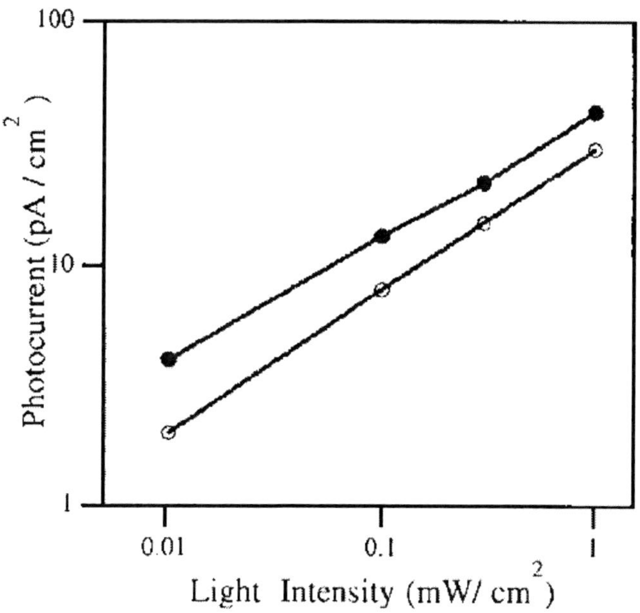

Figure 5. The photocurrent of bacteriorhodopsin LB films as the light intensity illuminated.

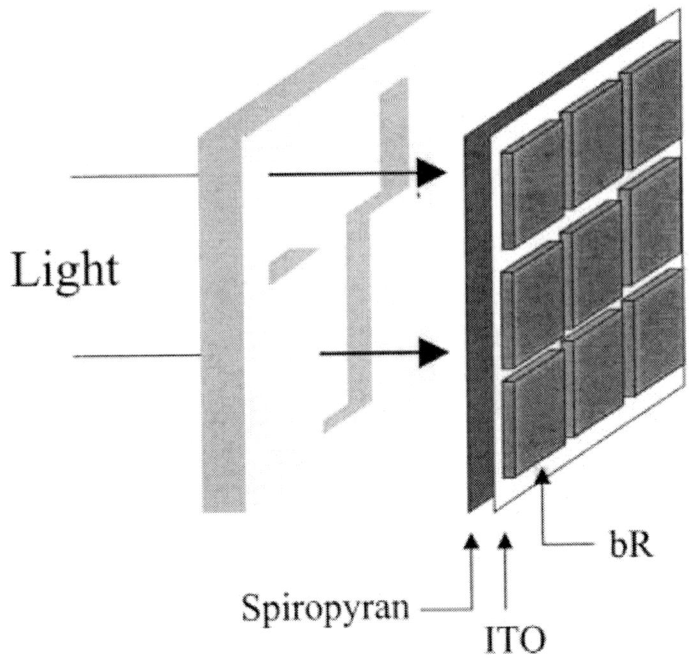

Figure 6. Schematic illustration of bacteriorhodopsin based photoreceptor for image extraction.

The photocurrent of bacteriorhodopsin film fabricated by LB technique and Electrophoretic sedimentation method was proportional to light intensity illuminated on bacteriorhodopsin layer as shown in Figure 5. This results represented the bacteriorhodopsin protein was suitable for artificial photoreceptor consisting of biomaterials.

The bioelectronic device composed of 9 photoreceptors in two-dimensional array was constructed for the enhancement of photosignal to noise ratio using spiropyran and bacteriorhodopsin as shown in Figure 6 [26].

Using bioelectronic device in Figure 6, the enhancement of target signals that are defined as slightly higher intensity than undesired signals (defined as noise) was verified using the T-shaped input pattern as shown in Figure. 6. The intensities of the input light transmitted through the spiropyran film are increased by the photochromic function of the spiropyran film. Over a period of time, the T-shaped image could be emphasized visually by the enhancement of the intensity ratio of the target signal to the noise.

Eventually, this T-shaped input image was extracted from the background because the photocurrent generation of bR films is directly proportional to the input light intensity, and the projected pixels can generate much more intensified photocurrent than those of the background. The enhancement efficiency of the target signal to noise ratio was 40% in 40 min as shown in Figure 7. This result implies that it is possible to apply a parallel processing function, one of the characteristics of the biosystem, without any software for information processing in bioelectronic devices for the elimination process of undesired optical/electrical signals [26].

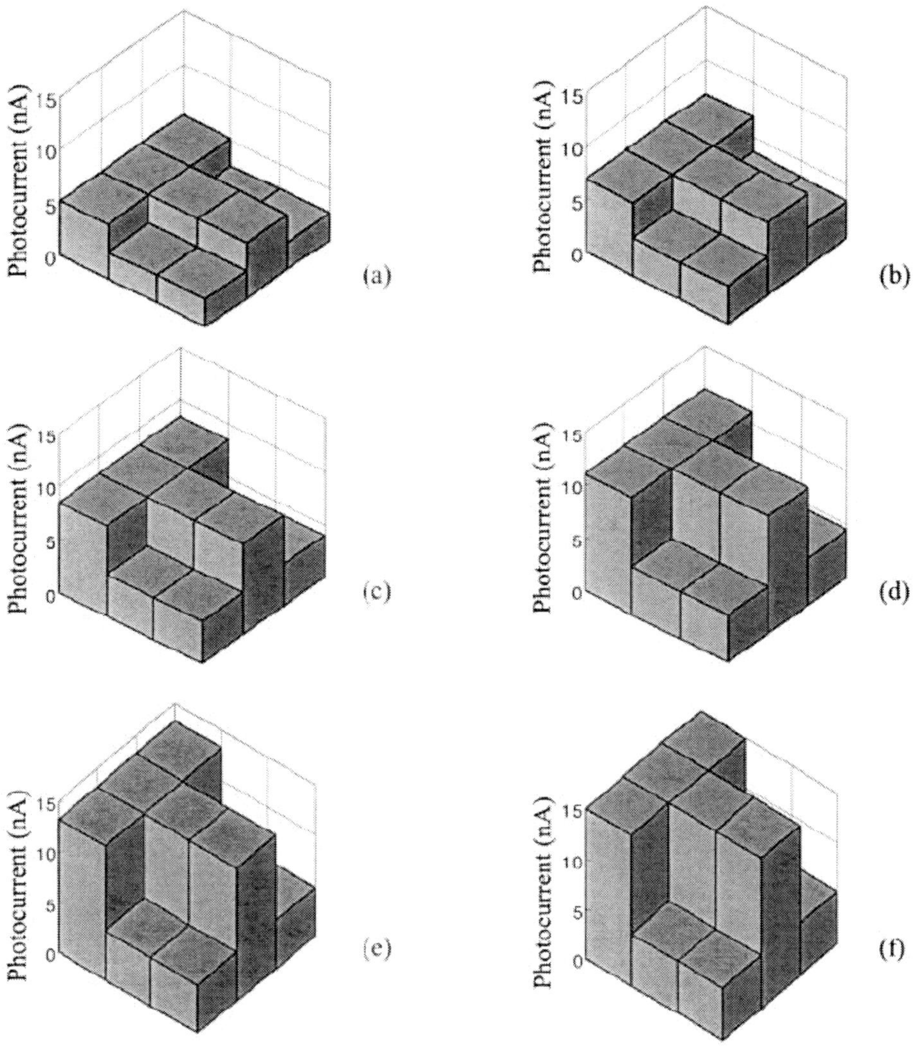

Figure 7. The process of image extraction using the proposed photoreceptor. In the initial process, indicated as (a), the target signals and undesired signals cannot be obviously distinguished from one another. However, as time goes by (a–f), the target signals can be much more intensified than those of background (undesired signals). Finally, the T-shaped input image can be clearly extracted from the background. The time taken for image extraction was as follows: (a) 1 min, (b) 9 min, (c) 18 min, (d) 25 min, (e) 32 min, and (f) 41 min.

The control of extraction efficiency of bioelectronic device is possible with simple method to control the extracting time and number of noise filter. But the extracting time determined by characteristics of the spiropyran film such as molecular interaction is not controllable. To reduce the extracting time of bioelectronic device proposed here, the design technique of new materials having faster photochromic property, the fabrication technique not to affect the molecular properties such as kinetic mechanism and the analysis of the effect of fabrication and operation condition might be developed.

2. Biomolecular Photoreceptor for Color Recognition

Among the sensory systems of human, the visual system is regarded as the most powerful source for obtaining information since it provides a way to interact flexibly with our environment, without any direct physical contact. The human visual system shows outstanding performance in pattern matching, color recognition, and 3-dimensional perception, and thereby the human visual system is often considered as the most powerful and efficient image pickup device in the world. By mimicking, the functions of human visual system, the image pickup devices have been developed, and some characteristics of the human eye have been successfully duplicated in current image pickup devices and machine vision algorithms.

The typical image pickup device would be a camera, especially video camera. The basic operation of a video camera begins within the camera, where light in the televised scene converts to video signals, such as electrical variations in charge, current, televised scene converts to video signals, such as electrical variations in charge, current, or resistance. In this process, the optical image focuses onto a photosensitive target in the camera, and the conversion of the entire scene into a video signal is accomplished by horizontal and vertical scanning. As the scanning continues in sequential order, the light values for every point in the scene are converted to signal output. The process is basically the same for color or monochrome TV, whereas the separate video signals are produced for the scene's reds, greens and blues in color TV.

With the invention of the charge-coupled devices (CCDs), it has become reality to use image pickup devices in our daily life for personal uses as well as in the industries or hospitals for specific ones [29-31]. The image capturing function of CCD is performed by a columnar grid array of very small photodiodes. Each diode is a light-sensitive semiconductor, thus the diodes transfer photoelectrons to the output stage of the device, proportional to the light energy they absorb. When an optical image is focused onto the diode array, the diodes produce a charge image corresponding to the picture. For color CCDs, the oblique color striped filter has been used to divide the light into red, green, and blue signals.

Another challenge in the development of artificial visual system would be a biomolecular photoreceptor using light-sensitive biological pigments. With growing interests in the artificial visual system and the behavioral neuroscience, the biomolecular photoreceptor will be one of the most challenging fields, and can provide a possible way to revolutionize the current technology of image pickup devices in the near future. Here, a biomolecular photoreceptor for color recognition by mimicking the light wave detection functions of visual receptors was discussed.

The most important intrinsic property of the materials to be used is the photoelectric conversion, the transformation of light energy to an easily sensible signal, such as current or voltage. Bacteriorhodopsin and flavin were chosen to organize the complex thin films to be used as a cardinal component of the biomolecular photoreceptor for color recognition because these have different photosignal shapes as well as different excitation wavelengths as shown in Figure 8.

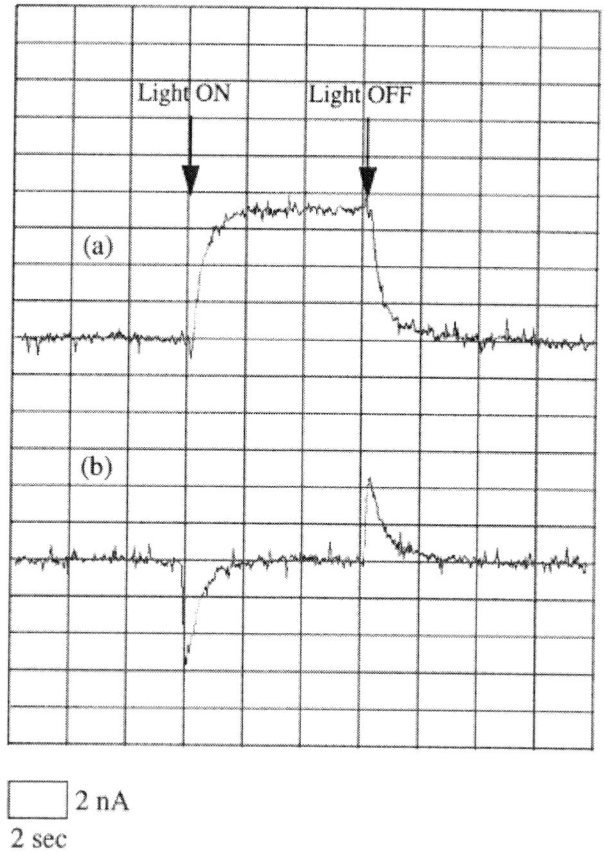

Figure 8. Typical photoelectrical responses of flavin and bR. (a) Flavin LB films (400 nm); (b) bacteriorhodopsin LB films (550 nm).

As a photoreceptor, bacteriorhodopsin has an advantage due to its nature of broad action spectrum, but the workable wavelength is restricted in the range of 500–680 nm. Using a biological or an organic molecule that produces photocurrent according to the light illumination of the shorter wavelength around 400–450 nm, together with bacteriorhodopsin, it is possible to expand the workable wavelength of the bacteriorhodopsin-based photoreceptor to the range of 400–680 nm [37].

Flavins, coenzymes for biological oxidoreductions, have been extensively studied since its discovery; at that time, flavin was known as vitamin B_2. It is now known that flavin is involved as either flavin adenine nucleotide (FAD) or flavin mononucleotide (FMN) in the biological electron transport systems, such as mitrochondrial electron transfer chain [29-35]. It has been reported that there are more than 20 flavoproteins covalently bound to flavin coenzymes. The versatility of flavoenzymes comes from at least the three redox states; oxidized, semioxidized, and reduced states. In the visible light region, the photoexcitation of flavin molecules is induced by the light illumination of 400–500 nm wavelength, and then the fast electron transfer takes place in the neighboring molecular assembly by the energy difference resulted from the charge separation. Therefore, in the excited state, flavin behaves like a conducting material.

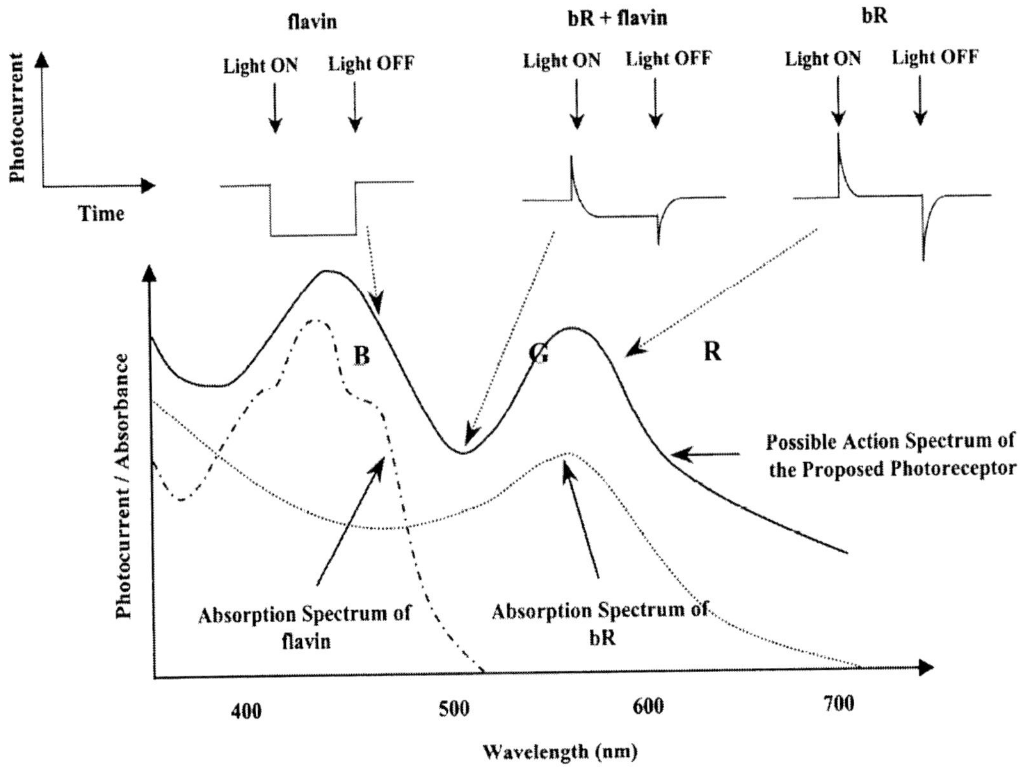

Figure 9. A schematic illustration of the principle of color recognition by the proposed biomolecular photoreceptor.

By mimicking the biological electron transfer system, such as photosynthetic reaction center, the artificial molecular photodiodes using flavin films as an electron-sensitizing component have been demonstrated.

A schematic illustration of the conceptual principle of color recognition using the proposed biomolecular photoreceptor was shown in Figure 9 [37]. As mentioned previously, the photocurrent of flavin can be generated by the photoexcitation according to the light illumination, and its action can be obtained in the range of 300~500nm, with two peaks at around 350 and 450 nm, which is slightly varied depending upon the side chains and the functional groups bound to the flavin molecule. One the other hand, the photocurrent of bacteriorhodopsin can be generated by the charge displacement in the bacteriorhodopsin molecules initiated by the rapid photoisomerization of the chromophore retinal, and its action spectrum can be obtained in the range of 500~600 nm with the peak at around 560nm. If the photoelectric conversion property of bacteriorhodopsin is combined with tat of flavin by appropriately aligning two molecules onto the single photoreceptor, it is possible to realize the biomolecular photoreceptor capable of sensing almost all visible light components, ranging from 400nm to 680 nm. It is expected that the action spectrum of the proposed biomolecular photoreceptor composed of bacteriorhodopsin/flavin complex thin films can be obtained in the range of 400–680 nm, which is corresponding to the almost all visible light components. In the range of 400–500 nm, flavin signal is dominant, and the only thing to be changed is the magnitude, peak value, of the photocurrent with changing the wavelength of the illuminated light. In the range of 500–680 nm, on the other hand, the dominant signal is of bacteriorhodopsin, and the flavin signal is rarely detected. Furthermore, since the signal shape of bacteriorhodopsin is impulse-like whereas that of flavin is pulse-like one, the signals of these two molecules can be unequivocally distinguished from each other. Therefore, it is possible that, in the visible light region, the illuminated light can be detected by the proposed biomolecular photoreceptor, and the color of the incident light can also be analyzed by comparing the signal shapes and the magnitude of the photocurrent to be generated, with a simplified knowledge-based algorithm to be obtained from the correlation between the incident light wavelength and the resulting photoelectric responses [36,37].

This preliminary processing of optical image is then linked with a higher-level processing network, including optic nerves, lateral geniculate nucleus, and visual cortex in cerebrum. The higher-level perception processes heavily rely on the mechanism called perceptual learning, which make it possible to recognize and distinguish between objects that have been seen in the past. For example, babies cannot recognize the shape or color of objects that he or she sees, but just can feel the difference in shape or color. As growing up, the or she learns a lot of information from their parents, teachers, friends, and even from books, and after learning, then he or she can recognize and discriminate the shape and color of objects by comparing with those accumulated in their brain through the learning processes [37]. The initial learning was simply performed by illuminating 400 nm monochromatic light, which produces a maximum photocurrent in visible region under the given light intensity, and the peak value of generated photocurrent was used as a standard. Then, the each value of measured photocurrent upon the light illumination with arbitrary wavelength is normalized to this standard value obtained upon the 400 nm light illumination, and the normalized value was applied to the knowledge based algorithm [37]. To validate color recognition function of bioelectronics, the simple color images were projected onto the 9 pixels of biomolecular photoreceptor consisting of bacteriorhodopsin/flavin complex LB films using a 300 W Xenon arc lamp with optically transparent color mask. The response profiles were obtained from each pixel upon the different color image projections as shown in Figure 10. The signal shapes and the magnitude of photocurrent were unambiguously distinguished from one another. The photocurrent measured from each pixel in the photoreceptor was interpreted by the simplified knowledge-based algorithm, and the processed image was then visualized onto the color monitor.

Using these distinctive photosignal as the wavelength of light input and knowledge-based algorithm, the several color images were projected onto the photoreceptor, and the processed images were visualized as shown in Figure. 11. The primary colors in visible light region, red, green, and blue, were clearly recognized, and additionally, cyan, yellow, black, and white colors were also discriminated form the other colors. The color images in Figure 11 were fairly well reproduced by the proposed photoreceptor with the simplified knowledge-based algorithm. However, there is some ambiguity in distinguishing the secondary colors, especially greeny-blue (at around 520–540 nm). since the signal shape obtained in the range of 520–540 nm is impulse-like and the magnitude of photocurrent show similar values to that obtained in the range of 570–620 nm.

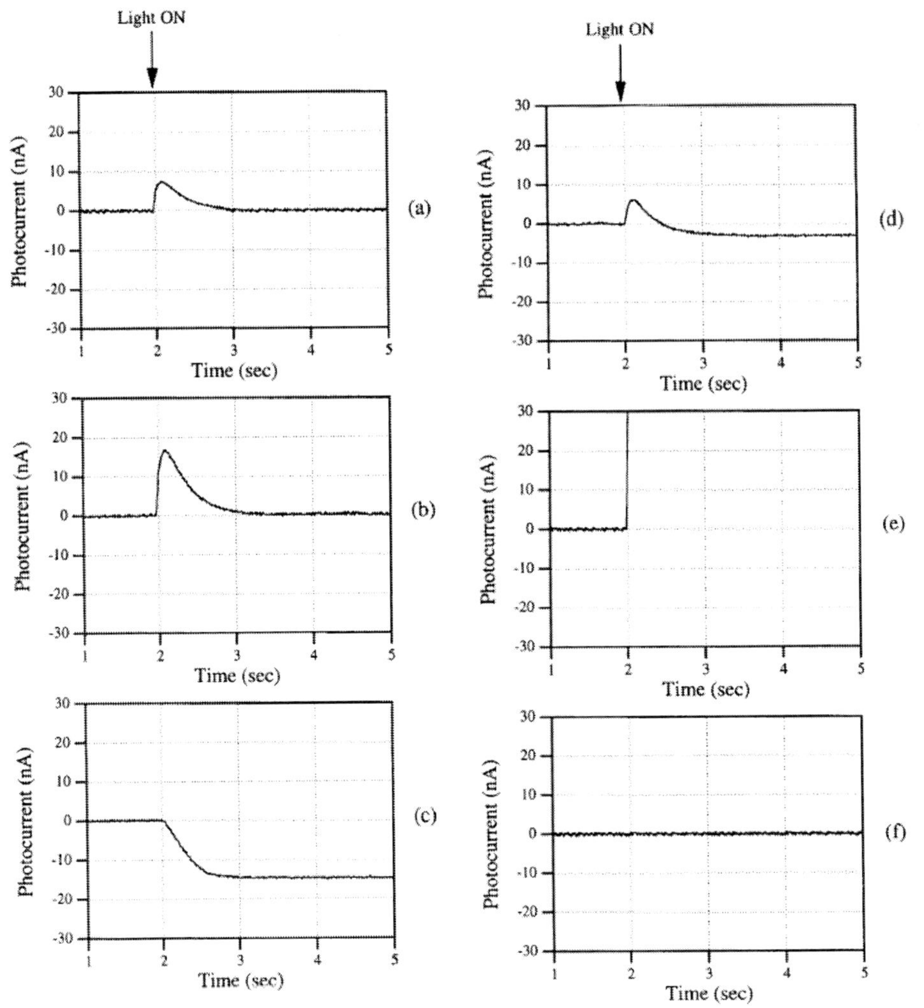

Figure 10. The photoelectric response behavior of the proposed biomolecular photoreceptor upon the several color projection. (a) red, (b) green, (c)blue, (d) cyan, (e) white, and (f) black.

3. Molecular Electronic Device for the Information Storage Based on the Photosystems

Since 1985, various artificial molecular devices have been fabricated by mimicking the electron transport function of biological photosynthesis. Some researchers investigated a molecular photodiode composed of multi-biomolecules complex films and its optical and electrical characteristics [17,22,38-42]. They use flavin, Cytochrome C, viologen, GFP (green fluorescence protein), TCNQ, and so on as a sensitizer (S) and electron acceptor (A). These molecules were assembled on solid electrode with their orientation by using LB method, which is one of the best techniques for aligning the molecules.

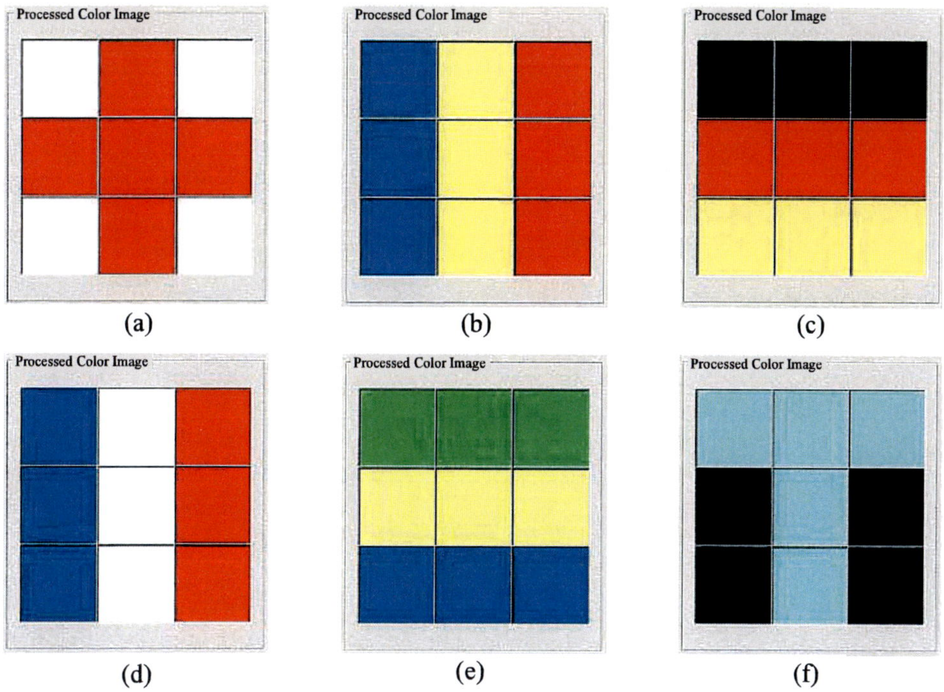

Figure 11. The reproduced color images by the biomolecular photoreceptor consisting of bacteriorhodopsin and flavin with a simplified knowledge based algorithm. (a) red-cross, (b) the national flag of Chad, (c) the national flag of Germany, (d) the national flag of France, (e) the national flag of Gabon, and (f) the alphabet "T".

Fujihira have investigated the electrochemical photodiode consisted of LB films of three functional molecules or an aligned triad on the electrode, which worked in electrolyte solution [43]. Ferrocene, pyrene and viologen were used as electron donor (D), S and A, respectively in biomolecule systems. The metal/insulator/metal(MIM) structured device consisted of hetroLB films of D, S and A was fabricated and photoinduced electron transfer was investigated [19]. Recently the molecular photodiode composed of electron D/S/relay (R)/A type 4 component MIM devices has been investigated as shown in Figure 12 [44]. Development of molecular photodiode is important in molecular electronics area since it can be applied to the molecular memory devices due to the photoswitching and rectifying characteristics. Molecular photodiode consisted of LB films of D, S, R, A units was designed based on the photoinduced electron transport in the natural system. By using two acceptor molecules (R and A), the time for separated charge area (A-/R/S/D+) can be sustained longer than that of A-/S+ hetero system. Charge recombination from R and A to the ground state S can be reduced due to the fast electron transport from S* to A via R and due to the increase distance between S and A with the presence of R. Based on these effects, the molecular photodiode composed of D/S/R/A hetero LB films is expected to show the better diode an switching properties than that of S/A and D/S/A hetero LB films. However, main reaction of photo induced bioelectronics diode is S/A system. GFP has been used as a sensitizer in biomolecular photodiode by Choi [39,40]. The GFP is the final light emitting protein in the bioluminescent jellyfish Aequorea victoria.

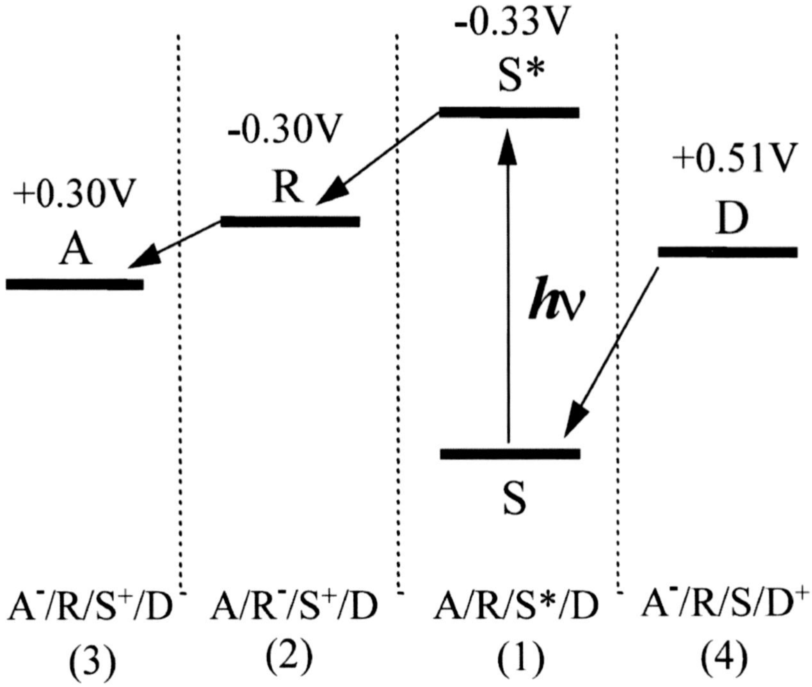

Figure 12. Energy level diagram of the molecules involved in the photosynthetic reaction center (number, the sequence of electron transfer reaction); A, N-docosilquinolinium tetracyanodimethan (TCNQ); R, N-Allyl-N`-[3-propylamido-N``,N``-di(n-octadecyl)]-4,4`-bipyridium dibromide (Viologen); S, 7,8-dimethyl-10-dodecyl isoalloxazine (Flavin); D, ferrocene octadecyl amine (Ferrocene).

The purified GFP, a protein of 238 amino acids, absorbs blue light and emits green light (peak emission at 510 nm). This fluorescence is very stable, and virtually no photobleaching is observed [22]. Since GFP shows very highly fluorescence quantum yield, approximately 80%, it is very reasonable approach to use GFP as an electron sensitizer in the development of molecular electronic device by mimicking biological photosynthesis mechanism. GFP and viologen/cytochrome C were utilized as a sensitizer and relay/acceptor, respectively as shown in Figure 13, which shows the energy diagram of the molecular array consisting of S/M/A hetero film. In this molecular photodiode, the photoinduced electric signal was generated by light irradiation, so that the light transparency and electrical resistance of the device were considered. The ITO coated glass substrate is adequate for the bottom electrode of the molecular photodiode [39].

The photocurrent of the biomolecular photodiode with the irradiation of a 460 nm monochromic light was shown in Figure 14(a). The photocurrent of this device consisted of S, GFP/M, viologen/A, cytochrome C heterofilm was generated by light excitation in Figure 14(a). When a forward bias was applied according to the energy level profile in the MIM device, the photocurrent was generated. With step illumination, the reproducible photocurrent was generated as shown in Figure 14(a). This result implied that the photoswitching function of the MIM device consisting of biomolecules was available by mimicking photosystem in biosystem.

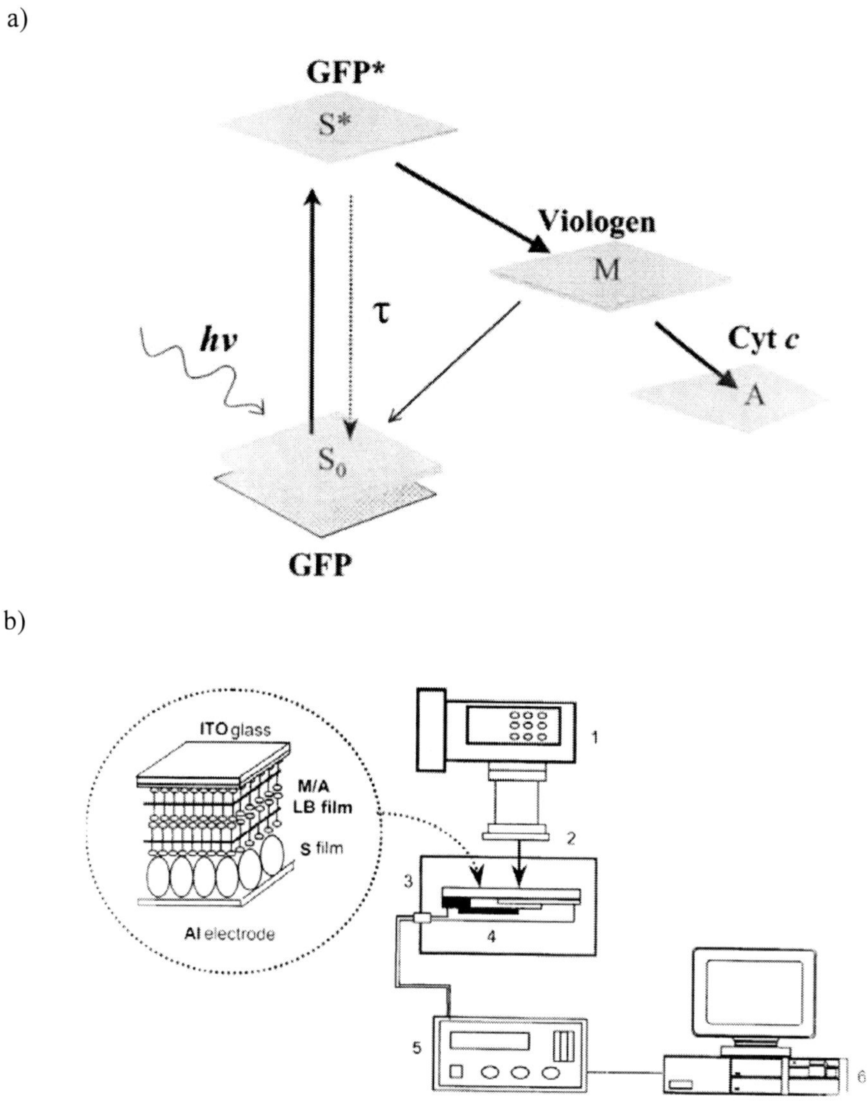

Figure 13. The Schematic diagram (a) of energy transfer of GFP(S)/Viologen(M)/Cytocrome C(A), (b) of light irradiation and photo signal detection system (1, 500 W xenon lamp; 2, 460 nm filter; 3, shield box; 4, sample device; 5, signal measurement unit (Keithley, SMU Model 236); 6, PC.

In the proposed molecular device, the photoinduced unidirectional flow of electrons could be achieved due to the redox potential difference as well as electronic coupling between the functional molecules as shown in Figure. 13(a). It was also observed that the intensity of the photocurrent was dependent on the external bias voltage. As the external bias voltage increased, higher photocurrents were generated. As shown in Figure. 14(b), the rectifying characteristic was also observed from the measurement of current–voltage characteristics. When a 460 nm monochromatic light was irradiated, the photocurrent was generated with the appropriate bias voltage. On the other hand, the current was less generated in the dark environment (without light irradiation) even though the bias voltage was increased. In photo state, the photocurrent was generated with 5mA when a forward bias was applied to the molecular photodiode whereas the photocurrent was 0 mA when reverse bias was applied to the molecular photodiode. This property was called as the rectification of photodiode. These results imply that the photodiode characteristics of molecular electronic device were verified and can be usefully applied as a model system for the development of the biomolecular photodiode by the mimicking of biological photosynthetic system [39].

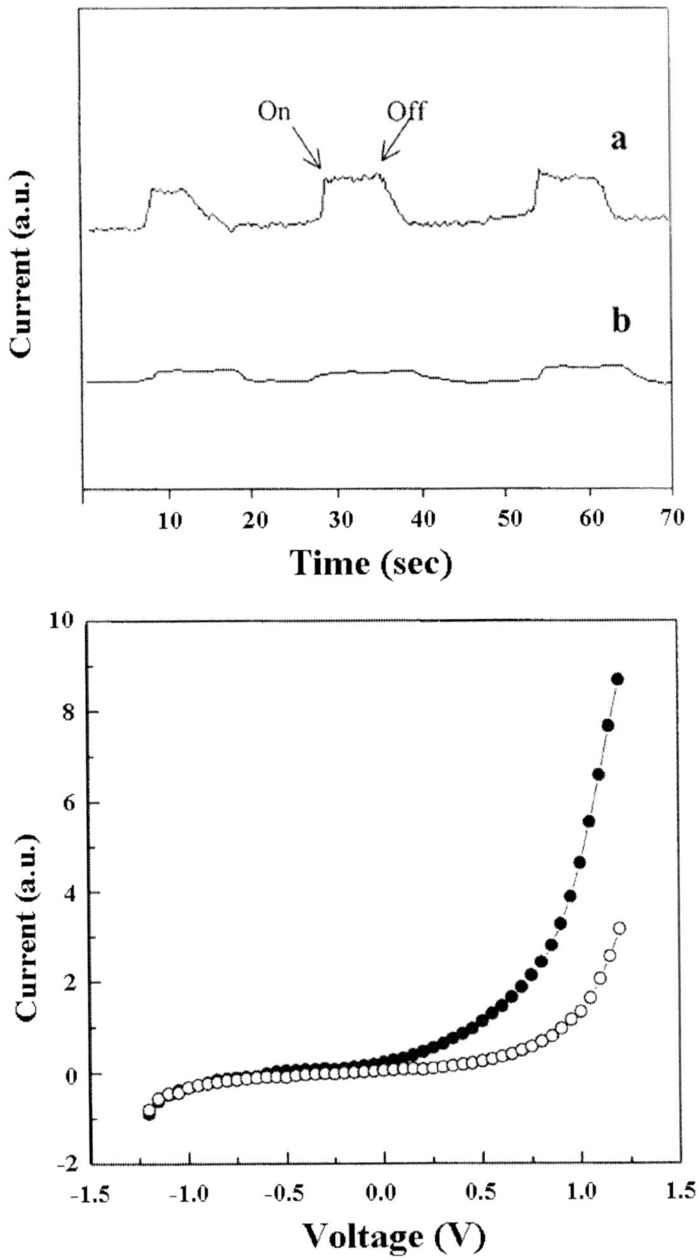

Figure 14. (a) Photo induced response of the MIM device: a, 2 V; b, 0 V. (b) I-V characteristics of the MIM device: (•) photocurrent; (○) darkcurrent.

Conclusion

The simply reviewed molecular electronic device which mimics the biological photosynthesis, biological photo receptor, biological image processor, can be usefully applied as a new electronic devices such as the molecular diode, molecular memory, Electroluminescence device, and solar cell in near future. Next challenge is the real application and commercialization of the molecular electronic device. As described here, the basic molecular electronic devices have been developed, and now it should be extended to make the advanced electronic device. These achievements lay the groundwork for further research on molecular electronics and bioelectronics, which can help to overcome the limit of current electronic device.

All of these application areas can be combined and extended to the development of biorobots with the aids of brain/neuroscience, materials science, and mechanics, and these joint activities will create a new field of biomechatronics. With this whole new world of electronics, the concept of a biocomputer or a biorobot will no longer appeared as science fiction, but will be realized in this century.

References

[1] Allison, J., Electronic engineering semiconductors and devices (2nd Ed.), McGraw-Hill, New York, 1990.

[2] Boyestad, R. and L. Nashelsky, Electronics: A survey of electrical engineering principles (4th Ed.), Prentice-Hall International, Inc., New Jersey, 1996.

[3] Kaminuma, T. and G. Matsumoto, *Tomorrow's computer, in biocomputer: the next generation from Japan* (T. Kaminuma and G. Matsumoto Eds., translated by N. Cook), Chapman and Hall, New York, pp 15-33, 1991.

[4] Beatty, J., *Principles of Behavioral Neuroscience*, Brown and Benchmark Publishers, Dubuque, pp.146-189, 1995.

[5] Birge, R. R., Nature of primary photochemical events in rhodopsin and bacteriorhodopsin, *Biochem. Biophys. Acta.*, 1016, 293-327 (1990a)

[6] Nicolini, C., *Towards the biochip: what, how and why*, in *Towards the Biochip* (C. Nicolini Ed.), World Scientific, Singapore, pp1-21,1990.

[7] Aizawa, M., *Biodevice computers, in Biocomputers: the next generation from Japan* (T. Kaminuma an G. Matsumoto Eds., translated by N. Cook), Chapman and Hall, New York, pp.107-119, 1991.

[8] Cass, A.E.G., Biosensors: A practical approach, IRL Press, New York, 1990.

[9] Turner, A.P.F., I. Karube, G.S. Wilson (Eds.), *Biosensors: Fundamentals and applications*, Oxford University Press, New York, 1987.

[10] Mosbach, K. (Ed.), *Methods in Enzymology*, Academic Press, New York, Vol. 137, 1988.

[11] Eggins, B.R., *Biosensors: An Introduction*, John Wiley and Sons Ltd., New York, 1996.

[12] Miyasaka, T., K. Koyama, and I. Itoh, Quantum conversion and image detection by a bacteriorhodopsin-based artificial photoreceptor, *Science*, 255, 342-344 (1992).

[13] Isoda, S., S. Nishikawa, S. Ueyama, Y. Hanazato, H. Kawakubo, and M. Maeda, Photo-induced electron transfer in molecular heterojunctions using flavin-porphyrin Langmuir-Blodgett multilayers, *Thin Solid Films*, 210/211, 290-292 (1992).

[14] Miyasaka, T. and K. Koyama, Image sensing and processing by a bacteriorhodopsin-based artificial photoreceptor, *Appl. Opt.*, 32, 6371-6379 (1993).

[15] Min, J., J. W. Choi, W. H. Lee and U. R. Kim, Photoreceptor consisting of spiropyran-bacteriorhodopsin films for photosignal enhancement, *Biosensors and Bioelectronics*, 13, 1151-1155 (1998).

[16] Fujihira, M., K. Nichiyama, and H. Yamada, Photoelectrochemical responses of optically transparent electrodes modified with Langmuir-blodgett films consisting of surfactant derivatives of electron donor, acceptor, and sensitizer molecules, *Thin Solid Films*, 132, 77-82 (1985).

[17] Isoda, S., Bioelectronic devices based on biological electron transport system using flavin-porphyrin molecular heterojunction, *FED J.*, 2, 59-65 (1992).

[18] Isoda, S., M. Maeda, H. Miyasaka, and N. Mataga, Picosecond laser photolysis studies on the electron transfer and charge shift processes in flavin mononucleotide-cytochrome c-EDTA molecular organization, Chem. *Phys. Lett.* 182, 379-383 (1991).

[19] Choi, J. W., G. Y. Jung, S. Y. Oh, W. H. Lee, and D. M. Shin, Photoinduced electron transfer in a MIM device composed of TCNQ-pyrene-ferrocene LB films, *Thin Solid Films*, 284/285, 876-878 (1996).

[20] Choi, J. W., M. J. Kim, S. W. Chung, S. Y. Oh, D. M. Shin, and W. H. Lee, Molecular photodiode consisting of flavin-viologen-TCNQ LB multilayers, *Mol. Cryst. Liq. Cryst.*, 294, 217-220 (1997).

[21] Choi, J. W., S. W. Chung, S. Y. Oh, D. M. Shin, and W. H. Lee, Photoinduced electron transfer in a MIM device composed of ferrocene-flavin-viologen-TCNQ molecular hetrojunctions, *Thin Solid Films*, 327/329, 671-675 (1998).

[22] Choi, J. W., Y. S. Nam, B. K. Oh, W. H. Lee and M. Fujihira, Artificial molecular photodiode consisting of GFP/viologen hetero-thin films, *Synth. Met.*, 117, 241-243 (2001).

[23] Yamazaki, I., S. Okazaki, T. Minami and N. Ohta, Optically switching parallel processors by means of Langmuir-Blodgett films, *Proc. 13th Symp. Futher Electron Devices (Tokyo, October, FED, Japan)*, 186-191 (1994).

[24] Miyasaka, T., K. Koyama, Recitified photocurrents from purple membrane Langmuir-blodgett films at the electrode-electrolyte interface, *Thin solid films*, 210/211, 146-149 (1992).

[25] Miyata,I., Y. Miuno, H. Arai, T. Ide, and Y. Unuma, Development of optical neural devices using molecular aggregates, *Proc. 13th Symp. Futher Electron Devices (Tokyo, October, FED, Japan)*, 152-157 (1994).

[26] Min, J., H.-G. Choi, B.-K. Oh, W. H. Lee, S.-H. Paek, J. W. Choi, Visual information processing using bacteriorhodopsin-based complex LB films, *Biosensors and Bioelectronics*, 16, 917-923 (2001).

[27] Yamaguchi, N., Y. Jinbo, M. Arai and K. Koyama, Visualization of the morphology of purple membrane surfaces by monoclonal antibody techniques, *FEBS letters*, 324, 287-292 (1993)

[28] McIntosh, A. R., and F. Boucher, On the action spectrum of the photoelectric transients of bacteriorhodopsin in solid-state films, *Biochim. Biophys. Acta.*, 1056, 149-158 (1991).

[29] Sweedler, J. V., K. L. Ratzlaff, and M. B. Benton (Eds.), *Charge-Transfer Devices in Spectroscopy*, VCH Publishers, Inc., New York, 1994.

[30] Shoenbeck, R. J., Electronic *Video systems: Diagnositics and Maintenance*, Prentice Hall, Inc., Singapore, 1996.

[31] Grob, B. and C. E. Herndon, *Basic Television and Video Systems*, McGraw-Hill, New York, 1999.

[32] hatefi, Y., Preparation and properties of NADH: Ubiquinone oxidoreductase (Complex I), EC 1.6.5.3., *Methods Enzymol.*, 53, 11-14 (1978)

[33] Galante, Y. M. and Y. Hatefi, Resolution of complex I and isolation of NADH dehydrogenase and an iron-sulfur protein, *Methods Enzymol.*, 53, 15-21 (1978)

[34] Heron, C., S. Smith, and C. I. Ragan, An analysis of the polypeptide composition of bovine heart mitochondrial NADH-ubiquinone oxidoreductase by two-dimensional polyacrylamide-gel electrophoresis, *Biochem. J.*, 181, 435-443 (1979)

[35] Lippard, S.J. and J. M. Berg, *Principles of bioinorganic Chemistry*, University Science Book, Mill Valley, 1994.

[36] Choi, H. G., J. Min, W. H. Lee, J.W. Choi, Molecular photoreceptor consisting of bacteriorhodopsin/flavin complex langmuir-blodgett films, *Biosensors and Bioelectronics*, 13, 1069-1075 (1998).

[37] Choi, H. G., W. C. Jung, J. Min, W. H. Lee, J.W. Choi, Color image detection by biomolecular photoreceptor using bacteriorhodopsin-based complex LB films, *Biosensors and Bioelectronics*, 16, 925-935 (2001).

[38] Akiyama, K., S. Nishikawa, S. Ueyama, and S. Isoda, Japanses Journal of *Applied Physics*, 34, 3942-3946 (1995)

[39] Choi, J. W., Y. S. Nam, S. J. Park, W. H. Lee, D. Kim and M. Fujihira, Rectified photocurrent of molecular photodiode consisting of cytochrome c/GFP hetero thin films, *Biosensors and Bioelectronics*, 16, 819-825 (2001)

[40] Choi, J.W., Y. S. Nam, S. C. Jeong, W. H. Lee, M. C. Petty, Molecular rectifier consisting of cytochrom c/GFP heterolayer by using metal coated optical fiber tip, *Current Applied Physics*, 6, 839-843 (2006)

[41] Nam, Y. S., K. W. Park, W. H. Lee, and J. W. Choi, Fabrication and electrical characteristics of ferredoxin adsorbed poly-L-lysin/11MUA hetero self assembly layer, *Materials Science and Engineering C*, 24, 35-38 (2004)

[42] K. S. Cho, Y. S. Nam, D. Kim, W. H. Lee, and J. W. Choi, Investigation of photoinduced electron transport in electron sensitizer/relay/acceptor hetero-LB films using transient absorption technique, *Synthetic Metals*, 129, 157-163 (2002).

[43] Sakomura, M., S. Lin, T. A. Moore, A. L. Moore, D. Gust, and M. Fujihira, Journal of Physical Chemistry A, 106, 2218-2226 (2002)

[44] Cho, K. S., Y. S. Nam, J. W. Choi, W. H. Lee, J. Choi, N. W. Song, and D. Kim, Investigation of charge transport in the molecular photodiode consisting of ferrocene/flavin/viologen/TCNQ molecular heterojunction, *Colloids and Surfaces A: Physicochemical and Engineering Aspects*, 202, 93-99, (2002)

In: Biomimetic and Supramolecular Systems Research
Editor: Arturo H. Lima
ISBN: 978-1-60456-405-1

Chapter XII

Fundamentals of the Nanocrystal Formation from Supersaturated Solutions

***Carlos Luna*[1] *and Raquel Mendoza-Reséndez*[2]**
1) Facultad de Ciencias Físico-Matemáticas. Universidad Autónoma de Nuevo León. Av. Pedro de Alba s/n, San Nicolás de los Garza, 66450 Nuevo León, México. carloslunacriado@hotmail.com
2) Facultad de Ingeniería Mecánica y Eléctrica. Universidad Autónoma de Nuevo León. Av. Pedro de Alba s/n, San Nicolás de los Garza, 66450 Nuevo León, México. mendozaresendez@hotmail.com

Abstract

The study of the linked processes related to the chemical precipitation from supersaturated solutions is required to understand the mechanisms involved in numerous natural processes, such as the biomineralization of biofunctional structures and the precipitation of chemical sedimentary rocks. Also, it offers us an exceptional opportunity to investigate how the matter self-assemblies into hierarchy ordered structures. On the other hand, the phenomenon of the chemical precipitation can be applied in the laboratory as an efficient and versatile synthetic route to prepare novel materials of both fundamental and practical interest.

In this chapter, we present an overview of the fundamentals of the crystallization from solutions, drawing especial attention to their application in the production of uniform nanocrystalline materials with engineered features.

1. Introduction

Crystallization from supersaturated solutions is a complex phenomenon broadly present in nature, which has played an essential role in the development of life and the formation of the environments in which life evolved. Consequently, comprehend the mechanisms that govern the precipitation from solutions is an essential challenge required for the understanding of a huge variety of natural processes, such as the formation of soils and chemical sedimentary rocks [1, 2], and the biomineralization of biofunctional materials (intracellular nanocrystals [3-5], teeth, bones, shells [6, 7], etc.).

For that reason, the research efforts of a growing number of scientific investigations are devoted to study the different aspects of this phenomenon. For example, the geochemical implications of the crystallization from solutions are currently being revisited to determine what changes have experienced the Martian environments during the course of time [8-11]. This information is using to evaluate the potential of such environments for the emergence and development of life taking into consideration the possible metabolic strategies that could have been originated [8, 12, 13].

On the other hand, it is generally accepted that the chemical precipitation from supersaturated solutions is an efficient and versatile synthetic route to produce ultra-fine colloidal crystals and ordered nanocrystalline structures with engineered properties [14-17]. The study of these materials offers us an exceptional opportunity to scrutinize cooperative behaviors which appear combined with quantum, surface and size effects to generate properties unobserved in other systems [18, 19], and that cover a large range of technological applications, such as energy [20] and information [15, 21, 22] storage, photonics [23], catalysis [24], biosensors [25], electronics [26] and the rapidly expanding nanomedicine [27].

The great success of this preparation technique is due to its ability to synthesize materials over a large range of compositions and characters without the necessity of expensive and complicated experimental setups, and the well level of control that it provides over the kinetics of crystal formation processes and the interactions that are present in the system. These redeeming features are attained by means of the reproducible and uncomplicated modification of the synthesis parameters (concentration and nature of the reactive compounds, reaction temperature, pH of the solution, presence of ligands, etc.). Such factors make possible the production of crystals and sophisticated structures whose characteristics (uniformity, size, morphology, composition, inner structure, crystalline phase, surface nature and aggregation state) can be tailored to satisfy specific requirements of numerous biological functionalities and fundamental and technological uses. However, despite the continuing upsurge of our knowledge about the crystallization mechanisms in solution, the prediction of the subtle balances between the kinetically and thermodynamically favored circumstances that lead to the precipitation of ultra-fine crystals with certain desired characteristics remains a difficult task. In fact, some related aspects have not been completely elucidated yet, and consequently, our understanding about this subject is rather limited. In this regard, we can learn much from nature about the control of the nucleation and growth of flaw-free crystals and about the self-assembly of the matter at molecular and supramolecular scale.

In this chapter, we present a review of the essential aspects of the crystallization in supersaturated solutions illustrated with some of our recent results. Particular attention is drawn to the intervention over the mechanisms of particle formation and the colloidal interactions to generate uniform nanocrystalline materials. In particular, the spontaneous nanocrystal aggregation in ordered structures, which can be considered like *a crystallization process on the nanometric scale* [28], is briefly examined. Some relevant mathematical expressions arising from the classical homogeneous nucleation theory and some crystal growth models are described to facilitate a further comprehension of this phenomenon.

The work is organized as follows. Firstly, in Section 2, a general classification of the mechanisms of nanomaterial formation is presented. In Section 3, the crystal precipitation by homogeneous nucleation is outlined taking into consideration the classical nucleation theory and its shortcomings as an approach to this phenomenon. In section 4, the induction of a heterogeneous nucleation is evaluated like a powerful route to control the crystal growth in solution. In Section 5, we give a very short description of the dependences of crystal growth by diffusion of the solute units toward the crystal surface. In Section 6 the Ostwald rule of stages is analyzed. The Section 7 deals with the fundaments of the colloidal nanocrystal stabilization and the nanocrystal self-assembly into ordered nanostructures. Then, in Section 8, the crucial roles played by the high-speed centrifugation in the purification of the crystallized materials, the improvement of the colloidal stability and particle size selection is mentioned. Finally, we present the final remarks of the work.

2. Mechanisms of Nanomaterial Formation

It is convenient to begin our discussions by considering the types of formation mechanisms on which all the existing methods of nanomaterial production are based. A very general classification of them can be established bearing in mind the fashion in which the nanometric scale is attained [14]. In this way, three kinds of mechanisms can be considered. In a first group, the desired dimensions are reached by means of the division or fragmentation of a macroscopic material. These splitting-up mechanisms can be produced through mechanical processes, usually by ball-milling [29], or dividing the material in tiny drops after its fusion, which can be solidified by a sudden cooling [30]. The techniques of preparation associated to these mechanisms, which go forward the nanometric scale in an *up-bottom* approach, are frequently used, albeit the resulting materials have usually a wide size distribution.

Another family of the formation mechanisms of nanomaterials consists in the condensation of atoms, ions or molecules that constitute them. In these mechanisms, the advance toward the nanometric size evolves in a *bottom-up* approach. It occurs in two steps: nucleation and diffusional growth, and it can be happen in gaseous phase by means of the chemical or physical deposition of the material in form of vapor [31], or in liquid phase, as it is the case of the chemical precipitation in supersaturated solutions, which will be studied in detail in the present contribution.

Finally, in a third group of these formation mechanisms, precursor nanometric materials (generated by a mechanism ascribed to one of the two previous categories) are transformed into other nanomaterials by means of the increment of the average volume of the system (by the aggregation, coalescence or sinterization of the initial nanometric units) and/or the induction of a phase transition by heat treatments, oxidation or reduction processes, and/or the application of high pressures.

3. Homogeneous Nucleation

3.1. Processes of the Crystallization in Supersaturated Solutions and the Classical Theory of the Homogeneous Nucleation

The formation of crystals from a supersaturated solution involves the attachment of the solute species (ions, atoms or molecules) that are going to constitute the new crystals. Usually, the term *monomer* is used to refer to these precursor units.

In order to characterize the supersaturation level of the system, it is suitable to introduce the *supersaturation ratio* $\beta = a/a_0$, defined as the relation between the actual activity of the solute and the equilibrium activity of the solute in the solution. For sufficiently dilute solutions, a and a_0 can be substituted by the solute concentration, C , and the critical concentration of solubility, C_{Sol}, respectively, and thus $\beta = C/C_{Sol}$. In this way, if $\beta < 1$ the solution is *undersaturated*, if $\beta = 1$ the solution would have reached its saturation and if $\beta > 1$ the solution would be *supersaturated*. Please observe that this dimensionless parameter facilitates the comparison between different supersaturated solutions. Also, the solution supersaturation is often characterized by the *relative supersaturation* σ defined as $\sigma = \beta - 1$.

The supersaturation of a solution can be induced by varying the equilibrium solubility of the solute modifying the solvent nature, the temperature or the pH of the solution, applying high pressures or adding another solute [32, 33], or well, through the increase of the solute concentration by evaporing the solvent or generating the solute monomers by chemical reactions [32, 33]. In the latter case, which is the most used for the preparation of nanoscale crystals and the one that we will consider in our subsequent discussions, the supersaturation ratio and its evolution with the course of time can be controlled with the modulation of the chemical reaction kinetic by, for example, the change of the concentrations and nature of the reagents, the temperature or pH of the solution, or the presence of impurities, ligands or other foreign substances.

Figure 1 shows a diagram of the linked processes involved in the crystallization from a supersaturated solution. Firstly, the precursor salts are dissolved and the chemical reactions take place to lead to the formation of the solute species. In consequence, β increases up to reach a critical value of supersaturation. At this moment the species in solution are effectively condensed generating new particles with the formation of embryos or nuclei. Such nucleus formation requires an expenditure of the free energy and occurs due to kinetic effects. In fact, the nucleus formation is traditionally visualized as a spontaneous process in which the solute monomers are associated when they collide between them with certain energies and orientations. This incorporation is reversible and the monomers can return to solution describing a *Szilard mechanism* [34]. Therefore, the chain reaction of nucleus formation can be represented as:

$$N_1 \rightleftarrows N_2 \rightleftarrows \ldots \rightleftarrows N_{n-1} \rightleftarrows N_n \rightleftarrows \qquad (1)$$

where N_i denotes a nucleus constituted by i (= 1, 2, ..., n-1, n , ...) attached monomers.

As we will analyze below, if the supersaturation level of the solution is low, the number of the successful encounters is rather limited, and the new produced nuclei attain very small sizes and their redissolution is energetically favored. However, if the increment of the supersaturation level is enough high, the large number of collisions that take place per unit of time and volume yields to the irreversible growth of the nucleus beyond a certain critical size, overcoming the tendency of the nucleus to be redissolved, see Figure 2.

Although there are other approaches more appropriate to analyze the crystal nucleation from supersaturated solutions [35-37], it is instructive to consider the conventional derivation of the *classical nucleation theory* (CNT) [34, 35, 38-40] to make a first analysis of this phenomenon and introduce some crucial concepts like the *critical nucleus size*. This theory has been mainly developed to explain the condensation of droplets from vapor, however, it can be extended for the study of crystallization from solutions like an oversimplified, but illustrative, approach. In this theory, the event of homogeneous nucleation is analyzed. This event occurs in the earliest stage of the transformation of an unstable phase to another more stable by the spontaneous generation, within the mother phase, of the first traces of the emergent phase in form of embryos or nuclei. During the subsequent course of the transformation, these nuclei are preferred centers of growth of the new phase. In the case of the supersaturated solutions, the mother phase is the monomers in solution and the new phase is the precipitated solid.

According to the classical nucleation theory (CNT), the change of the Gibbs free energy, ΔG, related to the formation of a nucleus by the condensation of the solute units, which is the work to form a nucleus, is given by the sum of two terms: the free energy needed to the formation of the nucleus volume V, ΔG_V, which is a negative amount when the solution is supersaturated, and the free energy required to generate the nucleus surface with an area a, ΔG_S, which is a positive amount. With these considerations, when the growth medium is a solution, we can write:

$$\Delta G = \Delta G_V + \Delta G_S = -\frac{V}{\Omega} k_B T \ln \beta + \gamma a \qquad (2)$$

where Ω is the molecular volume of the precipitated embryo, k_B is the constant of Boltzmann, T is the temperature and γ is the surface free energy by unit of area that characterizes the interface tension between the surface of the crystal in formation and the supersaturated solution in which is the nucleus. It is worth mentioning that we have considered only a value of γ to describe the free energy in all the surface of the nucleus. If the nucleus exhibits nonequivalent crystallographic faces, we would have to consider in the expression (2) the respective values of γ for each face and their contributions to the total surface, or well, consider effective values of the nucleus surface tension and surface area.

The equation (2) is illustrated by Figure 3. Obviously, this equation entails strong simplifications that originate the principal drawbacks of the CNT. For example, with the introduction of the parameter γ in the formulation we are assuming that the nuclei have similar properties than that the bulk crystals. This assumption is known as *capillarity approximation*. However, the definitions of superficial area, ΔG_S and ΔG_V become ambiguous in the earliest and crucial moment when the size of the nuclei is close to the dimensions of the elementary units that constitute them. Additionally, among other

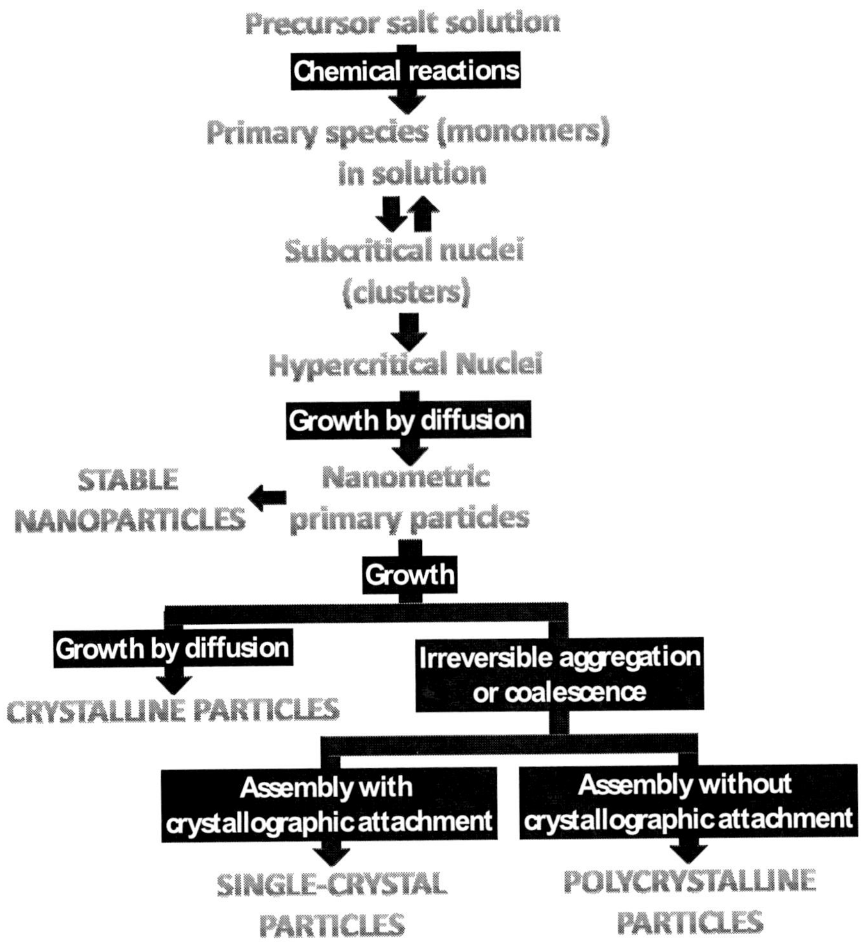

Figure 1. Stages of particle formation by the chemical precipitation from homogenous solutions.

difficulties [35], in this theory it is not contemplated how the incorporating units vary their positions during their attachment to build up a crystalline lattice of a certain phase. Therefore, it is important to emphasize the oversimplified character of this approach.

For simplicity, we will assume in our analysis that the nuclei are spherical with radii r. Therefore, the equation (2) can be written as:

$$\Delta G = -\frac{4\pi r^3}{3\Omega} k_B T \ln\beta + 4\pi\gamma r^2 \tag{3}$$

In the case of nuclei with polyhedral shapes, similar analyses could be realized considering a generalized particle radius r_g, defined as the ratio 3V/a.

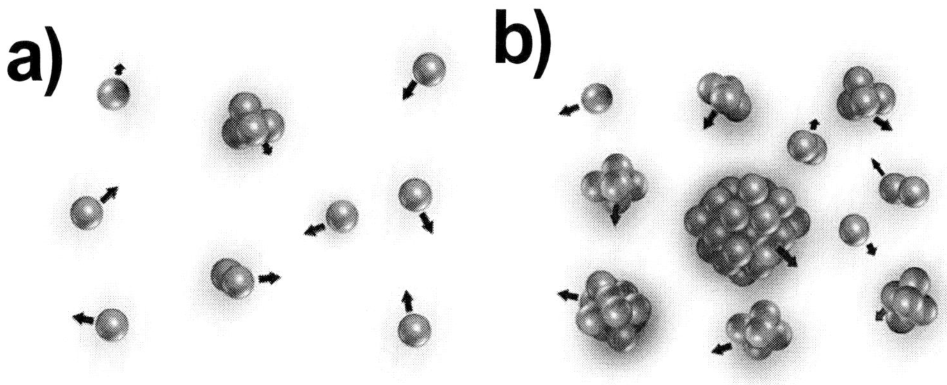

Figure 2. This sketch represents the construction mechanism that yields to the formation of crystal nuclei from supersaturated solutions: a) at low levels of supersaturation, the number of successful collisions per unit of time and volume is small and tiny short-living nuclei are formed. b) Increasing the level of supersaturation, the number of the effective collisions increases and bigger nuclei are generated.

In agreement with the equation (3), in non-supersaturated solutions, $\beta<1$, ΔG is always positive and it increases with r. Thus, in these circumstances, the chemical precipitation does not occur spontaneously. However, when the solution is supersaturated, $\beta>1$, the competition between the volume and surface terms yields to the exhibition of a positive maximum of ΔG, that we denote by ΔG_{max}, for a certain critical radius $r=r_C$ (Figure 3). This relative maximum is the work of critical nucleus formation and it represents the *activation energy of the nucleation*.

The value r_C, which is a thermodynamic property of the system, can be easily calculated considering the equation (3) and setting the condition of critical point, $d(\Delta G)/dr = 0$. Thus

$$r_C = \frac{2\gamma\Omega}{k_B T \ln \beta} \tag{4}$$

If β is replaced in the expression (4) by the ratio S_r/S_∞, where S_r is the solubility of particles of radius r and S_∞ is the equilibrium solubility of the crystallizer substance, (i.e., the solubility of particles of radius $r\to\infty$) the expression (4) is exactly transformed to the relationship between particle size and solubility originally derived by Thomson (before became Lord Kelvin) for a vapor-liquid system. Since Gibbs utilized this equation, and Ostwald and Freundlich extended its application for the crystal nucleation in solutions, this equation has been called in the literature by the names *Kelvin*, *Gibbs-Thomson*, *Gibbs-Kelvin* and *Ostwald-Freundlich equation* [41].

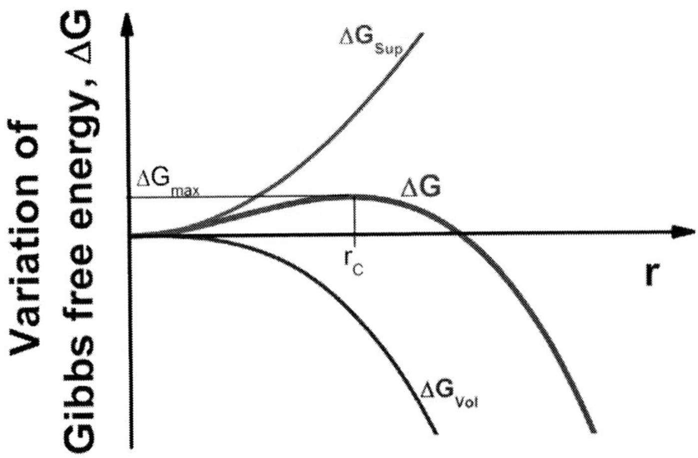

Figure 3. Free energy change for nucleation as a function of the nucleus size according to the CNT. The combination of the surface and volume terms gives the existence of a critical nucleus size. Please note that we have considered continuous values of r in this representation, despite the nuclei grow with the addition of an integer number of monomers. The supersaturation ratio β has been considered constant in this representation.

From equations (3) and (4) we get

$$\Delta G_{\max} = \frac{16\pi\Omega^2\gamma^3}{3[k_B T \ln \beta]^2} = \frac{4\pi\gamma r_C^2}{3} \tag{5}$$

This relationship indicates that the required energy to generate a nucleus with a radius equal to r_C for a given value of the supersaturation ratio β, is a third of the energy necessary to generate the nucleus surface.

Based on the above analysis, the stability of the new nucleus depends on its size. Once generated a nucleus of radius r due to the successive successful collisions of several solute monomers, the nucleus is in a state with an excess of energy and the system must evolve in order to deplete its free energy. Therefore, if the nucleus has dimensions smaller than the critical size, $r < r_C$, its dissolution will be energetically favorable and the nucleus shrinks and disappear. In contrast, if the nucleus size is bigger than the critical volume, $r > r_C$, it will spontaneously and irreversibly grow. The terms *cluster* and *nuclei* are often used for designating both types of nuclei respectively. Herein, we will use the terms *subcritical* and *hypercritical nuclei*. If the size of the nucleus is exactly the critical one the nucleus is in an unstable state. If one of the attached monomer leaves the crystal nuclei and it is not immediately replaced by other one, the nucleus dissolves spontaneously. In contrast, if one monomer is added to a critical nucleus, it tends to grow. This size-selection process is known in the literature as *Ostwald ripening* or sometimes *coarsening*.

The rate of nucleation, J_N, is defined as the number of nuclei that appear by unit of volume and time. Since the process of the nucleation can be analyzed, in approach, within the framework of the theory of collisions [42], it seems reasonable write J_N like a function of ΔG_{max} in terms of an Arrhenius equation:

$$J = J_0 \exp\left(-\Delta G_{\max} / k_B T\right) \tag{6}$$

where J_0 is the pre-exponential factor. This expresion can be more exhaustively derived assuming that the number of nuclei per unit volume follows a Boltzmann distribution with respect to the nucleus size (or well, with respect to the number g of nucleating units that the nuclei contain), see for example ref. [33]. Consequently, the rate of nucleation depends mainly on parameters T, γ and β. However, it is worth pointing out that there are other dependencies that does not explicitly appear in the previous expressions and that they should be considered. For instead, the interface energy between the crystal and the solution, γ, is affected by the variation of the solubility of the material. Thereby, if we change the nature of the solvent and the solubility of the solute increases, the interface tension γ will decrease [32].

The cases in which the nuclei are formed containing two or more components are considerably more complex and their molecular mechanisms are rather unknown. In this respect, the majority of the research efforts have been devoted to study the condensation of binary liquid drops in the atmosphere, and much research have relied on the extension of the above considerations of the unary homogeneous nucleation [43-45]. For the cases of binary vapor-to-liquid nucleation, the Gibbs free energy change is given by

$$\Delta G = -n_a k_B T \ln \beta_a - n_b k_B T \ln \beta_b + 4\pi \gamma r^2 \tag{7}$$

where n_a and n_b are the number of molecules of the incorporating substances a and b, respectively, and β_i, with $i = a, b$, are the supersaturation ratios defined as the ratio of the ambient partial pressure of free molecules of substance i to the partial vapor pressure of this substance over a flat surface of the solution. Of course, r depends on n_a and n_b. Therefore, the Gibbs free energy change associated to the generation of a binary nucleus is a function of n_a and n_b, whose saddle point in the multidimensional composition space determines the critical nucleus composition and size, and the activation energy of the nucleation. However, this generalization of the CNT for the *heteromolecular* nucleation presents some inconsistencies, as discussed in section 3.3.

3.2. Temporal Evolution of the Supersaturation Ratio during the Crystallization

In the great majority of the theoretical investigations about nucleation process, it is not considered the changes that the supersaturated mother phase experiences during the phase transformation. Therefore, due to the real systems are finite, the direct comparison of the experimental results with the estimations of these approaches could be erroneous. Hence, it is essential to note that both processes of nucleation and diffusional growth considerably diminish the level of the medium supersaturation. Consequently, the nucleus stability and the kinetics of the crystal nucleation and growth evolve with the course of time during the transformation, and events as nucleation and growth can be occur simultaneously.

LaMer and Dinegar studied in 1950 the mechanisms of chemical precipitation from supersaturated aqueous solutions of hydrochloric acid and sodium thiosulfate [46]. In their study, these authors proposed their famous model, represented in Figure 4, where the concentration of dissolved sulfur is described as a function of time. Three stages of the crystallization can be distinguished in this model. In the first step, called *pre-nucleation* or *induction step*, the precursor salts are dissolved and the chemical reactions start to occur to generate the primary species of solute. Therefore, β increases with time until reach a critical value of supersaturation β_{Min}, necessary to the formation of nuclei with sizes lager than the critical one. Since this moment, so-called *induction* or *delay time of the nucleation*, the solution experiences a transition from a metastable to an unstable state, and the process of nucleation begins [47]. As this stage occurs, β begins to diminish. When β decreases below β_{Min} the nucleation comes to an end and the nuclei already generated grow by diffusion of the monomers toward the nuclei surface until the concentration C is reduced close to the critical value of solubility, C_{Sol}, and $\beta=1$ (Figure 4).

Habitually, a general condition has to be satisfied to obtain final particles with uniform morphologic characteristics: *nucleation and growth must be two steps completely separated* [46]. If the nucleation rate is not enough high, the generation of new nuclei can be accompanied with the growth of the nuclei earlier produced. This situation usually yields to the obtaining of particles with wide size distributions. On the contrary, if the nucleation occurs suddenly, the concentration of the species dissolved is quickly reduced below the nucleation level, β_{Min}, being the nucleation a temporarily short stage which is immediately followed by the growth of the nuclei already generated, as it is described by the model of LaMer and Dinegar (Figure 4). As a consequence, the interdependence of these two stages can be used to accurately modulate the size of the obtained particles, the uniformity of the system and its crystalline structure. For instead, as it is deduced from the model of LaMer and Dinegar, a short induction time implies the formation of a large number of nuclei and a subsequent steep decay of β. Then, the crystal growth (by diffusion) should be very limited, see curve (a) of Figure 5. Conversely, if the induction time of the nucleation is large, a smaller number of nuclei will be generated and the solute concentration will slowly decrease until reach the value of the critical solubility concentration of the system, C_{Sol} (i.e., $\beta=1$). It leads to a longer period of time for the particle growth, and then, bigger particles should be produced, see curve (b) of Figure 5. On the other hand, a too fast nucleation rate can give rise to an excessive amount of nuclei and it can induce secondary growth processes by irreversible aggregation or coalescence (Figure 6), depending on the steric character of the solvent and the presence of additional stabilizing agents in the solution. Likewise, when the supersaturation is extremely high, the precipitated phase usually is poorly crystalline or even amorphous.

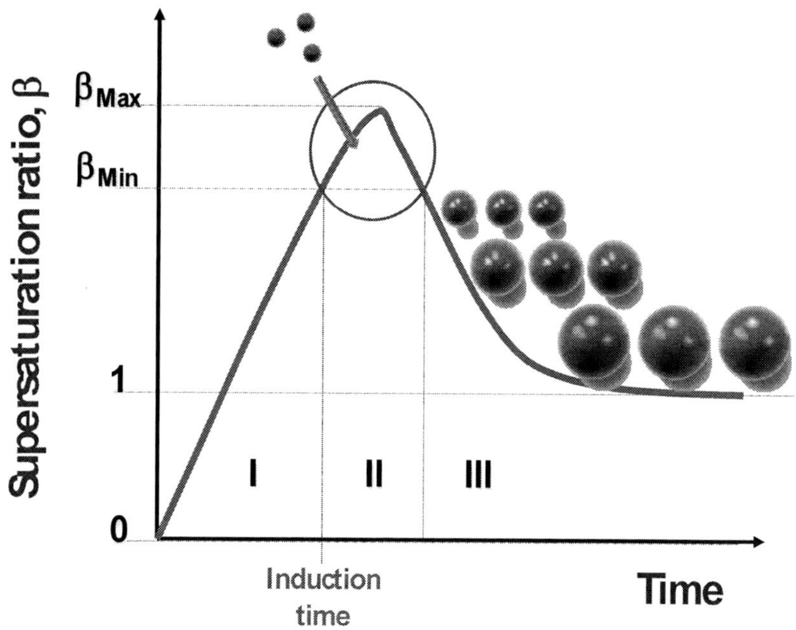

Figure 4. LaMer-Dinegar model [46]: Dependence of the supersaturation ratio with the course of time during the chemical precipitation from a solution. Three stages can be distinguished: I Pre-nucleation, II Nucleation and III Diffusional growth.

Since β passes through a maximum at a certain time (that we will denote by t^*) during the nucleation stage, one would expect from equation (4) that the critical nucleus size, which determines when the tendency of the nuclei to be redissolved is overcome, passes through a minimum at t^*, as it is illustrated in Figure 7. Therefore, during the earliest moments of the nucleation in which β increases due to the monomer generation by the chemical reactions, the critical radius substantially decreases, whereas when β is depleted down to values close to the unity, firstly due to the nucleation stage and then due to the growth one, the critical size increases and the Ostwald ripening should be rather present. Of course, there are several factors that could yield to a more intricate temporal dependence of the critical nucleus size (and the evolution of the crystal stability in solution) than that is represented in Figure 7. For example, instead γ is usually considered like a constant, γ could significantly changes with the crystal size. In fact, as we discussed above, the notion of surface tension causes important deficiencies in the theory when the nucleus is constituted by few atoms, ions or molecules. Also, in many synthesis the temperature of the reaction media is increased up to arise the boiling temperature of the solution. This change produces pronounced changes in the kinetics of the nucleation and growth rates that are very dificult to predice because the temperature is a very influental factor in several interdependent processes, such as the dissolution of the salt precursor, the kinetic of the chemical reactions, the critical solubility level and the mass transfer in solution.

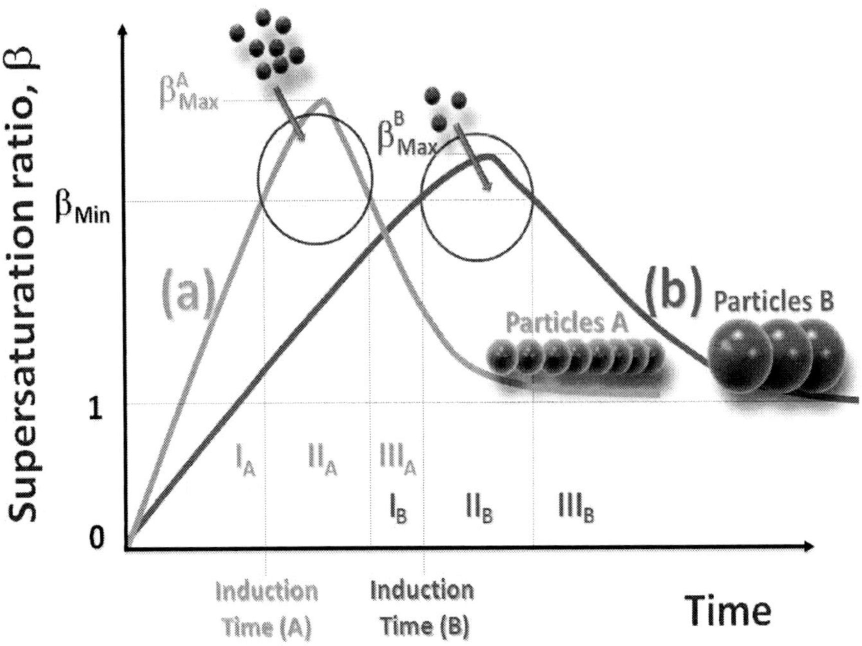

Figure 5. Effect of the variation of the nucleation rate on the particle size according to the LaMer-Dinegar model.

3.3. Refinements and Modifications of the Classical Nucleation Theory and Other Approximations to the Homogeneous Nucleation Process

Since nucleation is an essential process that participates in countless phenomena with important practical implications in science and technology, great efforts have been devoted to develop, test, and improve theories that describe its kinetic. In this way, the CNT approach has experienced several modifications during the last five decades. For instead, the classical expression of the homogeneous nucleation rate has been corrected with the addition of the factor e^{θ}/β where $\theta = \gamma s_1/k_B T$ and s_1 is the monomer surface area [48-50]. This temperature-dependent correction is derived when the equilibrium nucleus distribution, which is a Boltzmann distribution, is self-consistently written [49]. Also, thermodynamic inconsistencies of the multicomponent nucleation in the large nucleus limit have been amended [51]. Other corrections are focused on the modifications of the nucleation rates predicted by binary nucleation in the capillarity approach, which are not reduced to the expected unary nucleation rate when one component of the binary system is vanishing [52].

Other research efforts have been devoted to avoid the drawbacks of the CNT attributed to the nucleus interface tension through empirical approaches that provide effective values of γ, which yield to predictions of the nucleation rates more close to the experimental data [45], or by assuming alternative expressions for the Gibbs free energy of nucleus formation, retaining the basic procedure of the original CNT [53, 54]. In other approaches, the theory has been reconstructed without invoking the existence of a nucleus size distribution constrained to be in equilibrium with the supersaturated medium [55].

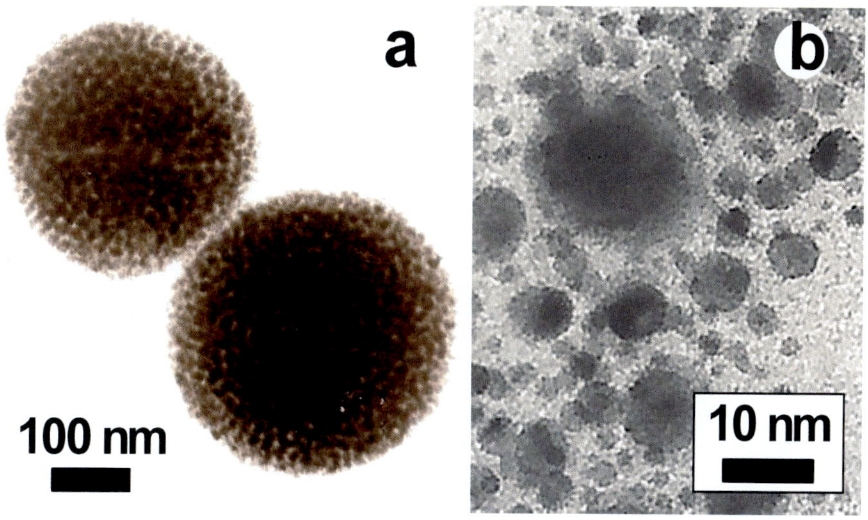

Figure 6. Transmission Electron Microscopy (TEM) images of: a) spherical aggregates of lauric acid-capped cobalt nanocrystals and b) cobalt nanoparticles that have experienced a secondary growth by coalescence.

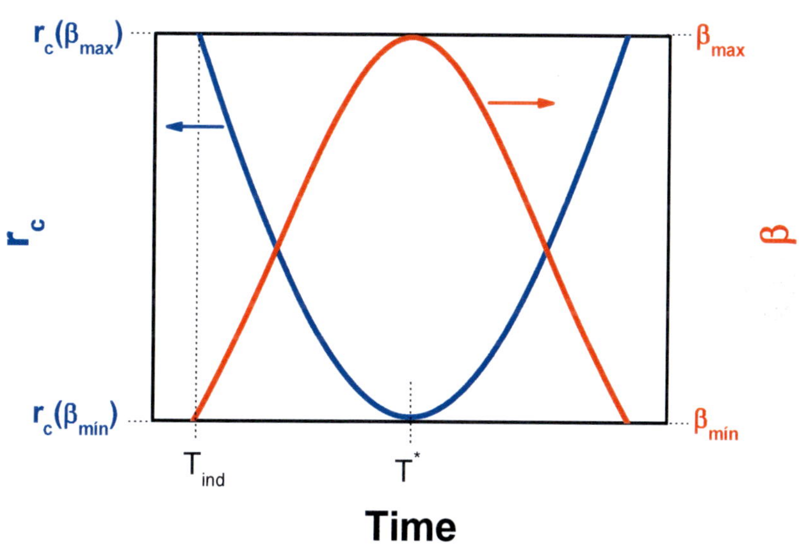

Figure 7. Evolution of the critical nucleus radius and the supersaturation ratio with the course of time during the nucleation stage, where the nucleation induction time has been denoted by t_{ind}. In this representation, γ is considered constant.

Other studies of the nucleation kinetic are based on Monte Carlo (MC) [56] and molecular dynamics (MD) [57] simulations, or the density-functional theory (DFT), where the interface between the nuclei of the new phase and the mother phase have a finite width that is characterized by density profiles [58, 59]. Also, a very relevant approach based on molecular interactions has been developed for both liquid-to-solid and vapor-to-liquid phase transitions [37, 60, 61]. In this approach, the Fokker-Plank equation is used like the master equation that governs the motion of a single molecule within a potential well generated in the around the nucleus surface. The obtained results are in agreement with estimations of classical nucleation theory (CNT) in the limit of large nucleus (for instead, an expression for the surface tension and the Kelvin equation are retrieved). On the other hand, it predicts higher nucleation rates than CNT for small critical nuclei [37]. In the case of liquid-to-solid phase transitions, the model describes how the nucleation rate can vary depending on the crystalline structure (it has been studied for three structures: amorphous, fcc and icosahedral).

Despite the real systems are finite, the majority of the above-mentioned theoretical approach do not considered the variations of the supersaturation due to the nucleation and diffusional growth, called by some authors as the *depletion* of the medium [62]. Beside the variation of the critical size and nucleation and growth rates mentioned in Section 3.2, this depletion yields to the existence of an additional stable state, as it is represented in Figure 8. In this way, the thermodinamical study realized by Vogelsberger shows that the change of the Helmholtz free energy due to the formation of a single droplet in a finite vapour phase under constant temperature, volume and total number of monomers exhibits a minimum after that the system passes the critical state [63]. In these calculations, the surface tension of the liquid was considered size dependent. The existence of the minimum point in a finite system has been also supported by Monte Carlo and molecular dynamics computer simulations [64].

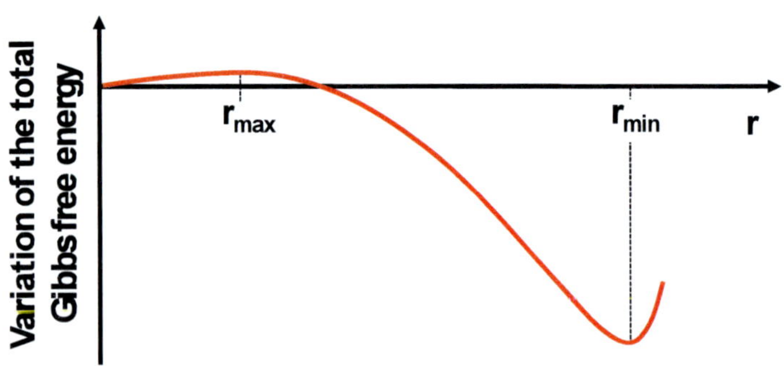

Figure 8. Representation of the Gibbs free energy change due to the formation of a constant number of nuclei in a finite system as a function of the nucleus size. It is observe that a minimum appears in addition to the maximum point ascribed to the formation of a critical nucleus. Please compare this curve with the curve represented in Figure 3.

Similarly, Schmelzer and Ulbricht found that a minimum appears in the change of the Helmholtz free energy due to the formation of a constant number of drops with the same radius under isochoric and isothermal conditions when the depletion of the medium is considered [62]. According to this investigation, the maximum point ascribed to the formation of a critical nucleus and such minimum depend on the depletion of the surrounding medium, and when the number of drops is increased the former critical state decreases whereas the latter increases

In a very recent work, Wasai et al. have reanalyzed the classical homogeneous nucleation theory taking into consideration the change of the Gibbs free energy in the mother phase obtaining analogous results [65, 66]. Thus, Wasai et al. have proposed that the physical meaning of the Kelvin equation should be re-interpreted due to such equation (now derived from the free energy change of the whole system) should correspond to the minimum of such curve, where the crystals would be stable, rather than its maximum which represents a barrier to nucleation.

The existence of this additional equilibrium state and how it changes with the variables of the system has an obvious impact on the first-order phase transition theories, material sciences and nanotechnology due to it indicates that when the nuclei attain the related critical size, which could be in the nanometric or micrometric scale depending on the synthesis conditions, they are stable within the supersaturated mother phase [65]. Please note that, if the solution is supersaturated, a continuous growth without stabilization for the hypercritical nuclei is expectable according to the equation (3) (see Figure 3). However, if an additional stable state must be considered in finite systems (Figure 8) according to the above-mentioned works, this crystal growth could come to an end before that the solution leaves its supersaturated state.

4. Heterogeneous Nucleation in Supersaturated Solutions

4.1. Nucleation with Seeds

A complete separation between the nucleation and growth stages can be attained with the replacement of the above described *homogeneous* nucleation, by a *heterogeneous* nucleation [67]. The induction of this latter kind of nucleation is an efficient tool to accurately control the crystallization mechanisms in solutions [68, 69]. It can be performed by *seeding* the initial solution with fine nanoparticles as *seeds*, which play the role of nuclei. These seeds act as active centers for nucleation due to catalytic effects, thus, the precipitating solute units trend to be accumulated onto their surfaces producing uniform final particles with tailored sizes and controllable core/shell structures [70, 71]. This seeding clearly reduces the times required for the chemical precipitation. Furthermore, certain reactions that do not occur under homogeneous conditions can take place in the presence of foreign nuclei or substrate surfaces.

The seeding can be induced with the inoculation of the seeds in the liquid reaction medium before that the homogeneous nucleation sets in, or with the formation of the foreign nuclei in situ [68]. Several illustrative examples of how the seeded nucleation enables the control of the final particle size can be found in the preparation of metallic particles by the so-called *polyol process* [68-71]. For instead, Co_xNi_{100-x} particles can crystallize by means of the chemical precipitation of metallic Co and Ni species from boiling polyol solutions, which are generated by the reduction of metallic ions (arising from the dissolution of salts such as hidroxides or tetrahydrated acetates) to their state of valence zero, where the reducing agent is the polyol itself. To seed the solution, foreign nuclei of Ag or Pt are prepared *in situ* in the initial solution [68]. These seeds are generated via the reduction of noble metallic salts that play the role of nucleating agents because the reduction of the ions Ag^+ and Pt^{2+} to his state of valence zero occurs at lower temperatures than that required for the reduction of the Co^{2+} and

Ni^{2+} ions to their metallic state, whose potentials standard of reduction are very similar (-0,28 and -0,26 V, respectively [72]).

Since the number of nuclei formed in situ is proportional to the mass of the added nucleating agent, it is feasible to modulate the crystallized particle size by the variation of the noble metal salt concentration [68]. In this fashion, for a fixed concentration of precursors of Co and Ni, a small amount of nucleating agent will give rise to the generation of a reduced number of nuclei, whereupon, the Co and Ni material resulting from the reduction reactions will precipitate over them and relatively big particles will be produced (see image a of Figure 9). Nevertheless, if the concentration of the nucleating agent is increased, a larger number of nuclei will be generated, giving rise to smaller final particles (see image b of Figure 9). Obviously, this size selection only can be attained if the following conditions are satisfied [69]:

i. The nuclei size must be roughly independent of the added nucleating agent concentration.
ii. Each Co_xNi_{100-x} particle should have only one nucleus of Ag or Pt.

From these hypotheses it is deduced that the average diameter of particles holds the following relation [69]:

$$d_m \propto \left(\frac{[Nucleating\,agent]}{[\Pr ecursor\,salts\,of\,Co\,and\,Ni]} \right)^{-1/3} \quad (8)$$

It is worth pointing out that the conditions i) and ii) will not be fulfilled if the number of nuclei is not enough to produce a complete replacement from the homogenous nucleation to a heterogeneous nucleation. In this situation the obtained crystals will be polydisperse because both types of nucleation will occur simultaneously. Also, since the steric stabilization effectiveness of the polyol diminishes according to the number of nuclei increases, when this latter is very large the nuclei experience a secondary growth by irreversible aggregation or coalescence.

The nature of the nucleating agent is also a crucial factor that affect to the size of the resulting particles. For example, it has been found that the Co_xNi_{100-x} particles prepared with Ag nuclei are considerably bigger than the particles prepared with Pt nuclei in similar conditions [68-71]. It is probably due to the Pt nuclei are smaller than the Ag nuclei (please note that the reduction potential for the reaction $Ag^+ \leftrightarrows Ag$ is 0.80 V and for the reaction $Pt^{2+} \leftrightarrows Pt$ is 1.18 V [72]). In this fashion, the variation of the nature and concentration of the nucleating agent have enabled us to vary the average particle size of the precipitated solid in a wide range (~4-400 nm) [70, 71], see Figure 10.

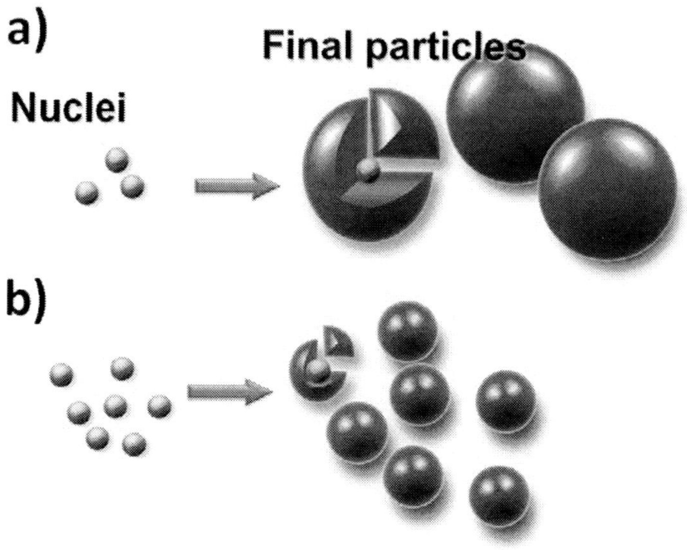

Figure 9. Schematic representation of the controlled particle size selection through the seeding of a supersaturated solution with small particles. As a consequence of this heterogeneous nucleation, the resulting particles have a core/shell structure.

4.2. Heterogeneous Nucleation at Planar Solid Substrate Surface

The actual solutions are far to be completely pure and the presence of foreign substances different to the solvent and the crystallizing phase that can catalyze the nucleation is expectable. Consequently, the occurrence of a true homogeneous nucleation is, usually, very unattainable. Besides, the presence of solid substrate surfaces, as the walls of the vessel that contains the solution, can affect the nucleation process when the solute has some affinity for such surface. In fact, several works have shown that the growth of many of the biomineralized composite materials, which usually are highly oriented flaw-free structures [73], is mediated by organic surfaces [6]. The comprenhension of these biological processes, optimized during millions or even billions of years, is an excellent source of inspiration for the design of new synthetic processes for the preparation of novel biomimetic materials [73-75]. Evidently, the texture of the substrate surface and its porosity are decisive on the formation of crystal nuclei over the surface, and their control can be exploited to grow and organize nanocrystals with engineered properties [76].

The heterogeneous circumstances produced by the presence of plane surfaces during the condensation of supersaturated vapor have been investigated by molecular dynamics [77, 78] and density-functional theory (DFT) [79], nevertheless they remain rather poorly explored for the crystallization from supersaturated solutions. Herein, we will consider the following assumptions developed within the framework of the capillarity approximation to give an intuitive picture of this process [32, 80].

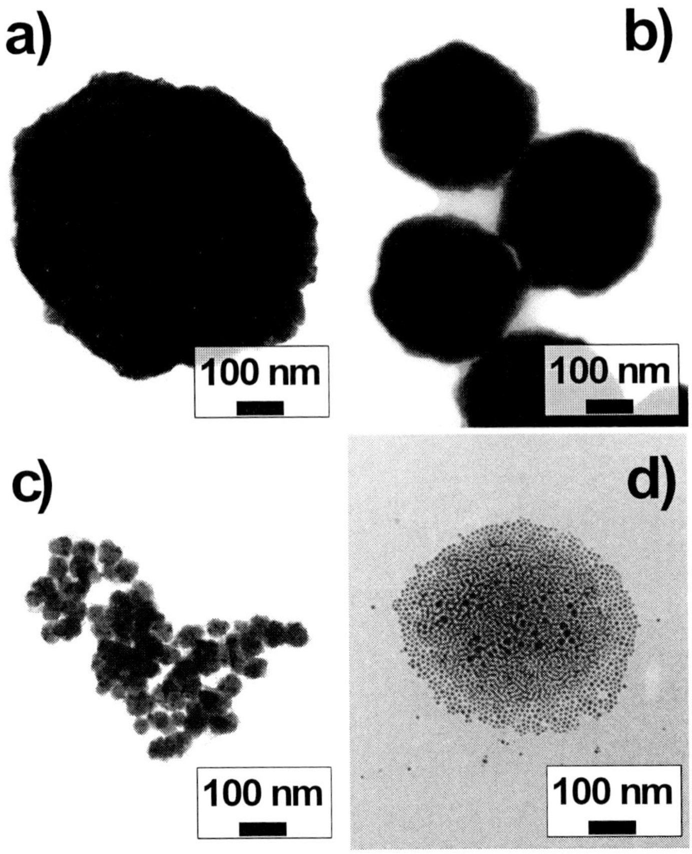

Figure 10. Transmission Electron Microscopy (TEM) images of $Co_{80}Ni_{20}$ particles prepared by polyol process inducing a heterogeneous nucleation: a) sample of particles of ~400 nm synthesized with Ag nuclei ($|Ag| / |Co + Ni| = 1.0 \cdot 10^{-5}$), b) particles of 175 ± 30 nm synthesized with Ag nuclei ($|Ag| / |Co + Ni| = 2.5 \cdot 10^{-5}$), c) particles of 47 ± 7 nm synthesized with Pt nuclei ($|Pt| / |Co + Ni| = 2.9 \cdot 10^{-4}$), and d) particles of 5.9 ± 0.8 nm synthesized with Pt nuclei ($|Pt| / |Co + Ni| = 1.0 \cdot 10^{-3}$) in presence of surfactant stabilizers (oleic acid).

In this approach it is obtained that the critical radius for the case of nuclei with a spherical cap shape of radius r forming onto a plane substrate (Figure 11), is determined by the same expression than that the formation of a spherical nucleus under homogeneous circumstances (i.e. equation (4)). However, the critical volume is smaller in this heterogeneous case and it is required less condensed solute units to form a *hypercritical nucleus*. Therefore, the value of the height of the activation barrier to nucleation in the presence of the substrate, $\Delta G_{\max}^{het}$, should be equal or less than the corresponding value associated with pure homogeneous nucleation, $\Delta G_{\max}^{\hom}$. Thus, we can write:

$$\Delta G_{\max}^{het} = \phi \Delta G_{\max}^{\hom} \tag{9}$$

where ϕ is a factor equal or less than unity.

Let consider three phases in contact: the precipitated crystal nucleus, a solid surface and the liquid solution. Then, three interfacial tensions can be defined: the interfacial tensions between the crystal and the surface, between the crystal and the liquid solution and between the surface and the liquid solution, denoted by γ_{cs}, γ_{cl} and γ_{sl}, respectively. Also, let consider the contact angle between the substrate surface and the crystal nucleus, θ, see Figure 11. Thus, the angle θ and the three interfacial tensions are related by the *Young equation*:

$$\cos\theta = \frac{\gamma_{sl} - \gamma_{cs}}{\gamma_{cl}} \tag{10}$$

On the other hand, it can be deduced that the dependence of the factor ϕ is a function of θ that satisfy the relation [32, 81]:

$$\phi = \frac{1}{2} - \frac{3}{4}\cos\theta + \frac{1}{4}\cos^3\theta \tag{11}$$

Therefore, for $\theta = 180°$, the nucleus has not affinity for the substrate surface and a homogeneous nucleation occurs (it is the *complete non-wetting case*). When $0 < \theta < 180°$, $\Delta G_{\max}^{het} < \Delta G_{\max}^{\hom}$ and therefore the substrate surface act as a catalyst of the nucleation process. For $\theta \to 0$, $\Delta G_{\max}^{het} \to 0$ (this limit corresponds to the *complete wetting case*).

5. Crystal Growth by Difussion

The mechanisms that are involved in the crystal growth depend on several internal factors (crystalline structure, nature of the bonds, crystal size, crystal defects, etc.) and external factors (presence of impurities, supersaturation ratio, temperature, pH of the solution, etc.) [32]. In this section we shall limit ourselves to briefly study some of these dependencies.

5.1. Some Mechanisms of Crystal Growth

There are a variety of kinds of crystal growth mechanisms that have different dependences. In this section we present a very restricted description of some of these mechanisms. For more exhaustive information on the crystal growth models, the reader is directed to the surveys presented elsewhere (see, for example, the references [32, 33, 82, 83]).

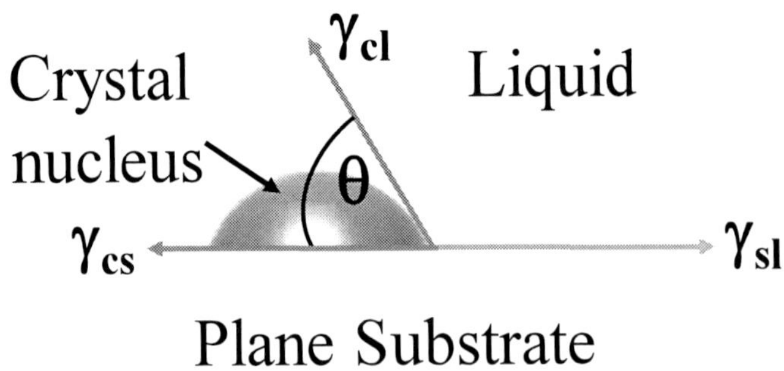

Figure 11. Formation of a nucleus with a spherical cap shape onto a plane solid surface.

As a crystal nucleus grows it exhibits more markedly corners and crystalline faces. The growth rates of this faces (and also their dissolution rates) depend on their crystallographic nature. In this manner, the relative growth rates of the crystallographic faces determine the growth habit of a crystal. Traditionally, three main types of crystal faces, represented in Figure 12, are considered: flat (F), stepped (S) and kinked (K) faces [84].

The growth of the flat faces occurs by two-dimensional nucleation events. Of course, it takes places when certain critical value of supersaturation is exceeded, as in the case of three-dimensional nucleation (see Section 3.2), nevertheless such critical value is lower in this case. In this fashion, the growth units are adsorbed to create two-dimensional clusters that diffuse to the crystal nucleus face until a monolayer is completed. During this process, the 2D nuclei can exhibit kink sites which are preferred growth places. If only one 2D nucleus is spreading across the crystal surface the growth is called *mononuclear* (Figure 13.a). In this case, similarly to the three-dimensional nucleation (Section 3.1.), the growth rate of the face can be written as an Arrhenius equation:

$$R = R_0 \exp\left(- \Delta G_{\max}^{2D} \big/ k_B T \right) \qquad (12)$$

where R_0 is the pre-exponential factor and $\Delta G_{\max}^{2D}$ is the activation energy of the two-dimensional nucleation. Expressions for $\Delta G_{\max}^{2D}$ and the critical size of the 2D nucleation can be derived in a similar manner to the case of the 3D homogeneous nucleation (Section 3.1). For example, for a circular disc of radius r and height h_d, the relationships of the critical radius and $\Delta G_{\max}^{2D}$ are, respectively

$$r_C = \frac{\gamma\Omega}{k_B T \ln \beta} \qquad (13)$$

and

$$\Delta G_{\max}^{2D} = \frac{\pi h_d \gamma^2 \Omega}{k_B T \ln \beta} = \pi h_d \gamma r_C \tag{14}$$

When several 2D nuclei are expanding at the same time across the crystal face the growth is called *polynuclear* (Figure 13.b), and the growth rate is given by

$$R = R_{'0} \exp\left(- \Delta G_{\max}^{2D} \middle/ 3k_B T \right) \tag{15}$$

whereas the 2D nucleation is the most expectable type of growth for nanocrystals, for crystals with larger sizes is habitually found other gorwth mechanisms because their faces often have a number of kinks and steps that are sites where is easier the incorporation of the solute species. For that reason, it is expectable that the crystal growth rate would be larger when S and K faces are growing. An expression of the growth rate when the crystal surface exhibit kinks is given by [85]

$$R = \Omega N_0 \sigma \frac{k_B T}{h} \exp\left(- W \middle/ k_B T \right) \tag{16}$$

where σ is the relative supersaturation, N_0 is the number of kinks per unit area of the surface, h is the Planck constant and W is the activation energy necessary to transport a monomer from the bulk solution to the crystal surface.

Another common type of the defects present in large crystal is the screw dislocations (Figure 14a). When a screw dislocation emerges from a flat crystal face, it can cause the formation of a succession of steps that extend over the surface describing a spiral pattern around the dislocation, as it is represented in Figure 14. This crystal growth pattern has been observed on many crystals [86-91] and when several screw dislocations appear, the interaction of two or more spirals has been encountered [87, 89].

Burton, Cabrera and Frank [92] develop a theory to estimate the rate of this complex kind of growth. The relationship of the crystal growth rate that they obtained for an Archimedean spiral can be written as

$$R = \xi \sigma^2 \tanh\left(\zeta \middle/ \sigma \right) \tag{17}$$

where ξ and ζ are functions of the temperature and the height and spacing of the steps. From this relationship, a parabolic dependence of the growth rate on the supersaturation σ at low supersaturations and a linear proportion at high supersaturations is expected for spiral growths.

Figure 12. Typical schematic representation of the three main types of the crystal faces: flat (F), stepped (S) and kinked (K) faces.

On the other hand, the dendritic growth is the most common form of crystal growth observed in the solidification of undercooled melts [93, 94] and the crystallization of supercooled salt solutions [95-97] in unstirred conditions. Also, it is very frequently found when a salt solution is deposited onto a plane substrate and the solvent is allowed to evaporate [98]. In fact, tree-like structures of unreacted salts or undesirable products are often observed when a solid obtained by chemical reactions is not purified and is deposited onto a substrate. An example of this situation is presented in Figure 15. The dendritic formations are produced by the rapid steady-state growth of the dendrite tip and the time-dependent crystallization of primary and secondary branches [99].

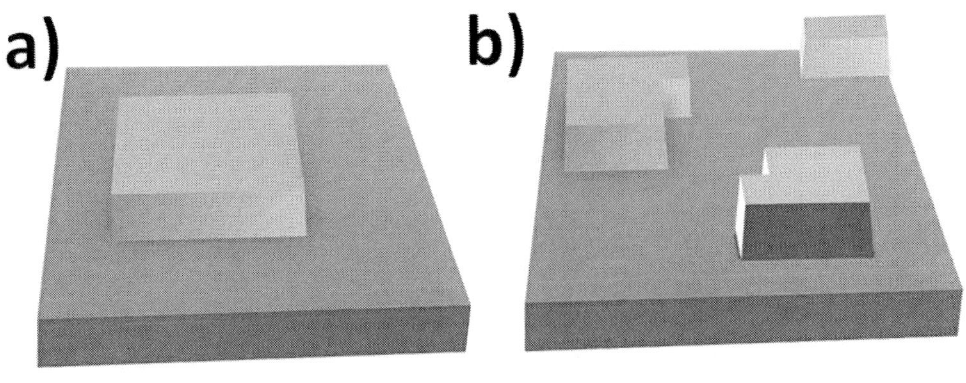

Figure 13. Growth by two-dimensional nucleation: a) mononuclear 2D nucleation and b) polynuclear 2D nucleation.

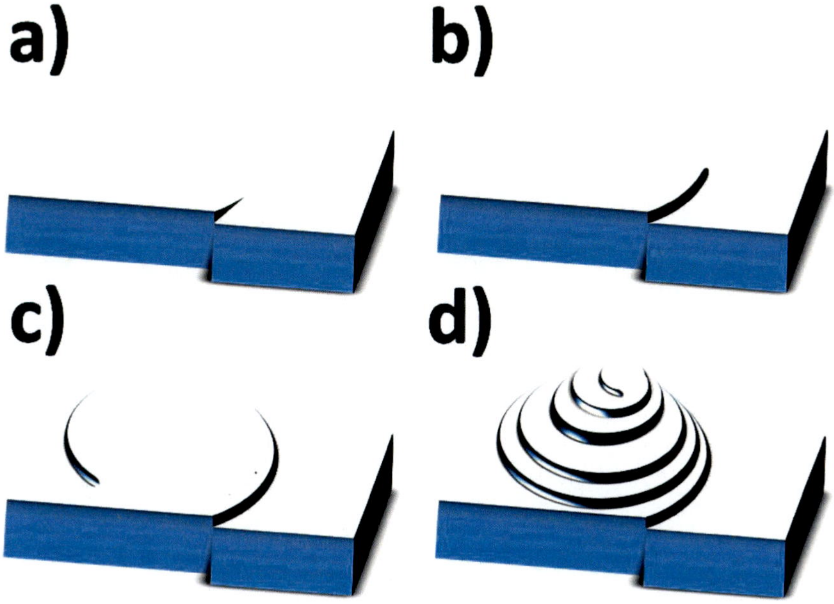

Figure 14. Schematic representation of a screw dislocation (a) and the spiral growth (b, c y d).

5.2. Size-Dependent Crystal Growth

An important factor that affects to the overall crystal growth rates in solution is the crystal size itself. This dependence has been studied for a long period of time, however it probably remains still not well understood. It is well-accepted that the size-dependence of the crystal growth varies with the size range under consideration. Whereas this size-dependence becomes negligible for macrocrystals, when the crystal sizes are in the microscopic and submicroscopic scales, the larger crystals usually grow faster than small [100-104]. This tendency could be related to the *solution velocity effect* on the crystal growth [105]. Due to the solute transfer to the crystal surface depends on the solution velocity through convection and advection processes, it could be expectable that the growth rate would be faster in the larger crystals than that in smaller ones due to the velocity of the supersaturated solution around a crystal increases in direct proportion to its diameter [105, 106]. Also, as a consequence of this solution velocity effect, the crystal growth rate substantially depends on the synthesis stirred conditions [105, 106]. Obviously, these size dependences of the crystal growth rate are decisive in the resultant crystal size distributions (CDSs). In fact, the most frequent way to experimentally study the dynamics of the crystal growth is analyzing the evolution of the crystal CDSs. Since there are several complex facts responsible of the CSDs, the interpretation of such experimental data have provoked controversial discussions [100-104, 106].

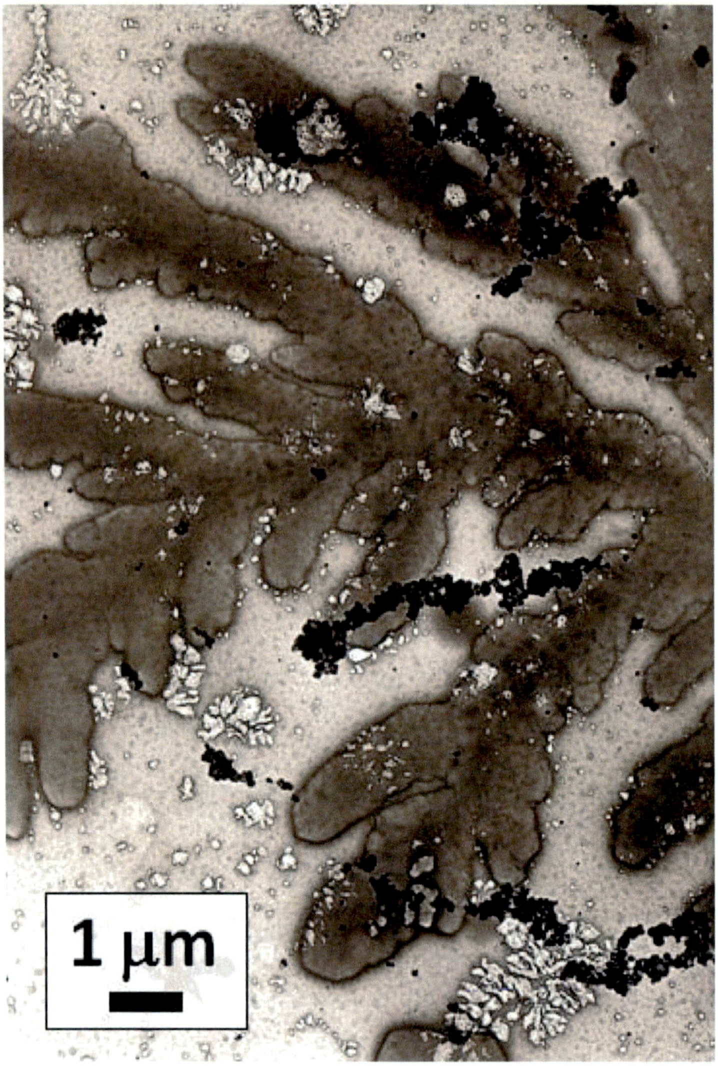

Figure 15. TEM image of an aliquot extracted from a solution where hematite nanocrystals were prepared by forced hydrolysis of an iron salt in presence of urea. This aliquot was directly deposited onto the carbon coated copper TEM grid. Together with the resulting hematite nanocrystals (small black dots) appear dendritic structures of undesirable salts (white and grey formations), which grow when the solvent is evaporated.

For nanometric crystals, the *solution velocity effect* is much less significant due to their tiny sizes are much smaller than that the turbulences present in the solution [105]. However, other effects appear in the nanoscale that yield to a complex size dependent crystal growth rate. Probably the most evident of them is the *coarsening effect,* described in Section 3.1 and which is schematically represented in Figure16. In this process, the crystals with size larger than the critical one grow to expenses of the subcritical crystals that are redissolved [107]. Important efforts have been concentrated on the description of this process. Among the more relevant contributions, a very important advance was established by the model developed by Lifshitz and Slyozov [108] and Wagner [109] (LSW theory), which is considered as the classical theory of the Ostwald ripening. This theory provides quantitative predictions of the evolution of coarsening systems without the necessity of numerical methods. However, the size distributions of nanocrystals experimentally observed are generally broader and more symmetric than the predictions of the classical LSW approach. In more recent works, other attempts have been proposed based on mean-field equations and numerical methods (for reviews, see for example ref. [110]). However, instead these considerable advances, our knowledge of the earlier transient stages of the nanocrystal growth probably remains rather restricted.

To provide an intuitive picture to the reader about the size-dependent crystal growth rate due to the Ostwald ripening, let us consider the model described by Sugimoto [111]. According to this model, the diffusion controlled growth rate of the spherical crystal with radius r is

$$\frac{dr}{dt} = K_D \frac{1}{r}\left(\frac{1}{r_C} - \frac{1}{r}\right) \qquad (18)$$

where the diffusion thickness is considered infinite, $K_D = \frac{2\gamma D C_S \Omega^2}{RT}$, D is the diffusion coefficient and N_A is the Avogadro number. The other variables that appear in equation (18) were introduced in Section 3.1. This model is derived from the *capillary assumption* and therefore contains the drawbacks of this approach. The expression (18), which is represented in Figure 17, indicates that the growth rates are negative for crystals smaller than the critical size. It means that these crystals shrink and disappear. In addition, the rate of this dissolution is more significant as the size of the subcritical crystal is smaller. Conversely, the hypercritical crystals have positive growth rates that decrease as the crystal size increases (Figure 17). Therefore, a narrowing of the crystal size distribution should be observed when all of the crystals have sizes exceeding the critical one [112].

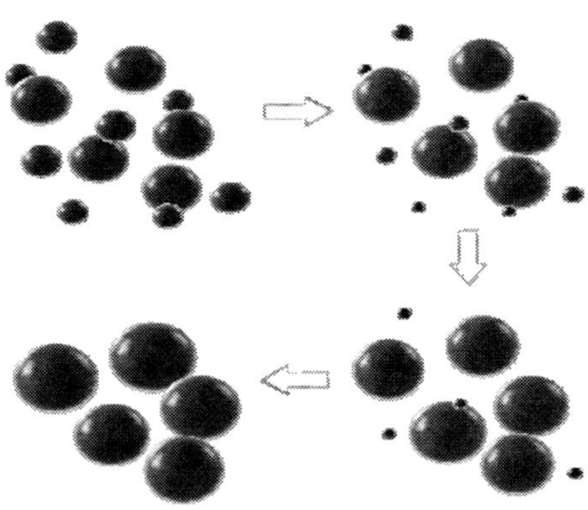

Figure 16. Schematic representation of the Ostwald ripening: the crystals with size larger than the critical one (represented by red spheres) grow to expenses of the subcritical crystals (represented by blue spheres) that shrink and disappear.

5.3. Effects of the Impurities on the Crystal Growth

The term *impurity* embraces all the foreign substances different to the crystallizing one that are present in the crystallizer medium. Thus, the solvent itself and any other compounds added in solution are impurities. The presence of such substances hugely affects to the crystal growth mechanisms in different ways depending on their nature and concentration, and the pH of the solution. These effects can be considerable even if the impurity concentration is very low (even lower than 1 part per million) and they are particularly notorious when the supersaturation is small. Some of such effects, which are often quite difficult to quantitatively estimate, are the alteration of the supersaturation ratio, the crystal solubility and the growth rate in the sites where they are adsorbed (usually in the steps and kinks of the crystal faces and in the branches of the dendritic structures). In the latter case, this adsorption can yield to the modification of the crystal growth habit. Also, the impurities can react with the solute species to form complex that delay the nucleation process. Some surveys about the mechanisms of impurity adsorption and their effects on the growth of crystals can be found in the following works and their references [113, 114].

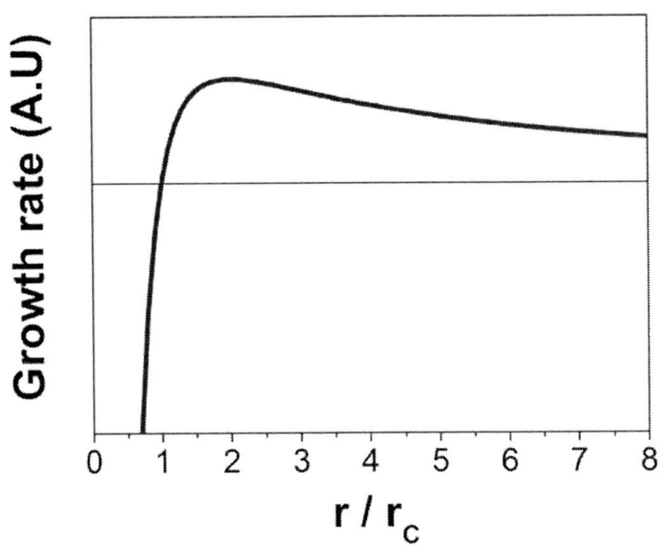

Figure 17. Growth rate as a function of crystal size for diffusion-controlled, according to equation (18). The temperature has been considered constant.

6. Precipitation of Several Phases

A good example of the complexity of the crystallization in supersaturated solutions and how the kinetic effects can take precedence over the thermodynamic factors is the precipitation of two or more phases until obtain the most stable one in solution. These precipitated phases can have the same chemical composition but different crystal structure, or they can differ by the number of solvent molecules present in their crystalline structure [32]. This behavior is described by the rule that Wilhelm Ostwald proposed in 1897 [115], where it is established that an unstable system not necessarily experiences a direct transition to the most stable state, if not that usually it previously pass through one or several intermediate states and, in each transformation, the smaller loss of possible free energy takes place.

In the attempt to find an theoretic explanation of the Ostwald rule, which cannot be considered like a general law since not always it is fulfilled, it has been encountered that it cannot be explained solely with thermodynamic arguments, and that the kinetics effects of the precipitation have to be considered to obtain some justification. In recent works, new explanations have been proposed based on the crystalline structural changes or the foundations of the irreversible thermodynamics [116, 117]. Also, computational simulations of the crystallization in supersaturated solutions have been carried out with results consistent with the Ostwald rule of stages [118].

In the present work, we will only recall the qualitative considerations described in the reference [32]. The stability of a phase in solution is determined by its solubility, therefore, as it is least soluble it is more stable. If a solution is supersaturated with respect to two phases at the same time, habitually, the phase with larger solubility, and consequently the less supersaturated, is the phase that firstly crystallizes, in agreement with the rule of Ostwald. As the formation of this phase occurs, the solute concentration decreases until reach the critical value of solubility of the crystallized phase. At this moment, the solution is solely supersaturated with respect to the phase of smaller solubility. While the crystal formation of this last phase has not occurred, the phase that precipitated firstly remains in a metastable state. However, when the first crystals of the phase with least solubility begin to be generated, the solute concentration decreases again, and the phase with greater solubility becomes unstable and it is dissolved to generate species in solution that will precipitate contributing to the growth of the most stable phase. Hence, the phase of larger solubility acts like a material reserve, and its precipitation mediates the growth of the most stable phase. An example of the fulfillment of the rule of Ostwald can be encountered in the formation of hematite nanocrystals by hydrolysis of perclorate of iron (III) in the presence of phosphate and urea. In this synthesis, lepidocrocite particles firstly precipitate. Later, the crystallization of hematite nanocrystals takes place, which tend to grow to expenses of the lepidocrocite particles that disappear, see Figure 18.

In some special cases, where two different phases precipitate, it has been observed that the phase of smaller solubility precipitate over the less stable phase, epitaxially growing over it [119].

Another consequence of the Ostwald rule is the above-studied *Ostwald ripening* process. In this case, the precipitated crystals have the same crystalline structure and composition and differ only by their size. Since the energy associated to atoms located in the particle surface is higher than that atoms placed in the particle inner, the bigger nuclei are more stable than the smallest, because the latter have a larger surface area/volume ratio. Therefore the bigger crystals tend to grow to expenses of the smallest, than they are dissolved (Figure 16) [107]. This process is the responsible of why, in certain occasions, it has been synthesized very uniform particles in spite of to the occurrence of several events of nucleation.

Also, it is worth pointing out that the Ostwald ripening process can be present in the secondary growths of particles by aggregation, coalescence or sinterization.

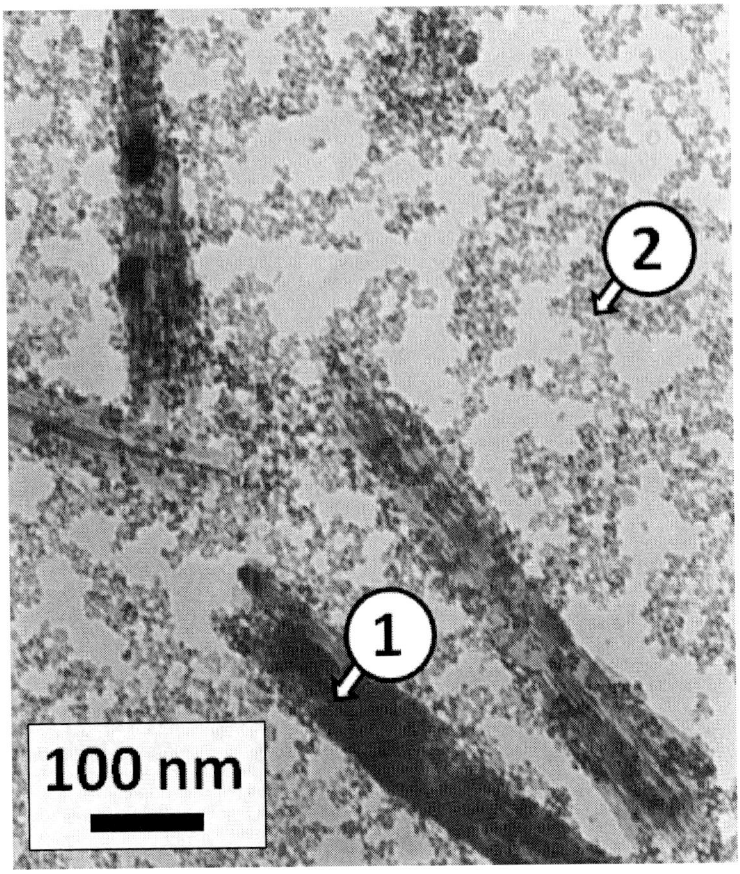

Figure 18. TEM image of a sample removed from the reaction media during the hydrolysis of perclorate of iron (III) in the presence of phosphate and urea. Two different phases are detected: a) lepidocrocite particles and b) hematite nanocystals.

7. Colloidal Satabilization and Nanocrystal Self-Assembly Into Ordered Structures

A decisive aspect of the formation of nanometric crystal in supersaturated solutions is their stabilization in the disperser media. As we mentioned in the previous discussions, several kinds of attractive interactions between particles exist in a colloidal dispersion, including the forces of van der Waals and the magnetostatic interactions. These interactions have to be balanced by other additional forces with a repulsive character to avoid the flocculation of the nanocrystals and to prevent secondary growths by irreversible aggregation or coalescence. In general, the stabilization of particles in a colloidal suspension consists in the covering or encapsulation of each particle by means of the adsorption of ions or molecules, called *ligands*, which play the role of stabilizing agents.

The ligands more suitable to stabilize nanoparticles in polar solvents are ions (or sometimes very polar molecules), whose adsorption provides to the nanocrystal surface an electric charge distribution. These surface charges attract to the opposite charges that are in the surroundings of the crystal. This double layer of electrical charges generates a potential of Coulomb which can be enough intense to avoid the agglomeration of particles, as it is represented in Figure 19a. On the other hand, if the solvent is not polar, the agglomeration and coalescence of the nanocrystals can be prevented with the adsorption of long molecules of organic acids, thiols, polymers or amines. These stabilizers reduce the high superficial tension of the nanocrystals and provide them a steric barrier that separates between them, allowing the formation of crystals with dimensions within the nanometric scale, see Figure 19b.

As a result of the use of this protective agents, it is possible act over the interactions in the dispersion and, therefore, it is possible to take part on the state of aggregation of the colloidal crystals and the symmetry of their assembly by means of the variation of the nature of the stabilizing agent, and the properties of the nanocrystals (size, shape, crystallinity, magnetic behavior, etc.).

When the nanocrystal size distribution is considerably narrow, with a standard deviation around 10%, the colloidal stabilization promotes the spontaneous organization of the nanocrystals into ordered structures known like *superstructures*, *supercrystals*, *colloidal crystals* or sometimes *crystals of nanoparticles*. In these circumstances, given to the uniformity of the system and the great mobility of nanoparticles in the colloidal suspension, the nanometric crystals tend to occupy certain positions of equilibrium, which are defined by the competition of the attractive interactions and the electrostatic or steric forces. This self-organization can occur in solution [14, 120-124], or during the evaporation of a colloid deposited onto a plane substrate [15-17]. The constituted nanometric architectures, in which the nanocrystals plays the role of the building blocks, can adopt a great variety of possible configurations, including chains and rings [16], spherical structures [120], acicular aggregates [121-124], close-packets lattices of nanocrystals [15-17] and branched aggregates [125]. Some examples of these types of ordered structures can be found in Figures 6a, 10d, 20a, and 21a.

An especially interesting situation is produced when the stabilizing agent is only adsorbed in certain crystalline surfaces. This selective adsorption can entail the inhibition of the growth in these surfaces, inducing an anisotropic growth that can be rather different from the growth habit of the crystal [16, 126]. This alteration of the crystal habit can be enough important to cause the formation of a new crystal phase, unstable in others circumstances [127, 128]. Also, the selective adsorption of ligands can promote the nanocrystal oriented attachment in nanometric architectures with monocrystalline characteristics [121-124]. For instead, hematite nanocrystals prepared by forced hydrolysis of iron perchlorate in presence of phosphate ions adsorb these latter on surface planes parallel to the c axis [129, 130]. Such adsorption produces a slight distortion in the c parameter of these crystals and alters the distribution of the electric charges in the nanocrystal surface, introducing additional anisotropic interactions in the system that determine the [001] crystallographic direction like a preferential direction of aggregation [121]. Consequently, the hematite nanocrystals self-assemble into flattened ellipsoidal architectures, as can be seen in Figure 21a. The well orientation of the nanocrystals into the aggregate is manifested in the monocrystalline properties that exhibit the secondary ellipsoidal particles. For instead, the electron diffraction analysis of one isolated ellipsoidal aggregate shows a monocrystalline pattern, see Figure 21b.

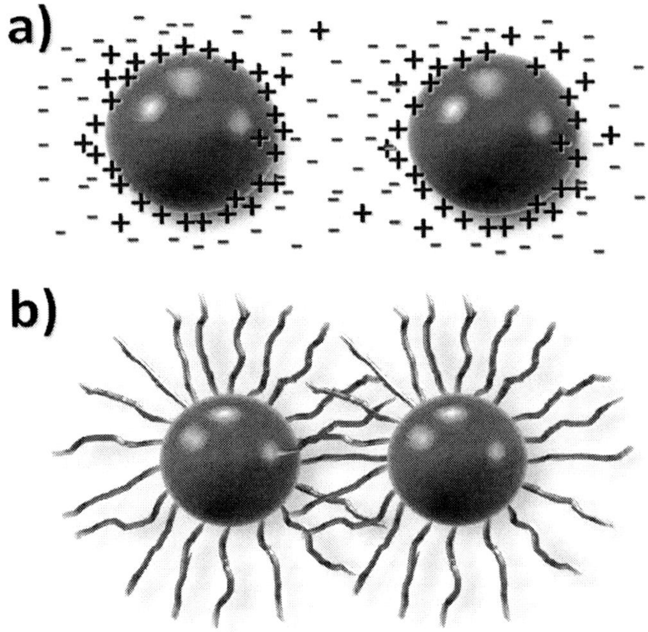

Figure 19. Schematic representation of the a) electrostatic and b) steric stabilization.

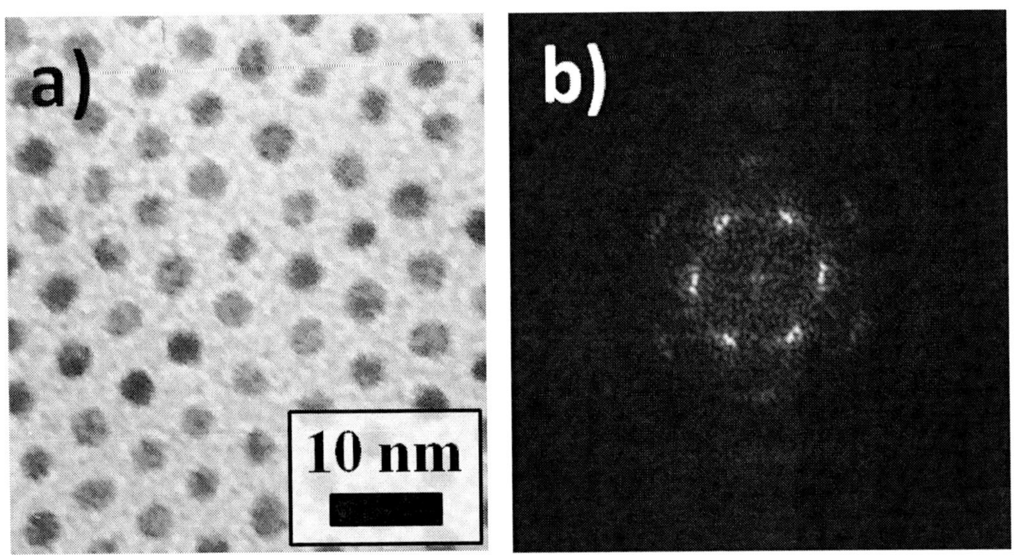

Figure 20. a) TEM image of a 2D self-assembly of 3.39±0.35 nm $Fe_{43}Pt_{57}$ nanocrystals coated with oleic acid. b) Fourier transform of image a) that confirms the well nanocrystal organization into the superlattice.

8. High-Speed Centrifugation

The high-speed centrifugation is a versatile technique that has a crucial role in the production of monodisperse nanocrystals and high-stable colloids.

The separation of the as-prepared nanocrystals from the solvent, unreacted matter and other undesired chemical products clearly is a decisive process necessary for the study and technological use of the solids synthesized from a supersaturated solution. It can be carry out by centrifugation or, in the case of magnetic crystals, by magnetic decantation. Habitually, the addition of other solvents to the initial dispersion is convenient to facilitate the precipitation of the solid. Then, the supernatant solution is removed and the precipitated solid is redispersed in an appropriate solvent (or a mixture of solvents) which should have the ability to dissolve the impurities, and at the same time, it should be incapable of stably dispersing the nanocystals. Some solvents often used are ethanol, which can dissolve both polar and non-polar impurities, water that is used to remove undesired salts, and acetone that is utilized to remove water. Then, the nanocrystals are isolated again by centrifugation or magnetic decantation. This process can be repeated several times to assure a good purification of the nanocrystals. Usually, the precipitated powder is dispersed again in a certain carrier solvent and stored under N_2.

The high-speed centrifugation also provides a simple route to improve the stability of a colloidal suspension. It is due to the possible particle aggregates and the surplus particles that exceed the maximum solubility of the solvent carrier can be precipitated by this technique. Afterwards, the supernatant is kept and the resulting precipitated solid is discarded. This process can be repeated several times to obtain a nanoparticle suspension with an optimized stability. The same procedure can be applied to realize a size selection in a polydisperse system. The aggregates and bigger particles of a colloidal suspension can be precipitated by high-speed centrifugation whereas the finest nanoparticles remain in suspension, see Figure 22.

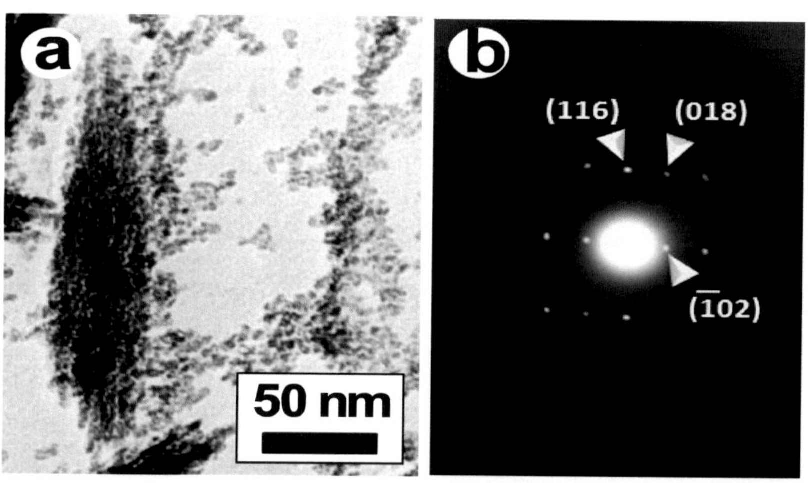

Figure 21. a) TEM image of hematite nanocrystals self-assembling into an ellipsoidal aggregate. b) Electron diffraction pattern of an isolated ellipsoidal hematite nanocrystal aggregate with reflections corresponding to the planes (116), (018) and (102) of a hematite phase.

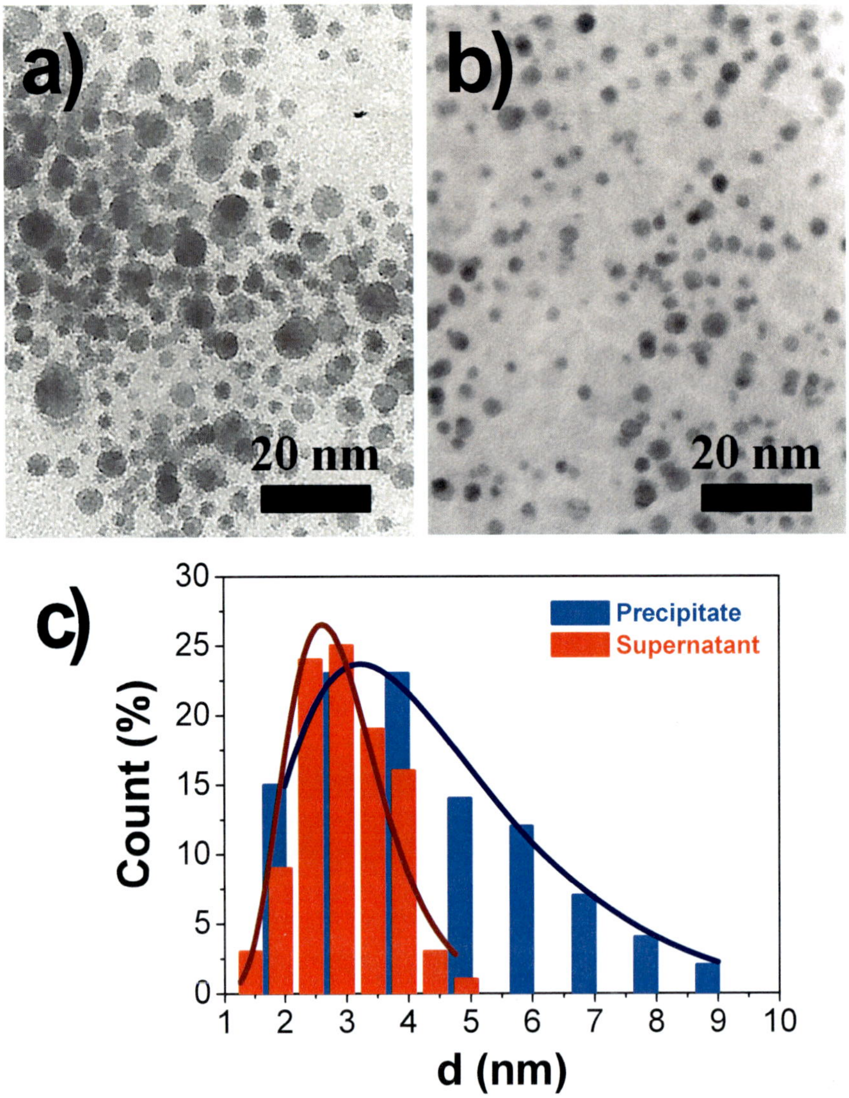

Figure 22. Particle size selection by high-speed centrifugation in a polydisperse sample of Co nanoparticles coated by oleic acid suspended in hexane: a) TEM image of the precipitated nanoparticles. b) TEM image of the supernatant nanoparticles. c) Size distributions of both precipitated and supernatant particles. The solid lines show fits to log-normal distributions.

9. Conclusion

A number of natural processes and some experiments realized in the laboratory show that the control of the interdependent processes occurring in supersaturated solutions allows the obtaining of very uniform colloidal nanocrystals and sophisticated nanocrystalline structures with well-defined characteristics (size, morphology, crystalline structure, etc.) that can be adjusted to satisfy a multitude of applications and functionalities. However, instead the remarkable advances in the field and the strenuous research efforts attracted during more than a century, there are many interesting questions concerning to the crystallization from supersaturated solutions which remain quite open. In particular, the stability of a small crystal in a fluid is, for the moment, imprecisely determined, and the theoretical approaches available do not provide us satisfactory predictions of the subtle conditions that promote the crystallization of nanostructures with the desired characteristics from a supersaturated medium. In fact, the majority of the principal experimental successes attained with this synthetic route are based on intuitive procedures and trial and error methods.

The great encountered difficulty to develop theoretical approaches more realistic is that the study of the crystallization from solutions implies resolve a many-body problem whose interactions between the bodies and also of them with their environment are not well-known. Thus, the complexity of the phenomenon force us to simplify the calculations considering important restrictive assumptions to obtain closed analytical approaches or realize numerical computer simulations based on idealized potential functions. Both routes yield to an incomplete understanding of the many processes involved in a real crystallization phenomenon and their complicated interdependencies. Thus, inquire into the molecular processes of nucleation to make available more detailed theoretical descriptions of the crystallization mechanisms (nucleation, Ostwald ripening, precipitation of several phases, etc.) and the factors that affect them (presence of substances different to the crystallizer phase, supersaturation ratio, temperature, depletion of the medium, etc.) is an exciting and very relevant current challenge that would allow us to gain a deeper insight of how the involved self-assembly mechanisms of the matter are hierarchically linked to generate complex structures with a diversity of engineered properties. A deep understanding of the crystal nucleation from solutions would be generalized to develop a more complete general theory of the first-order phase transition dynamics, and consequently, they would be extended to a variety of physical and chemical phenomena.

Acknowledgments

The authors are indebted to Prof. Manuel Vázquez, Prof. Carlos Serna and Dr. Maria del Puerto Morales for the collaboration that they have offered us during the last years.

References

[1] Thompson, G.R. and Turk, J. Introduction to physical geology; Sauders Collegge Publishing: 1997; 2th Edition.

[2] Tarbuck, E.J.; Lutgens, F.K. and Tasa, D. Earth: An Introduction to Physical Geology; Prestice Hall: 2007; 9th Edition.

[3] Dunin-Borkowski, R.E.; McCartney, M.R.; Frankel, R.B.; Bazylinski, D.A.; Pósfai, M.; and Buseck, P.R. *Science*, 1998, *282*, 1868-1870.

[4] Schüler, D. *J. Molec. Microbiol. Biotechnol.* 1999, 1, 79-86.

[5] Wiltschko, R. and Wiltschko, W. Magnetic Orientation in Animals; Springer-Verlag: Berlin, GE, 1995.

[6] Lowenstam, H.A. *Science*, 1981, 211, 1126-1131.

[7] Nancollas, G.H. and Wu, W. *J. Crys. Grow.* 2000, 211, 137-142.

[8] Squyres, S.W. and Knoll, A.H. *Earth and Planetary Science Letters*, 2005, 240, 1-10.

[9] Catling, D.C. and Moore, J.M. Sixth International Conference on Mars 2003.

[10] Squyres, S.W. and Knoll, A.H. *Earth and Planetary Science Letters*, 2005, 240, 1-10.

[11] Torrent, J. and Barrón, V. *Icarus*, 2000, 145, 645-647.

[12] Fernández-Remolar, D.C.; Morris, R.V.; Gruener, J.E.; Amils R.; and Knoll, A.H. 2005, 240, 149-167.

[13] Amils, R.; González-Toril, E.; Fernández-Remolara, D.; Gómez, F.; Aguilera, A.; Rodríguez, N.; Malkib, M.; García-Moyano, A.; Fairén, A.G.; de la Fuente, V.; and Sanz J.L. *Planetary and Space Science*, 2007, 55, 370-381.

[14] Goia, D.V. *J. Mater. Chem.* 2004, 14, 451-458.

[15] Sun, S.; Murray, C.B.; Weller, D.; Folks, L.; and Moser, A. 2000, 287, 1989-1992.

[16] Puntes, V.F.; Krishnan, K.M. and Alivisatos, A.P. *Science*, 2001, 291, 2115-2117.

[17] Wang, Z.L. *Adv. Mater.* 1998, 10, 13-30.

[18] Schmid, G. (Ed.). Nanoparticles; Wiley-VCH Verlag GmbG & Co. KGaA: Weinheim, GE, 2004.

[19] Rao, C.N.R.; Müller, A. and Cheetham, A.K. (Ed.) The chemistry of Nanomaterials; Wiley-VCH Verlag GmbG & Co. KGaA: Weinheim, GE, 2004.

[20] Raimondi, F.; Scherer, G.G.; Kötz, R.; and Wokaun, A. *Angew. Chem. Int. Edit.* 2005, 44, 2190-2209.

[21] O'Grady, K.; White, R.L. and Grundy, P.J. *J. Magn. Magn. Mater.* 1998, 177-181 Part 2, 886-891.

[22] Iwasaki, S.I. *J. Magn. Magn. Mater.* 2005, 287, 1-8.

[23] Zheludev, N.I. *J. Opt. A: Pure Appl. Opt.* 2006, 8, S1-S8.

[24] Hamilton, J.F. and Baetzold, R.C. *Science*, 1979, 205, 1213-1220.

[25] Kouassi, G.K.; Irudayaraj, *J. Anal. Chem.* 2006, 78, 3234-3241

[26] Nayfeh, M.H.; Therrien, J. and Smith, A.D. United States Patent US6410934, European Patent EP1412984. 2002. Available from: http://www.patentstorm.us/patents/ 6410934-fulltext.html.

[27] Mornet, S.; Vasseur, S.; Grasset, F.; and Duguet, E. *J. Mater. Chem.* 2004, 14, 2161-2175.

[28] Velev, O.D. *Science*, 2006, 312, 376-377.

[29] Sort, J.; Suriñach, S.; Muñoz, J.S.; Baró, M.D.; Nogués, J.; Chouteau, G.; Skumryev, V.; and Hadjipanayis, G.C. *Phys. Rev. B.* 2002, 65, 174420.

[30] Gurevich, S.A.; Kozhevin, V.M. and Yavsin, D.A. United States Patent 20060222780 2006. Available from: http://www.freshpatents.com/Method-for-obtaining-nanoparticles-dt20061005ptan20060222780.php.

[31] Tavakoli, A.; Sohrabi, M. and Kargari, A. *Chem. Pap.* 2007, 61, 151-170.

[32] Boistelle, R. and Astier, J.P. *J. Crys. Grow.* 1988, 90, 14-30.

[33] Dirksen, A. and Ring, T.A. *J. Chemical Engineering Science*, 1991, 46, 2389-2427.

[34] Farkas, L.Z. *Phys. Chem.* (Munich) 1927, 125, 236-242.

[35] Oxtoby, D.W. *J. Phys.*: Condens. Matter 1992, 4, 7627-7650.
[36] Cacciuto, A. and Frenkel, D. *J. Phys. Chem.* B 2005, 109, 6587-6594.
[37] Ruckenstein, E. and Djikaev, Y.S. *Adv. Colloid. Interface Sci.* 2005, 118, 51-72.
[38] Volmer, M. and Weber, A.Z. *Phys. Chem.* 1926, 119, 277-301.
[39] Becker, R. and Doring, W. *Ann. Physik.* 1935, 24, 719-752.
[40] Frenkel, J. The Kinetic Theory of Liquids; Clarendon Press: Oxford, 1946, p. 488
[41] Mullin, J.W. Crystallization; Butterworth-Heinemann: Oxfort, 2001, 4th Edition, p 108.
[42] Connors, K.A. Chemical Kinetics: The Study of Reaction Rates in Solution; VCH Publishers: New York, USA, 1990; p. 188.
[43] Reiss, H. *J. Chem. Phys.* 1950, 18, 840.
[44] Doyle, G.J. *J. Chem. Phys.* 1961, 35, 795-799.
[45] Hamill, P.; D'Auria, R. and Turco, R.P. *Annals of Geophysics*, 2003, 46, 331-340.
[46] LaMer, V.K. and Dinegar, R.H. *J. Am. Chem. Soc.* 1950, 72, 4847-4854.
[47] Volmer, M. Kinetik der Phasenbildung; Ann. Arbor. Edwards Bros.: Mich., 1945.
[48] Girshick, S.L. and Chiu, C.P. *J. Chem. Phys.* 1990, 93, 1273-1277
[49] Girshick, S.L. *J. Chem. Phys.* 1991, 94, 826-827.
[50] Kulmala, M.; Laaksonen, A. and Girshick, S.L. *J. Aerosol. Sci.* 1992, 23, 309-312.
[51] Wilemski, G. *J. Chem. Phys.* 1984, 80, 1370-1372.
[52] Zhuo, L.; Wu, D.T. *J. Chem. Phys.* 2006, 125, 104506.
[53] Lothe, J. and Pound, G.M. *J. Chem. Phys.* 1962, 36, 2080-2085.
[54] Reiss, H.; Katz, J.L. and Cohen, E.R. *J. Chem. Phys.* 1968, 48, 5553-5560.
[55] Katz, J.L. and Wiedersich, H. *J. Colloid. Interface Sci.* 1977, 61, 351-355.
[56] Chapela, G.A.; Saville, G.; Thopson, S.M.; and Rowlinson, J.S. *J. Chem. Soc. Faraday Trans.* II 1977,73, 1133-1144.
[57] Bai, X.M. and Li, M. *J. Chem. Phys.* 2005; 122, 224510.
[58] Talanquer, V. and Oxtoby, D.W. *J. Chem. Phys.* 1996, 104, 1993-1999.
[59] Laaksonen, A. and Napari, I. *J. Phys. Chem.* B 2001, 105, 11678-11682.
[60] Ruckenstein, E. and Nowakowski, B. *J. Colloid Interface Sci.* 1990, 137, 583-592.
[61] Nowakowski, B. and Ruckenstein, E. *J. Chem. Phys.* 1991, 94, 8487-8492.
[62] Schmelzer, J. and Ulbricht, H. *J. Coll. Inter. Sci.* 1987, 117, 325-338.
[63] Vogelsberger, W. *Chem. Phys. Lett.* 1979, 64, 601-604`.
[64] Rao, M.; Berne, B.J. and Kalos, M.H. *J. Chem. Phys.* 1978, 68, 1325-1336.
[65] Wasai, K.; Kaptay, G.; Mukai, K.; and Shinozaki, N. *Fluid Phase Equilibria*, 2007, 254, 67-74.
[66] Wasai, K.; Kaptay, G.; Mukai, K.; and Shinozaki, N. *Fluid Phase Equilibria*, 2007, 255, 55-61.
[67] Zsigmondy, R. and Thiessen, P.A. Das Kolloide Gold; Akad. Verlagsges: Leipzig, 1925.
[68] Fiévet, F.; Lagier, J.P.; Blin, B.; Beaudoin, B.; and Figlarz, M. *Solid State Ionics*, 1989, 32, 198-205.
[69] Toneguzzo, P.; Viau, G.; Acher, O., Fiévet-Vincent, F., and Fiévet, F. *Adv. Mater.* 1998, 10, 1032-1035.
[70] Luna, C.; Morales, M.P.; Serna, C.J.; and Vázquez, M. *Nanotechnology*, 2003, 14, 268-272.

[71] Luna, C.; Morales, M.P.; Serna, C.J.; and Vázquez, M. *Nanotechnology*, 2004, 15, S293-S297.

[72] Lide, D.R. *Handbook of Chemistry and Physics*, 2004-2005; CRC Press. 2004, 85th Edition.

[73] Bunker, B.C.; Rieke, P.C.; Tarasevich, B. J.; Campbell, A.A.; Fryxell, G.E.; Graff, G.L.; Song, L.; Liu, J.; Virden, J.W.; and McVay, G.L. *Science,* 1994, 264 48-55.

[74] Mann, S.; Heywood, B.R.; Rajam S.; and Birchall J.D.; Proc. R. Soc. Lond. A 1989, 423, 457-471.

[75] Aksay, I.A.; Trau, M.; Manne, S.; Honma, I.; Yao, N.; Zhou, L.; Fenter, P.; Eisenberger P.M.; and Gruner S.M. *Science*, 1996, 273, 892-898.

[76] Folch, B.; Guari, Y.; Larionova, J.; Luna, C.; Sangregorio, C.; Innocenti, C.; Caneschi, A.; and Guérin. C. *New. J. Chem.* 2008 (in press).

[77] Yasuoka, K., Gao, G.T. and Zeng, X.C. *J. Chem. Phys.* 2000, 112, 4279-4285 ;

[78] Toxvaerd. S. *J. Chem. Phys.* 2002, 117, 10303-10310.

[79] Bykov, T.V. and Zeng, X.C. *J. Chem. Phys.* 2002, 117, 1851-1868.

[80] Kashchiev, D. and van Rosmalen, G.M. *Cryst. Res. Technol.* 2003, 38, 555-574.

[81] Mullin, J.W. Crystallization; Butterworth-Heinemann: Oxfort, 2001, 4th Edition, p. 194.

[82] Mullin, J.W. Crystallization; Butterworth-Heinemann: Oxfort, 2001, 4th Edition, p. 216.

[83] Irisawa, T. In Crystal Growth Technology; Byrappa, K.; Ohachi, T.; (Ed.); William Andrew Inc. Springer-Verlag: 2003, pp 25-54.

[84] Elwell, D. and Scheel, H. J. Crystal Growth from High-tempperature Solutions; Academic Press: London, 1975; p. 206.

[85] Sangwal, K. *Prog. Crystal Growth and Charact.* 1996, 32, 3-43.

[86] Amelinckx, S. *Nature*, 1951, 167, 939-940

[87] Amelinckx, S. *Jour. de China. Phys.* 1952, 49, 411.

[88] Forty, A. *J. Phil. Mag.* 1951, 42, 670-672.

[89] Hendricks, S.B. *Amer. Min.* 1955, 40, 136-146.

[90] Wang, K.; Sun, D.; Zhang, J.; Yu, W.; Liu, H.; Hu, X.; Guo, S.; and Geng, Y. *J. Crystal Growth*, 2004, 261, 63-69.

[91] Ranganathan, M.; Dougherty, D.B.; Cullen, W.G.; Zhao, T.; Weeks, J.D.; and Williams, E.D. *Phys. Rev. Lett.* 2005, 95, 225505.

[92] Burton, W.K.; Cabrera, N. and Frank, F.C. *Phil. Trans. Roy. Soc.* 1951, A243, 299-2358.

[93] Boettinger, W.J.; Coriell, S.R.; Greer, A.L.; Karma, A.; Kurz, W.; Rappaz, M.; and Trivedi, R. *Acta mater.* 2000, 48, 43-70.

[94] Panofen, C. and Herlach, D.M. *Mat. Sci. Engin.* A 2007, 449-451, 699-703.

[95] Raz, E., Lipson, S.G. and Polturak, E. *Phys. Rev. A.* 1989, 40, 1088-1095.

[96] Aksel′rod, E.G.; Martyushev, L.M.; and Levkina, E.V. *Tech. Phys. Lett.* 1999, 25, 830-832.

[97] Dougherty, A. and Nunnally, T. *J. Crystal Growth*, 2007, 300, 467-472.

[98] Sarig, S. and Tartakovsky, F. *J. Crystal Growth*, 1975, 28, 300-305.

[99] Glicksman, M.E. and Lupulescu, A.O. *J. Crystal Growth*, 2004, 264, 541-549.

[100] Garside, J. and Jancic, S.J. *Chem. Engng. Sci.* 1978, 33, 1623-1630.
[101] Molnar, I.; Halasz, S. and Bickle, T. *Chem. Engng. Sci.* 1990, 45, 1243-1251.
[102] Mydlarz, J. and Briedis, D. *Computers Chem. Engng.* 1992, 16, 917-922.
[103] Mydlarz, J. and Jones, A.G. *Chem. Eng. J.* 1993, 53, 125-135.
[104] Choi, B.S. and Ring, T.A. *Fluid Phase Equilibria*, 2005, 228-229, 099-107.
[105] Mullin, J.W. Crystallization; Butterworth-Heinemann: Oxfort, 2001, (4th Edition), pp. 243, 253.
[106] Kile, D.E. and Eberl, D.D. *Amer. Min.* 2003, 88, 1514-1521.
[107] Ostwald, W. *Lehrbruck der Allgemeinen Chemie*. Vol2. Leipzig, GE, 1896.
[108] Lifshitz, I.M. and Slyozov, V.V. *J. Phys. Chem. Solids*, 1961, 19, 35-50.
[109] Wagner, C.Z. *Elektrochem.* 1961, 65, 581-591.
[110] Baldan, A. *J. Mater. Sci.* 2002, 37, 2171-2202.
[111] Sugimoto, T. *Adv. Colloid. Interface Sci.* 1987, 28, 65-108.
[112] Peng, X.; Wickham, J. and Alivisatos, A.P. *J. Am. Chem. Soc.* 1998, 120, 5343-5344.
[113] Sangwal, K. *Prog. Crystal Growth and Charact.* 1996, 32, 3-43.
[114] Sangwal, K.J. *Crystal Growth*, 1999, 203, 197-212.
[115] Ostwald, W.Z. *Phys. Chem.* 1897, 22, 289-302.
[116] Nývlt, J. *Cryst. Res. Technol.* 1995, 30, 445-451.
[117] Threlfall, T. *Organic Process Research & Development*, 2003, 7, 1017-1027.
[118] Anwar, J. and Boateng, P.K.J. *Am. Chem. Soc.* 1998, 120, 9600-9604.
[119] Barrón, V.; Gálvez, N.; Hochella, M. F.; Jr., and Torrent, *J. Amer. Min.* 1997, 82, 1091-1100.
[120] Luna, C; Morales, M.P.; Serna, C. J.; and Vázquez, M. *Mater. Sci. Eng.* C 2003; 23, 1129-1132.
[121] Ocaña, M.; Morales, M.P. and Serna, C.J. *J. Colloid. Interface. Sci.* 1995, 171, 85-91.
[122] Mendoza-Reséndez, R.; Bomati-Miguel, O.; Morales, M.P.; Bonville, P.; and Serna, C.J. *Nanotechnology*, 2004, 15, S254-S258.
[123] Mendoza-Reséndez, R.; Pozas, R.; Morales, M.P.; Bonville, P.; Ocaña, M.; and Serna, C.J. *Nanotechnology*, 2005, 16, 647-654.
[124] Zhang, Z.; Sun, H; Shao, X.; Li, D.; Yu H.; and Han, M. *Adv. Mater.* 2005, 17, 42-47.
[125] Guan, Z.S.; Zhang, Y.; Zhang, Q.; and Li, D.X. *J. Colloid. Interface Sci.* 2006, 302, 113-122.
[126] Peng, X.; Manna, L.; Yang, W.; Wickhan, J.; Scher, E.; Kadavanich, A.; and Alivisatos, A.P. *Nature*, 2000, 404, 59-61.
[127] Dinega, D.P. and Bawendi, M.G. *Angew. Chem. Int. Ed.* 1999, 38, 1788-1791.
[128] Sun, S.; Murray and C.B.J. *Appl. Phys.* 1999, 85, 4325-4330.
[129] Barron, V.; Herruzo, M. and Torrent, J. *Soil. Sci. Soc. Am. Proc.* 1988, 52, 647.
[130] Reeves, N.J. and Mann, S.J. *Chem. Soc.*, Faraday Trans., 1991, 87, 3875-3880.

In: Biomimetic and Supramolecular Systems Research
Editor: Arturo H. Lima
ISBN: 978-1-60456-405-1

Chapter XIII

Crystallization Effect on Bonding Strength and Failure Mechanism of Plasma-Sprayed Hydroxyapatite Coatings

***Chung-Wei Yang*[8] [a] *and Truan-Sheng Lui* [a]**

[a] Department of Materials Science and Engineering,
National Cheng Kung University, Tainan 701, Taiwan, R.O.C.

Abstract

The objective of present investigation is to systematically study the crystallization effect of autoclaving hydrothermal treatment at 100-200°C, vacuum and atmospheric heat treatments at 400-800°C on the microstructural evolution of plasma-sprayed HA coatings (HACs), especially focus on clarifying the effect of hydrothermal crystallization with lower heating temperatures. This study will also clarify the HA crystallization on the fluctuation of bonding strength and evaluate the failure mechanism of coatings through the Weibull model.

Post-heat treatments can promote the crystallization of HACs, and it is worth noting that just 150°C is required to see a significant rise in crystallinity for the hydrothermally-treated HACs compared with the other heat treatments. The ambient saturated steam atmosphere of hydrothermal treatment is favorable to eliminate the TCP, TP, CaO and amorphous calcium phosphate phases within the HACs. Low-temperature hydrothermal treatment shows a superior crystallization effect than high-temperature heat treatments in a vacuum or in an atmospheric environment.

According to the reaction kinetics of Arrhenius equation, the heating temperature, especially higher than 500°C, is a control factor for promoting the HA crystallization. Considering the hydrothermal treatment, experimental evidence confirmed that the saturated steam pressure of hydrothermal treatment can promote and accelerate the HA

[8] Corresponding author: C.W. Yang, No. 1 University Road, Tainan 701, Taiwan, R.O.C. Tel: +886-6-2757575 ext. 62931, Fax: +886-6-2380698, Rei@mse.ncku.edu.tw.

crystallization. The ambient saturated steam pressure is thought of as another control factor that not only reduces the activation energy of HA crystallization calculated from the Arrhenius equation, but also plays an important role in lowering crystallization temperatures. The finely-crystallized HA crystalline observed within hydrothermally-treated HACs is recognized as a nucleation and grain growth phenomenon by the replenishment of OH^- groups into the hydroxyl-deficient HA structure. A self-healing effect can be recognized to diminish the spraying defects with these crystalline HA.

Referring to the results of tensile adhesion tests, the HA crystallization and the thermal-induced microstructural degradation are concluded as two effects to dominate the bonding strength evolution and data fluctuation. Hydrothermally-treated HACs with higher bonding strength (38.9 MPa) than vacuum and atmospheric heat-treated ones (36.0 MPa and 37.1 MPa, respectively) can be recognized for its enhanced fracture toughness and much denser microstructure from the self-healing effect to reduce the strength degradation. Through the statistical analysis of Weibull model, the strengthening HACs are generally reliable materials with a wear-out failure model. It means that the crystallization effect promotes plasma-sprayed HACs to have a better reliability for further biological applications.

The bonding strength degradation occurred at heating temperatures above 600°C is mainly due to the effect of its crystallization-induced contraction through the thermal dilatometry analysis. This effect generates a thermal stress within the HACs because the mismatch of the thermal expansion coefficient between the crystallized HACs and Ti-6Al-4V substrate, and the HACs will be cracked in the form of significant contraction-induced cracks. The cracked microstructural feature will induce the bonding strength degradation and further increase the bonding strength data fluctuation. This phenomenon even induces a low Weibull modulus and a decreased reliability of the HACs. In contrast, the hydrothermally-treated HACs with low-temperature crystallization and a self-healing effect will be effective to prevent the above-mentioned microstructural defect and bonding strength degradation from the high-temperature crystallization process.

1. Introduction

The concept to design a surface biological fixation for orthopedic joint implant has recently achieved from bone ingrowth and bone apposition methods. With the same chemical and crystallographic structure as the apatite of living bone, hydroxyapatite ($Ca_{10}(PO_4)_6(OH)_2$, HA) is a widely preferred calcium phosphate ceramic as the biomaterial of choice in both dentistry and orthopaedics due to its favorable osteoconductive and bioactive properties [1,2]. The advantages including: (1) rapid fixation and stronger bonding between the host bone and the implant, and (2) increased uniform bone ingrowth and ongrow the bone-implant interface. HA is therefore considered as an excellent bone substitute [3,4], and HA are generally applied as coatings for the purpose of improving the bioactivity of metal implants. Investigations into calcium phosphate ceramic coatings on metallic implants started with the observation that HA in the pores of a metal implant with a porous coating would significantly affect the rate and vitality of bone ingrowth into the pores [5]. Compared with many other techniques, currently, plasma spraying has been the most extensively used method in the commercial product of HA-coated metal implants. Plasma-sprayed bioactive HA coatings (HACs) on metallic substrate exhibits enhanced interfacial strength and represents propensity of avoiding the inherent mechanical property limitations of HA without significant loss in biocompatibility [6,7]. Ti-6Al-4V alloy as a metallic substrate is more favorable compared to other alloys because of the compatibility between elastic moduli of bone and that of the Ti-alloy is important in order to avoid a difference in deformation between bone and the implant when stress is applied. Therefore, clinical application of plasma-sprayed HA coatings on Ti-6Al-4V alloy has high potential as an implant material not only owing to HA can bond physicochemically with surrounding bone [2,8,9] and promote bone growth onto its surface [10,11], but also due to the excellent corrosion resistance and lower elastic modulus of Ti-alloy compared to 316L stainless steel and cobalt chromium alloys [12-14].

However, the high temperature and enthalpy during the plasma spraying process will result in the dehydroxylation and decomposition of crystalline HA phase into tri-calcium phosphate ($Ca_3(PO_4)_2$, TCP), tetra-calcium phosphate ($Ca_4P_2O_9$, TP), calcium oxide (CaO), oxyhydroxyapatite and amorphous calcium phosphate within the sprayed coatings. HACs with a higher content of impurity phases and amorphous calcium phosphate will increase higher dissolution rate than crystalline HA in aqueous solutions and body fluids. It results in some problems with decreasing the structural homogeneity and the mechanical properties in the firm fixation between the implant and surrounding bone tissue [15,16]. Therefore, decreasing the impurity phases and amorphous calcium phosphate is very important for the long-term biological stability of plasma-sprayed HACs.

Referring to the reports of HA, many researchers have investigated the material and medical properties of plasma-sprayed HACs in the past ten years. Since amorphous and impurity calcium phosphate is thermodynamically meta-stable and undesirable for the HACs, previous studies pointed out that controlling spraying parameters [17] or performing appropriate thermal treatments are available methods to significantly promote the HA crystallization and improve the dissolution behaviors [18-21]. Furthermore, a better bonding strength should also be required with a higher HA crystallinity level and fewer coating defects as a result of HA crystallization. However, with different materials preparation, manufacturing and characterization, or different *in vitro* and *in vivo* examination methods, it is difficult to systematically conclude the relationship between the HA crystallization, biological responses and mechanical properties of coating layers. Additionally, reports have mainly focused on the changes of phase composition and microstructural feature of HACs [22-24], as well as the mechanism of crystallization of the coating layer [25]. The effect of HA crystallization on bonding strength still needs further investigation.

Therefore, the aim of this chapter will clarify the crystallization effect which influencing the microstructural characteristics, bonding strength and failures of the HACs through systematically performing post-thermal treatments within the ambient air, vacuum atmosphere and the autoclaving hydrothermal treatment. The benefits of low-temperature hydrothermal crystallization on the HACs compared with the other methods will also be clarified through the evaluation of reaction rate and activation energy by the Arrhenius kinetics. In addition, the reliability and the failure mechanism of as-sprayed and heat-treated HACs will be examined by the statistical analysis of the Weibull model with the observation of failure morphologies.

2. Microstructural Characteristics of Plasma-Sprayed and Crystallized HA Coatings with Various Heat Treatments

High purity HA powder with particle size ranging from 15 to 40 μm was used in the coating process, and Ti-6Al-4V alloy (ASTM F-136 ELI) was used as a substrate. The suitable plasma spraying parameters with a power of 40.2 kW was selected according to our relative studies [17,26]. The thickness of coatings was controlled at a range of 120 ± 10 μm. By x-ray diffraction (XRD) analysis, Figure 1 shows the diffractograms of the as-received HA powder (HAP) and the as-sprayed HACs. A fairly high content of amorphous and impurity calcium phosphate, including α-$Ca_3(PO_4)_2$ (α-TCP), β-$Ca_3(PO_4)_2$ (β-TCP), $Ca_4P_2O_9$ (TP) and CaO, can be recognized in the as-sprayed HACs besides the desired HA phase. The difference in the phase composition and crystallinity between HAP and as-sprayed HACs are evident.

In order to quantitatively evaluate the crystallinity of HACs, a commonly used index of crystallinity (IOC) which defined from the ratio of three strongest HA main peaks ((211), (112), and (300), JCPDS 9-432) integral intensity of the HACs (*Ic*) and the as-received HAP (*Ip*) according to the relationship IOC = $(Ic/Ip) \times 100\%$ [20,27] herein. As shown in Figure 1, the HAP represents a well-crystallized single phase of $Ca_{10}(PO_4)_6(OH)_2$, and this method supposes that the IOC of the as-received HAP is 100%. Figure 2(a) shows the variation of IOC values for the vacuum, atmospheric and hydrothermally heat-treated HACs at different heating temperatures. The crystallinity of the as-sprayed HACs is about 20.3%, and the crystallinity of HACs increases with increasing heating temperatures. In addition, it is worth noting that just low heating temperatures of 150-200°C are required to see a significant rise in crystallinity for the hydrothermally-treated HACs (the IOC is about 70-85%) compared with the 600-800°C heat treatments in the air or in vacuum (the IOC is about 71-80%).

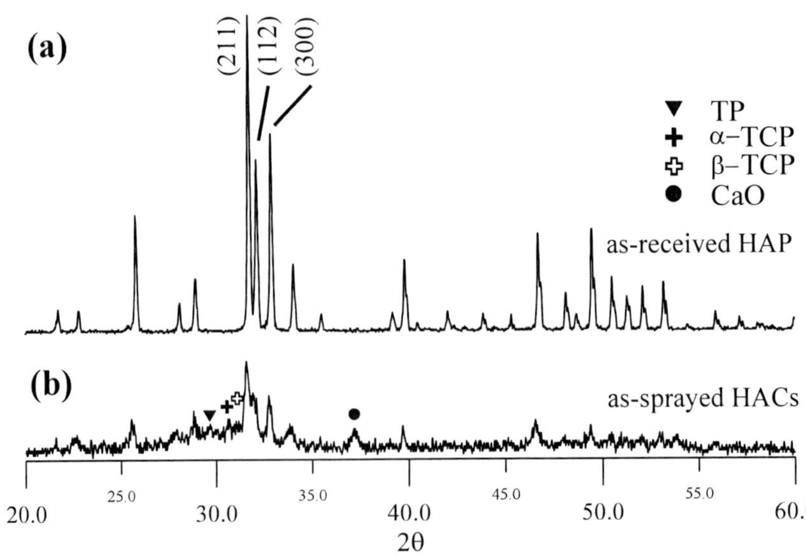

Figure 1. The diffractograms of (a) the as-received HA powder and (b) the as-sprayed HACs.

Moreover, the internal standard method [28] was used to quantitatively determine the fluctuation of phase composition within the heat-treated HACs. The integral intensity of known weight percent pure Si powder added in the specimens was taken as internal standard. The calibration curves for impurity phase content were quoted according to the studies by Wang et al. [29,30]. The main peak integral intensity ratio between TCP, TP, CaO and HA phase from various XRD patterns of heat-treated HACs were compared to the calibration curves and the concentrations (in wt. %) in the HACs were calculated. As shown in Figure 2(b), the impurity phases generally tend to decrease with increasing heating temperatures. However, when the heating temperature raises to higher than 600°C within vacuum (VH-series), a significant increasing trend of impurity phases content (TP and CaO phases are mostly) resulted from HA decomposition [31,32] can be recognized as the temperature increases further, resulting in greater total impurity phases about 26.1 wt. % at 800°C (VH800). According to the phase diagram of CaO-P_2O_5 [33], since there is a lack of ambient partial steam pressure during vacuum heat treatment, TCP, TP and CaO cannot be eliminated without the replenishment of hydroxyl groups (OH^-) and the increased impurity phase content in the temperature range of 600-800°C is resulted from the decomposition of HA [31,34]. For the air heat-treated HACs (AH-series), the partial steam pressure of atmospheric moisture can recover and promote the reconstitution of TCP, TP and amorphous calcium phosphate into crystalline HA. The residual impurity phase within 600-800°C air heat-treated HACs is recognized as CaO. It cannot easily be converted into HA if the ambient heating atmosphere without abundant H_2O molecules [31,35,36].

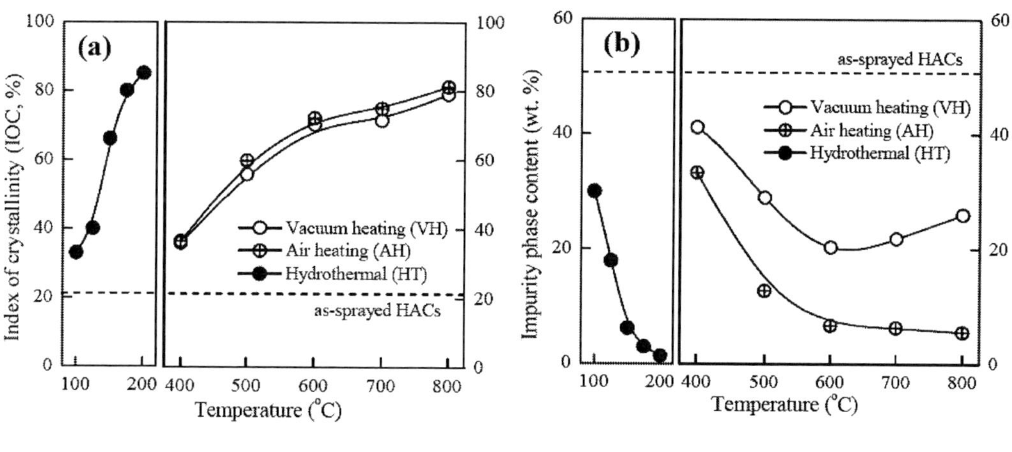

Figure 2. (a) The index of crystallinity (IOC), and (b) the impurity phase content of different post-heat treatments compared with the as-sprayed HACs.

Considering the HACs hydrothermally-treated in the autoclave with an ambient saturated steam pressure (HT-series), it is worth noting in Figure 2(b) that the impurity phases and amorphous calcium phosphate content significantly decrease with performing the hydrothermal treatment, and the HACs which hydrothermally-treated at 200°C (HT200) contains the least impurity phases content of about 1.4 wt. % among all the heat-treated specimens. Since the hydroxyl groups (OH^-) promote the reconstitution of amorphous calcium phosphate into crystalline HA [37], therefore, the saturated steam pressure atmosphere of the autoclaving hydrothermal treatment can significantly improve HA crystallization and effectively eliminate TCP, TP, CaO phases and amorphous component of plasma-sprayed HACs with the replenishment of OH^- groups.

Besides XRD analysis, FT-IR can further identify structural features such as crystallization and hydroxylation state of the HACs after post-heat treatments. Figure 3(a) shows the diffuse reflectance FT-IR (DRIFT) spectrum of plasma-sprayed HACs. No obvious hydroxyl stretching band (OH^-, at 3572 cm^{-1}) is observed, and the phosphate asymmetric stretching band (PO_4^{3-}, at 1087 cm^{-1}), the bending band (at 602 cm^{-1}) [38] are almost invisible because of the dehydroxylation and low crystallinity of plasma-sprayed HACs. HA attains a stable phase at room temperature only if the ambient atmosphere contains H_2O molecules [33], which phenomenon indicates that the decrease of OH^- groups has a negative effect on the stability of HA phase. Comparing the spectra for the HACs with a similar IOC (specimens VH800, AH800 and HT175), as shown in Figure 3(b)-(d), the OH^- band at 3572 cm^{-1} and the two PO_4^{3-} bands in HT175 spectrum is much sharper than VH800 and AH800. The prominence and sharpness of the OH^- and PO_4^{3-} peaks demonstrates the superior crystallized integrity of hydrothermally-treated HACs. However, The broad spectra at about 1400-3300 cm^{-1} in Figure 3(b) and 3(c) represent the various band energies of calcium phosphate compounds [38], and the appearance of a peak at 3642 cm^{-1} is assigned to another OH^- band which is related to the existence of $Ca(OH)_2$ [39]. For vacuum and air heat treatments, CaO is not easily eliminated without abundant environmental H_2O molecules, whereas the intensity of this band disappeared with eliminating the CaO by the hydrothermal treatment. To summary the above analysis, the hydrothermal treatment conducted with an abundant H_2O atmosphere is significantly help to minimize impurity phases and promote HA crystallization. This is a result of the replenishment of missing OH^- groups within the dehydroxylation HA. Well-crystallized HACs with pure phases through the autoclaving hydrothermal treatment are favorable to prevent dissolution problems for clinical application.

Figure 4 shows the representative surface morphologies with significant crystallization features of vacuum, air and hydrothermally heat-treated HACs. Considering the vacuum and air heat treatments, the broken features of VH600 and AH600 specimens, which displayed with obvious surface cracks as indicated by arrows in Figure 4(a) and 4(b), are resulted from HA crystallization-induced contraction effect [31,32] of coating layers during high-temperature heat treatments, especially in temperature interval of 600-800°C. However, the typical hydrothermally-treated HACs, similar with HT150 specimen as shown in Figure 4(c), display a structural feature without obvious surface cracking. High magnification SEM evidences as illustrated in Figure 4(d)-(f) represent the main differences in microscopic cracking features among the VH600, AH600 and HT150 specimens. It is worth noting that evident new-growth crystalline indicated by arrows in Figure 4(e) and 4(f) is observed in the vicinity of the cracks and on the coating surface. From previous XRD and FT-IR analysis results, these grains can be attributed to crystalline HA, which crystallized from the hydroxyl-deficient structure of plasma-sprayed HACs with the replenishment of hydroxyl groups. For hydrothermal treatment, the crystalline HA further grain growth with a larger crystal size and reduce microcracks with increasing the hydrothermal heating temperatures.

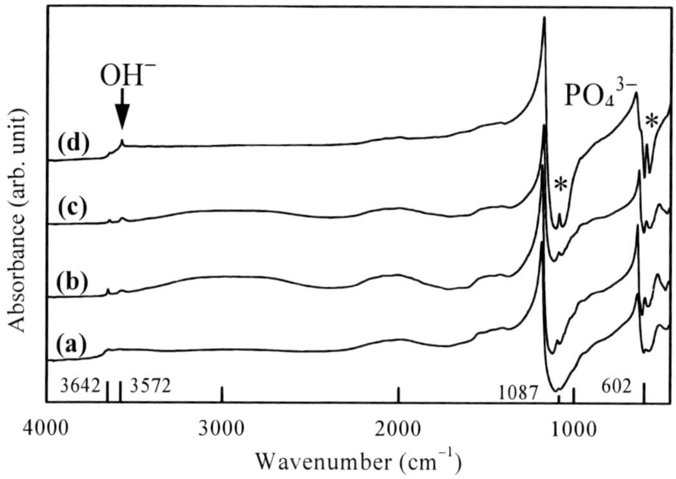

Figure 3. Representative FT-IR spectrum of (a) the as-sprayed HACs, (b) VH800, (c) AH800 and (d) HT175 specimens (The arrow indicates OH^- stretching band at 3572 cm^{-1}, the star indicates the PO_4^{3-} bands at 1087 cm^{-1} and 602 cm^{-1}).

The reduction of coating defects for hydrothermally-treated HACs can be recognized as the self-healing effect of the hydrothermal treatment [26,40]. As a result of crystallization during heat treatments, the contraction-induced cracking and the self-healing phenomena will significantly influence the bonding strength and the failure mechanism of HACs. The reason will be discussed later.

3. Kinetics of HA Crystallization

The above-mentioned results represent that an appropriate thermal treatment can actually promote significant crystallization to improve the phase purity and crystallinity of plasma-sprayed HACs. The heating temperature plays a role in changing the crystallinity. Referring to the past reports [25,29,32], however, the crystallization of hydroxyl-deficient HACs requires at least 600°C, and high heating temperatures tend to undermine the structural integrity and cause phase decomposition of crystalline HA. Therefore, a heating process conducted with a water vapor atmosphere was developed to minimize impurity phases and promote HA crystallization. This is a result of replenishment of missing OH^- groups with surrounding H_2O molecules [18,26,41]. The effect of the ambient heating atmosphere is another factor that should be considered to affect the reaction rate and the activation energy for HA crystallization. The purpose of the following discussion is to clarify the effect of the crystallization kinetics related to heating temperatures and ambient atmosphere. The influence of saturated steam pressure during the autoclaving hydrothermal treatment is studied in detail to understand the kinetics of the hydrothermal crystallization at lower heating temperatures.

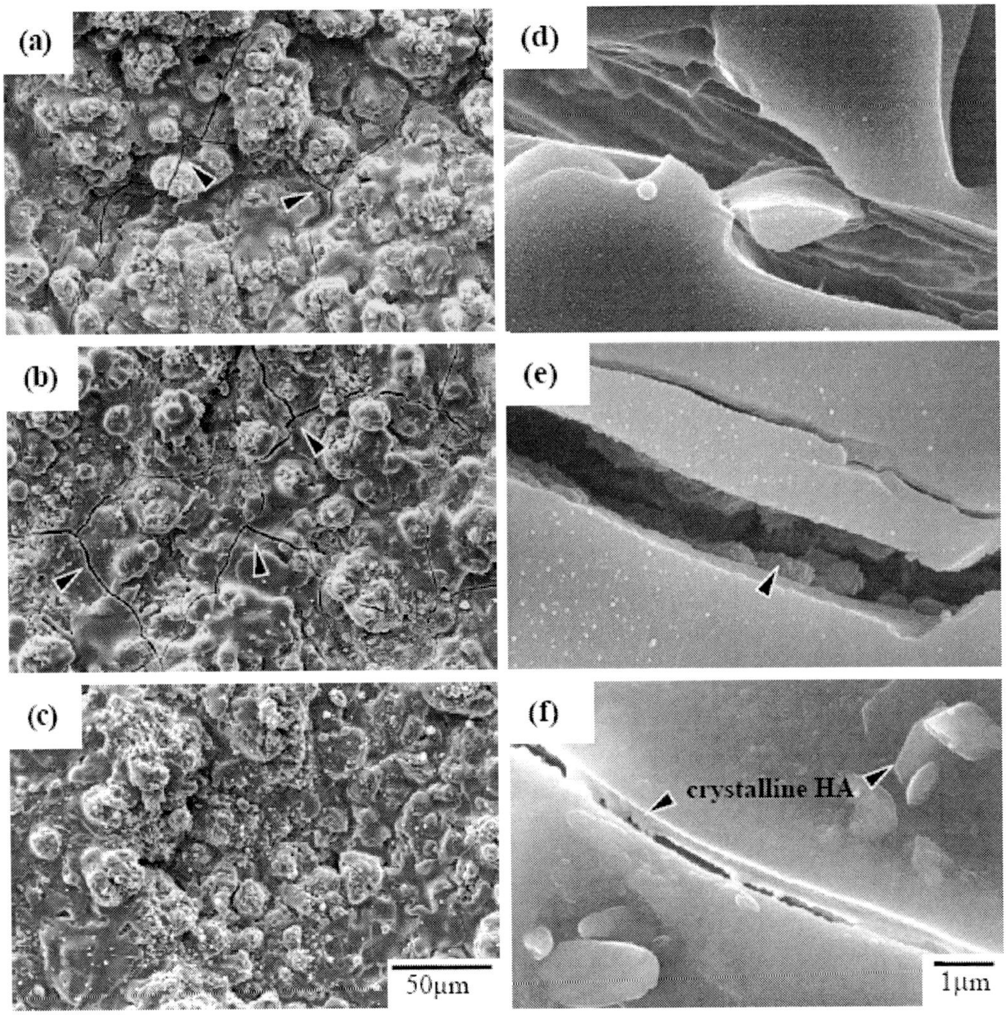

Figure 4. Surface morphologies of (a) VH600, (b) AH600 and HT150 specimens. The arrows indicate the crystallization-induced cracks during high-temperature heat treatments. Images (d), (e) and (f) are micro-features of VH600, AH600 and HT150, respectively. The arrows indicate the crystallized HA grains after air and hydrothermal treatments.

According to the theory of chemical reaction kinetics, it is recognized that the HA crystallization should follow the Arrhenius equation [42-44] as represented in Eq. (1):

$$r = \frac{d\alpha}{dt} = k(1-\alpha)^n \tag{1}$$

r is for the reaction rate, k is for the rate constant, α is the conversion ratio of HA from amorphous calcium phosphate to crystalline HA, and it represents the IOC value herein. Figure 5 shows the linear plots of ln($dIOC/dt$) *versus* ln($1-IOC$) from the natural logarithmic of Eq. (1). The slope represents the reaction order n, which calculated by the least-squares fitting method of Eq. (2) at a maximum coefficient of determination (R^2).

$$\ln\frac{dIOC}{dt} = n\ln(1 - IOC) + \ln k \tag{2}$$

The reaction order n is approximated from 1.8 to 2.1 for vacuum and air heat treatments as denoted in Figure 5(a) and 5(b). Therefore, it can be deduced that HA crystallization should follow the second-order reaction kinetics of Arrhenius equation, and the relationship of Eq. (1) can be converted into Eq. (3). Figure 6(a) and 6(b) shows the linear plots of $(1-IOC)^{-1}$ *versus* heat treatment time (t) from the integration of Eq. (3), and the slope represents the rate constant k, which calculated by the least-squares fitting method at a maximum R^2 value.

$$r = \frac{dIOC}{dt} = k(1 - IOC)^2 \tag{3}$$

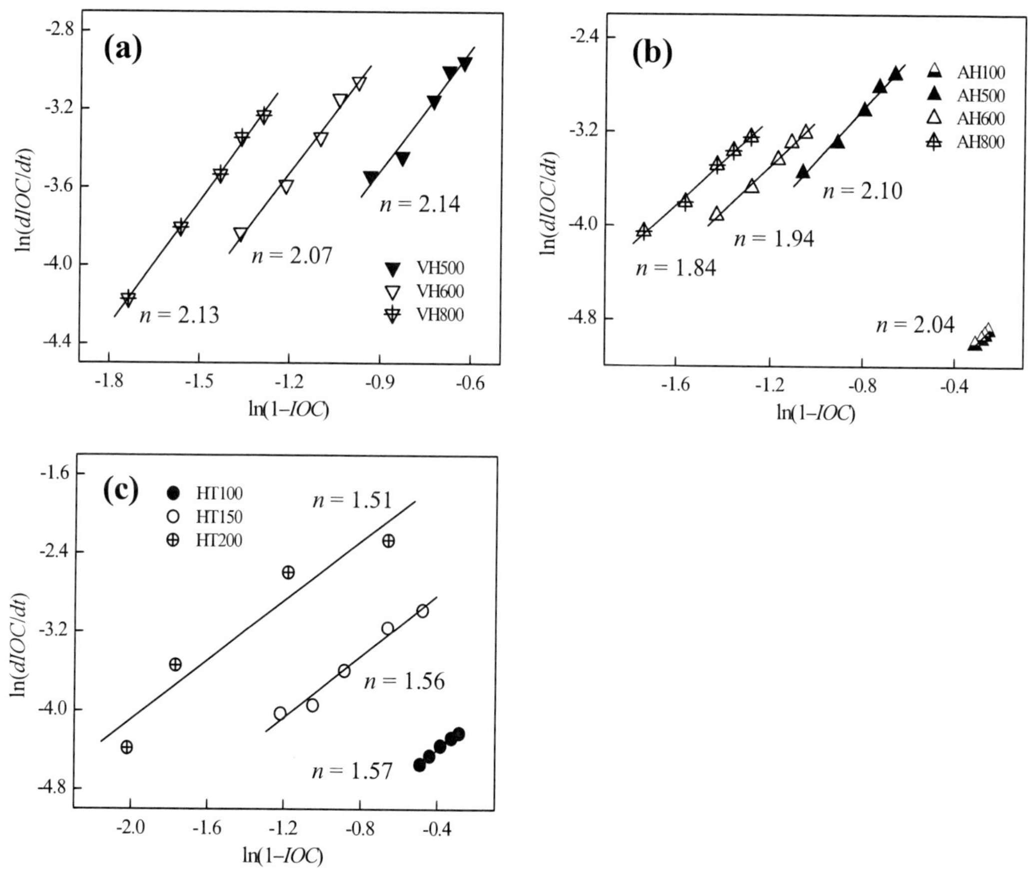

Figure 5. Plots of ln($dIOC/dt$) *versus* ln($1-IOC$) from the natural logarithmic of Eq. (1) for (a) the vacuum heat treatment (VH-series), (b) the atmospheric heat treatment (AH-series), and (c) the hydrothermal treatment (HT-series).

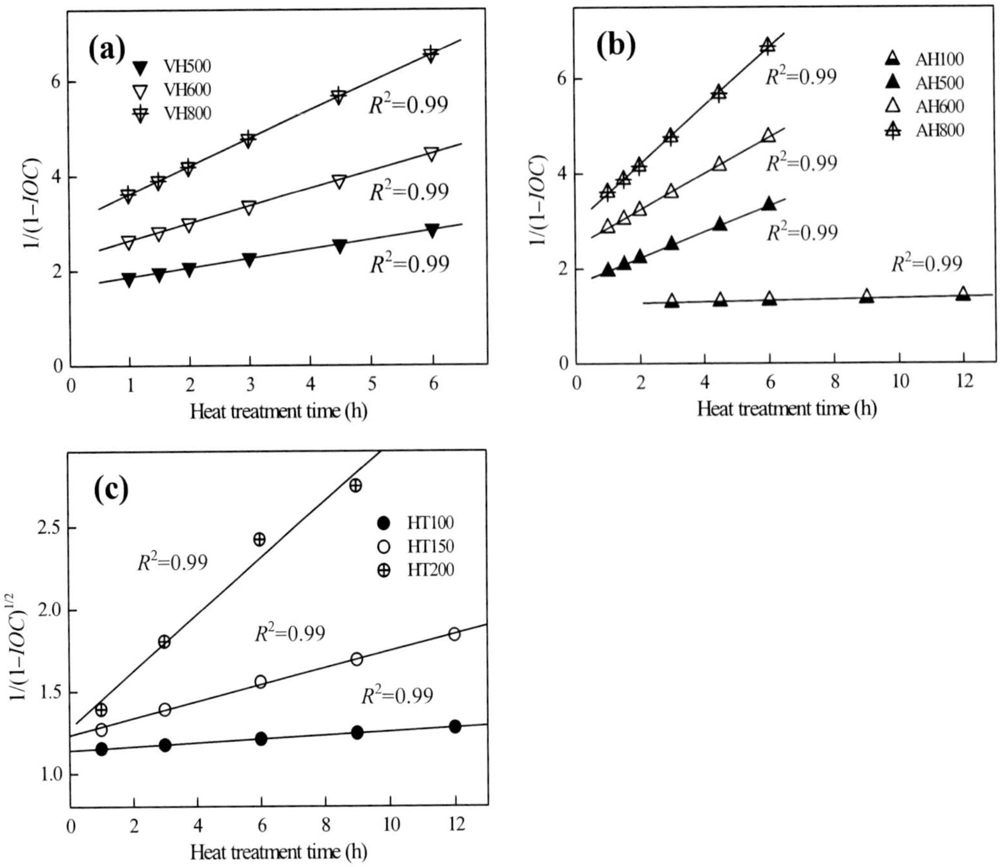

Figure 6. Plots of $(1-IOC)^{-1}$ *versus* heat treatment time (t, hours) from the integration of Eq. (3) for (a) the vacuum heat treatment, (b) the atmospheric heat treatment, and the plot of $(1-IOC)^{-1/2}$ *versus* t for (c) the hydrothermal treatment.

The reaction rate and the activation energy of HA crystallization within different ambient heating atmospheres in vacuum and in the air can be quantitatively evaluated by the IOC of each specimens. The activation energy (E_a) of HA crystallization can also be quantitatively evaluated from the Arrhenius plot of ln$k$ *versus* $1/T$ (heating temperatures). These results are finally concluded in Table 1. Through the kinetics calculation from Figure 6(b), the rate constant for 100°C atmospheric heat treatment (k_{100}) is about 0.011. According to the reaction kinetics of Arrhenius equation, the rate constant (k) represents the reaction rate, and it represents the crystallization rate during heat treatments herein. Compared the rate constant k_{100} (0.011) with any other k value for the vacuum and atmospheric heat treatments, therefore, it can be recognized that the heating temperature, especially higher than 500°C, is a controlling factor for promoting HA crystallization rate both in vacuum and in the air heat treatments. Besides, referring to the phase diagram of CaO-P_2O_5 under 500 mmHg partial water vapor pressure (P_{H_2O}) [33], HA is a stable phase, and P_{H_2O} is also a significant factor to promote HA crystallization.

Table 1. Results of crystallization kinetics of different heat-treated HACs

	Rate constant, $k = A\exp(-E_a/RT)$			Activation energy, E_a (kJ/mol)†
	k_{500}	k_{600}	k_{800}	
Vacuum heating	0.197	0.365	0.586	24.6
Atmospheric heating	0.272	0.379	0.612	19.1
	k_{100}	k_{150}	k_{200}	
Hydrothermal treatment	0.294	0.591	0.905	16.6

† Data were calculated from the slope of ln*k* vs. heating temperatures (1/*T*) graph.

As shown in Table 1, the experimental results prove that there is a faster crystallization rate (with a higher rate constant of k_{500}, k_{600} and k_{800}) and lower activation energy of HA crystallization for the heat treatment in an ambient atmosphere with a partial water vapor pressure than in a vacuum. Except for considering the oxidation problem of Ti-6Al-4V during high-temperature heating, the atmospheric heat treatment can be recognized as a better way to improve the HA phase purity and crystallization rate of plasma-sprayed HACs than vacuum heat treatment under the same heating conditions.

Since the dehydroxylation is a result of hydroxyl groups (OH^-) broken away from HA crystal structure during the high temperature and high enthalpy of plasma spraying process, the formation of $Ca_{10}(PO_4)_6(OH)_{2-2x}O_x{\bullet}V_x$ (oxyhydroxyapatite, V is for the lack of OH^- in HA lattice) [45] and amorphous calcium phosphate with a decreased crystallinity of HA occurred in the coating layers. When the autoclaving hydrothermal treatment is applied for promoting the crystallization of plasma-sprayed HACs, the water molecules are reduced and the hydroxylation of HA proceeded with the replenishment of hydroxyl groups according to Eq. (4) and Eq. (5), respectively.

$$H_2O \rightarrow H^+ + OH^- \tag{4}$$

$$Ca_{10}(PO_4)_6(OH)_{2-2x}O_x{\bullet}V_x + xH_2O \rightarrow Ca_{10}(PO_4)_6(OH)_2 \tag{5}$$

Figure 5(c) shows the linear plot of ln(*dIOC*/*dt*) *versus* ln(1−*IOC*) from the natural logarithmic of Eq. (1). The reaction order *n* calculated from the slope is 1.57, 1.56 and 1.51 for 100°C, 150°C and 200°C hydrothermal treatment, respectively. According to previous reports [42-44] and the above-mentioned calculation, HA crystallization is a second-order reaction kinetics, which depends on heat treatment time and temperature effects. However, the saturated steam pressures during autoclaving hydrothermal treatment are 0.10 MPa (100°C), 0.48 MPa (150°C) and 1.56 MPa (200°C), and therefore, the effect of the saturated steam pressure should be considered as another significant factor affecting HA crystallization. Equation (6) shows the modified form which involving a saturated steam pressure term. Figure 6(c) shows the $1/(1-IOC)^{1/2}$ *versus* heat treatment time (*t*) plot resulting from the integration of Eq. (6), P_{H_2O} can be thought of a constant at each hydrothermal heating temperature. Since the above-mentioned saturated steam pressure term (P_{H_2O}) is independent of time (t), the slope of the least-squares fitting curves in Figure 6(c) is deduced in a form of $(k P_{H_2O}{}^{1/2})/2$ and the hydrothermal crystallization rate constant k is obtained. Generally, HA crystallization requires at least 600°C in vacuum or in the air [23,25,29,46], however, high heating temperatures will cause phase decomposition of crystalline HA and structural deterioration of coating layers as denoted in the previous section. Through the analysis of Arrhenius kinetics, the HA crystallization under hydrothermal treatment is performed only at lower temperatures with a significant larger crystallization rate constant than other heat treatments as shown in Table 1. Furthermore, compared with the activation energy (E_a) of HA crystallization for the vacuum and atmospheric heat treatments (are 24.6 kJ/mol and 19.1 kJ/mol, respectively), the activation energy for the hydrothermal crystallization (16.6 kJ/mol) is significantly lower than the other heating process.

$$r = \frac{dIOC}{dt} = k(1 - IOC)^{3/2} P_{H_2O}{}^{1/2} \qquad (6)$$

On the basis of the previous discussion, the heating temperature is a controlling factor for HA crystallization. In addition, post-heat treatment in an atmosphere with abundant H_2O molecules can further promote and accelerate the HA crystallization, because H_2O molecules can replenish the missing OH^- groups within the dehydroxylation state to improve the structural integrity of plasma-sprayed HACs. Hydrothermal treatment can induce a low-temperature crystallization process, and the saturated steam pressure is thought to be a factor which reduces the activation energy and accelerates the HA crystallization (larger rate constant k, and lower activation energy E_a). Experimental evidence confirmed that the ambient saturated steam pressure plays an important role in lowering heating temperatures and promoting HA crystallization.

4. Effect of HA Crystallization on the Bonding Strength and Failures of Coatings

The bonding strength of the as-sprayed and different heat-treated HACs measured by the standard tensile adhesion test (ASTM C633-01) [47] is shown in Fig 7. For both the vacuum and atmospheric heat treatments, the bonding strength tends to increase with increasing the heating temperature up to 600°C, whereas the bonding strength markedly decreases with further increasing the heating temperature up to 800°C. Considering the cross-sectional features also shown in Figure 7, both the 600°C vacuum and atmospheric heat-treated HACs represent dense microstructure. However, it is to be noted that obvious cracking features (indicated by arrows) and an interfacial TiO_2 layer (a Rutile phase) [35] can be recognized under a high-temperature heating in vacuum and in the air.

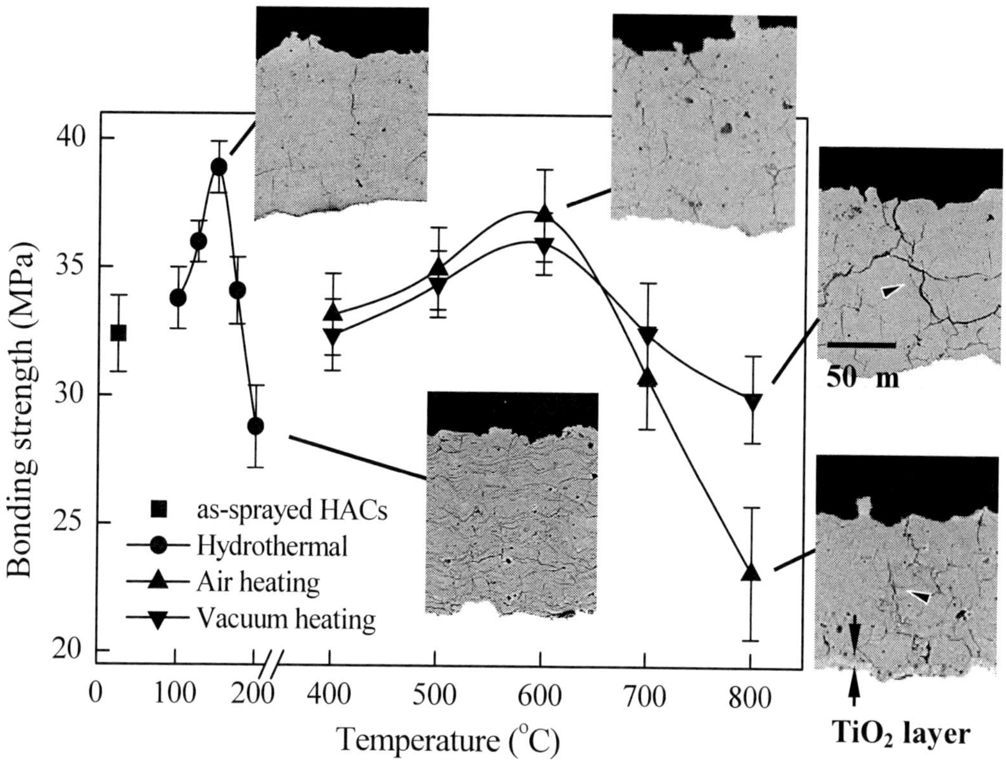

Figure 7. The variation of bonding strength that correlated with the microstructural features after vacuum, atmospheric and hydrothermal heat treatment at different temperatures.

The cracking feature is resulted from the crystallization-induced contraction during high-temperature crystallization. For the hydrothermal treatment, the bonding strength is also significantly improved with HA crystallization, and 150°C hydrothermally-treated HACs has a higher bonding strength (38.9 MPa for HT150) than both of the optimal condition (at 600°C) for atmospheric and vacuum heat-treated HACs (37.1 MPa for AH600 and 36.0 MPa for VH600, respectively). But the bonding strength of hydrothermally-treated HACs tends to decrease with the appearance of induced inter-lamellar microcracks (at the lamellar-splat boundaries) in the heating temperature interval of 150-200°C. The bonding strength evolution can be recognized as a result of HA crystallization, and the deterioration of bonding strength should depend on the above-mentioned detrimental crystallization-induced defects.

Since the Young's modulus is a measure of the inter-atomic binding forces, the evolution of Young's modulus for the HACs depends on the extent of crystallization [35]. Besides, the microstructural features such as defects, lamellar structure and the contact between splats boundaries should be also considered as other influencing factors on the Young's modulus of a coating [48,49]. Through the Young's modulus measurements, it can help to clarify the effect of crystallization on the bonding strength and the failure mechanism. To measure the Young's modulus, the 1.2 mm-thick HA coating was coated on the Ti substrate, and a standard three-point bending test was applied to evaluate the Young's modulus [50]. The HAC test pieces with dimensions of 50 (l) × 5 (w) × 1 (t) mm^3 were carefully cut from Ti substrates by a low-speed diamond saw along the coating/substrate interface, and then they were hydrothermally and vacuum heat treated at the above-mentioned temperatures, respectively.

Figure 8 shows a linear relationship between the Young's modulus and the crystallinity (IOC) of the hydrothermal and vacuum heat-treated HACs. The SEM/BEI images of vacuum heat-treated HAC test pieces represent that there is no obvious microstructural difference between the HAC test pieces with different crystallinity. Thus, it is reasonable to suggest that the crystallinity is a main controlling factor in increasing the Young's modulus of HACs. According to Griffith-Orowan-Irwin theory, the fracture toughness of a ceramic is correlated with the ability of impeding crack propagation (γ) and the Young's modulus (E) from the relationship: $K_{IC} = \sqrt{2\gamma E}$ [51]. The fracture toughness of the heat-treated HACs is further enhanced about 5-20% owing to the improvement of Young's modulus which resulted from the crystallization of HACs [20,52,53]. Moreover, the crystallization can also help to enhance the cohesion between splats boundaries within the coatings [48,49,52]. Therefore, the enhanced bonding strength of the heat-treated HACs is owing to the improvement of Young's modulus and fracture toughness resulting from the crystallization effect of HA coatings.

As illustrated in Figure 8, it should be worth noting that the significant improvement of the Young's modulus for hydrothermally-treated HACs (HT-HAC) compared with the vacuum heat-treated HACs can be recognized due to the self-healing effect of hydrothermal crystallization. The evidences can not only be observed from surface morphologies as shown in Figure 4(f), but also demonstrated from the fracture specimens of HT-HAC test pieces after three-point bending tests. Figure 9(a) represents a significant reduction of pores and cracks within HT-HAC coatings, and the lamellar boundaries are healed with the nano-crystalline as indicated by arrows in Figure 9(b). The microstructural homogeneity and the contact between splats boundaries are significantly improved after hydrothermal treatment.

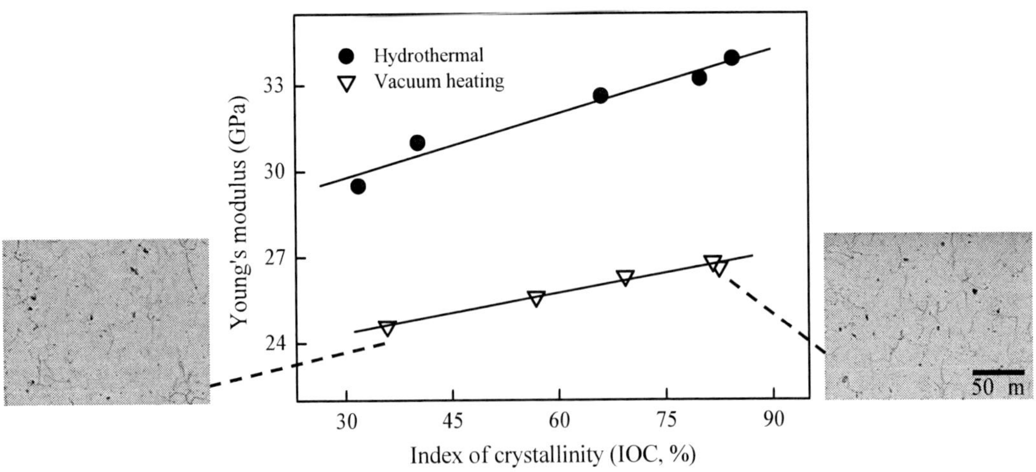

Figure 8. A linear correlation between the Young's modulus and the crystallinity (IOC) of the vacuum and hydrothermally heat-treated HACs. The SEM/BEI images show the microstructural features of Young's modulus test pieces with different crystallinity.

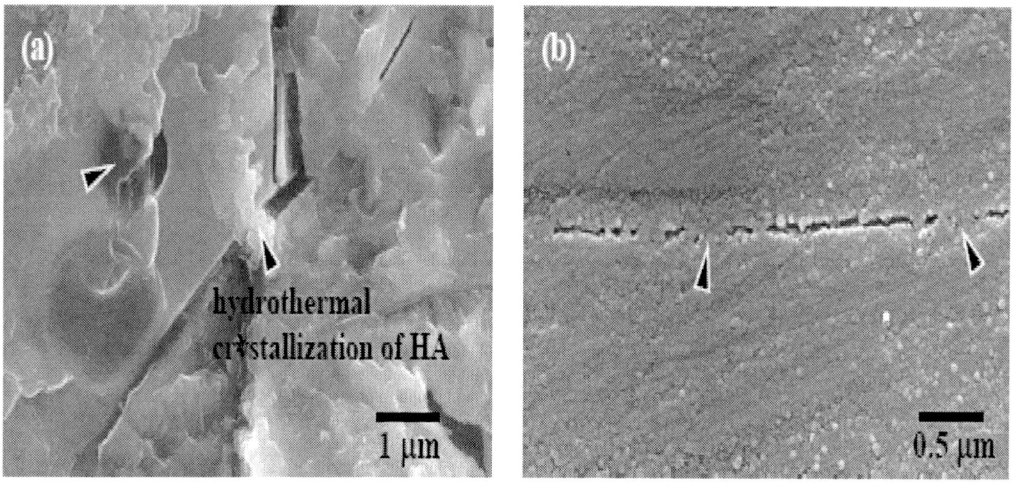

Figure 9. The self-healing effect of hydrothermal crystallization (as indicated by arrows) on plasma-sprayed HACs observed at (a) the coating fracture surface after three point bending test, and (b) the lamellar boundaries of the coatings.

Thus, besides the HA crystalline observed on the coating surface (Figure 4), the hydrothermal crystallization with a self-healing effect occurred throughout the whole HA coating layers under an abundant saturated steam environment. Since the reduction of coating defects, lamellar structure and the contact between splats boundaries are effectively improved by the self-healing effect of hydrothermal crystallization, therefore, the bonding strength of HT150 specimen is significantly further enhanced and higher than AH600 and VH600 specimens.

Although the bonding strength is improved with the crystallization of coating layers during heat treatments, however, it can be recognized that a detrimental effect accompanied with HA crystallization may result in the deterioration of bonding strength as shown in Figure 7. The dilatation of HACs during high-temperature crystallization is help to clarify the degradation of coating layers. Therefore, plasma-sprayed HAC test pieces with dimensions of 20 (l) × 3 (w) × 3 (t) mm^3 without Ti substrate were used for thermal dilatometry analysis by a dilatometer. The heating condition were 400-800°C in vacuum with a heating rate of 10°C/min, followed by furnace cooling after holding for 3h.

Figure 10 illustrates the thermal dilatometry curves for the HACs without Ti-6Al-4V substrate, heating at 400°C, 500°C, 600°C, 700°C and 800°C, respectively. They all show an approximately linear thermal expansion when the heating temperature is in the range from 100°C to 400°C. The coefficient of thermal expansion (CTE, α) of a HAC within this temperature range can be evaluated about 15.6×10^{-6} $°C^{-1}$ by measuring the slope of the dilatation curves at above-mentioned temperature interval. In addition, it is worth noting that the nonlinear region of the dilatation curves represents a consequence of greater crystallization of plasma-sprayed HACs at 500°C to 800°C temperature interval. Table 2 lists the extent of dilatation values, which calculated from the dilatation curves cooling to about 100°C, and a significant crystallization-induced contraction after heat treatments is recognized herein. The CTE data of crystallized HACs are calculated from the slope of cooling curves, and the change in the CTE value also means that the effective crystallization temperature of the HA coating layer by post-heat treatment should be over 600°C.

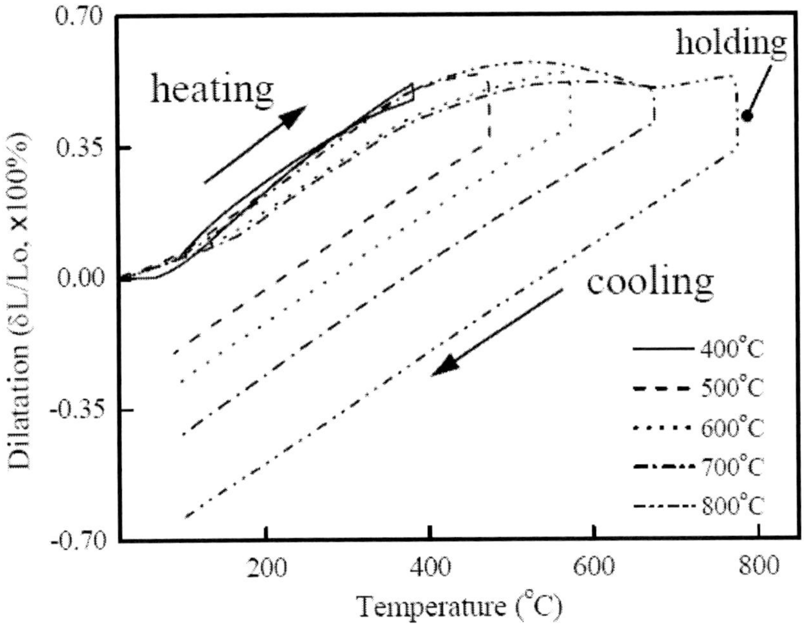

Figure 10. The dilatation curves of HACs obtained through thermal dilatometry.

A significant crystallization-induced contraction has been demonstrated from thermal dilatometry, then Figure 11 shows an obvious SEM/SEI observation nearby the coating contraction-induced cracks of the crystallized HACs in the heating temperature interval of 600-800°C, and there is an obvious contrast within this area. Through the semi-quantitative analysis of SEM/EDS results (each value is an average of five EDS spot-checks), the light gray contrast denoted by "C" with a Ca/P ratio of 1.61 represents the occurrence of HA crystallization close to the contraction-induced cracks after heat treatment, and the dark gray region "N" with a low Ca/P atomic ratio of 1.35 represents a non-stoichiometric calcium phosphate region away from the "C" region. By incorporating of the IOC and the thermal dilatometry results, therefore, it is recognized that the formation and the propagation of cracks are due to the coating contraction during crystallization. Raising the heating temperature will induce significant degradation in the microstructural feature and bonding strength of the HACs. Furthermore, since the formation of titanium oxide often causes serious interfacial fractures at elevated heating temperatures [54], Figure 7 reveals that the lowest bonding strength for atmospheric heating specimens at 700-800°C not only resulted from the above-mentioned HA crystallization-induced cracks, but also the formation of detrimental interfacial TiO_2 layer.

Table 2 The coefficient of thermal expansion (CTE, α) and the contraction of the crystallized HACs

Heating temperatures	α ($\times 10^{-6}$ °C^{-1}) †	Contraction (%) ‡
400°C	15.1	0.05
500°C	14.8	0.20
600°C	13.6	0.28
700°C	13.9	0.43
800°C	14.3	0.65

† Values were measured from the slope of cooling curves.
‡ Data obtained from total dilatation of the sample dimension after cooling.

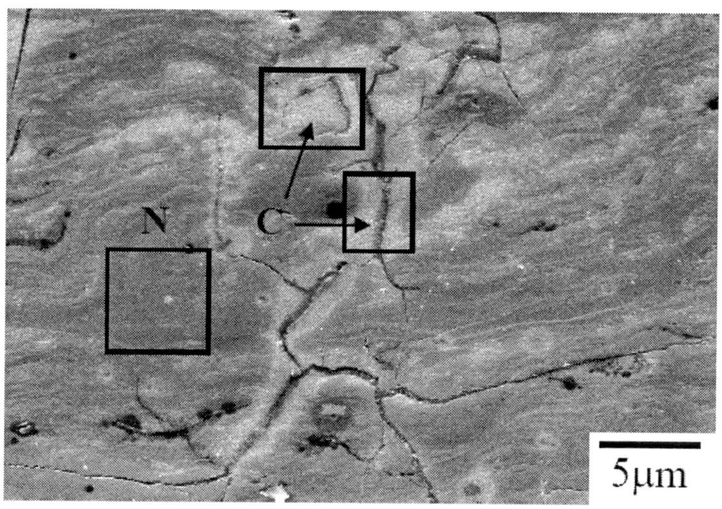

Figure 11. Enlarged cracks of the crystallized HACs at heating temperatures in the range of 600°C to 800°C, which showing the presence of the crystallized region (denoted by "C") and the non-stoichiometric calcium phosphate region (denoted by "N") that can be recognized by different contrast.

According to the previous discussion, the ambient saturated steam pressure plays an important role for the hydrothermal crystallization, which promoting a significant effect of microstructural self-healing through the grain growth of crystalline HA. However, the bonding strength degradation of hydrothermally-treated HACs within 150°C to 200°C is also resulted from this crystallization effect. During the hydrothermal treatment, the crystalline HA further grain growth with a larger crystal size, and the reduction of pores and microcracks proceeded with increasing the hydrothermal heating temperatures. This phenomenon results in a compressive stress applied within HA coating layers when the specimens cooling to room temperature. The compressive stress will induce serious inter-lamellar cracks, which is recognized as the strain relaxation effect, as shown in Figure 7. Generally, the inter-lamellar cracks are harmful for the plasma-sprayed coatings. Thus, the bonding strength degradation can be recognized as a result of the lamella decoherence in HT175 and HT200 coatings.

Figure 12 depicts the overall dependence of bonding strength on heating temperature that can be generalized into two characteristic regions. According to the previous discussion, plasma-sprayed HA coating is strongly affected by crystallization, which acted as a main controlling factor for improving the bonding strength. Additionally, the inter-lamellar and contraction-induced cracks resulted from the crystallization of HACs may change the fracture behavior and cause the deterioration of bonding strength. The relationship between the heating temperature and the bonding strength is found to be related to the HA crystallization.

The representative failure morphologies of these coatings are shown in Figure 13. According to the criterion of ASTM C633-01, it is recognized that the variation of bonding strength in situ is suggested to be governed by the cohesive strength within the coating layers themselves and the adhesive strength of a coating to a metal substrate. The cohesive failure (co) is dominated by the microstructural feature such as crystallinity, defects, lamellar texture, and a large area fraction of cohesive failure can be commonly observed at high strength coatings [20,35]. The representative failure morphology which resulted from the crystallization-induced cracks and the interfacial TiO_2 can be clarified through the high magnitude SEM observation of AH800 specimen as shown in Figure 13(a). The interfacial TiO_2 layer results in a serious debonding of the coating layer from the substrate can be observed as indicated by the arrow, and the bared Ti-substrate is found by SEM/BEI obsertion. The failure morphology of VH600 specimen displays a larger cohesive failure area without interfacial debonding as shown in Figure 13(b).

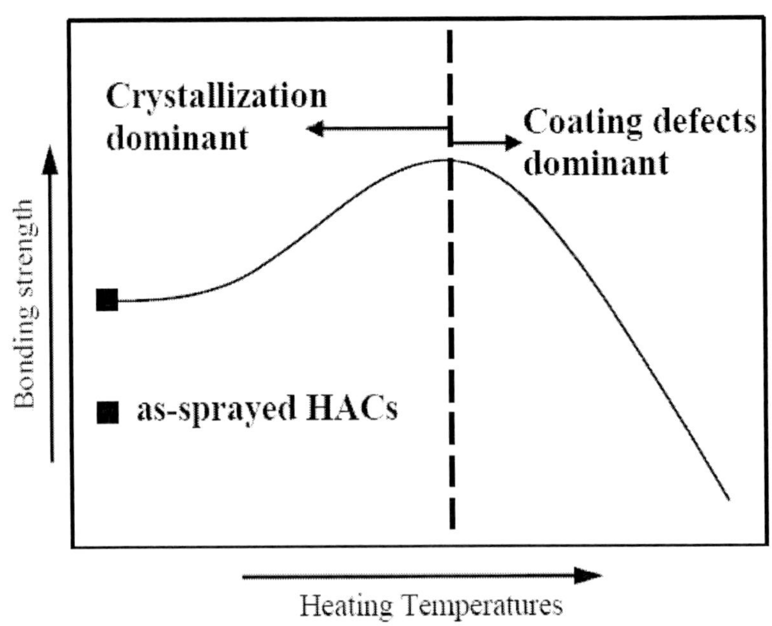

Figure 12. The correlation of the crystallization and the microstructural integrity on the evolution of bonding strength with various heating temperatures.

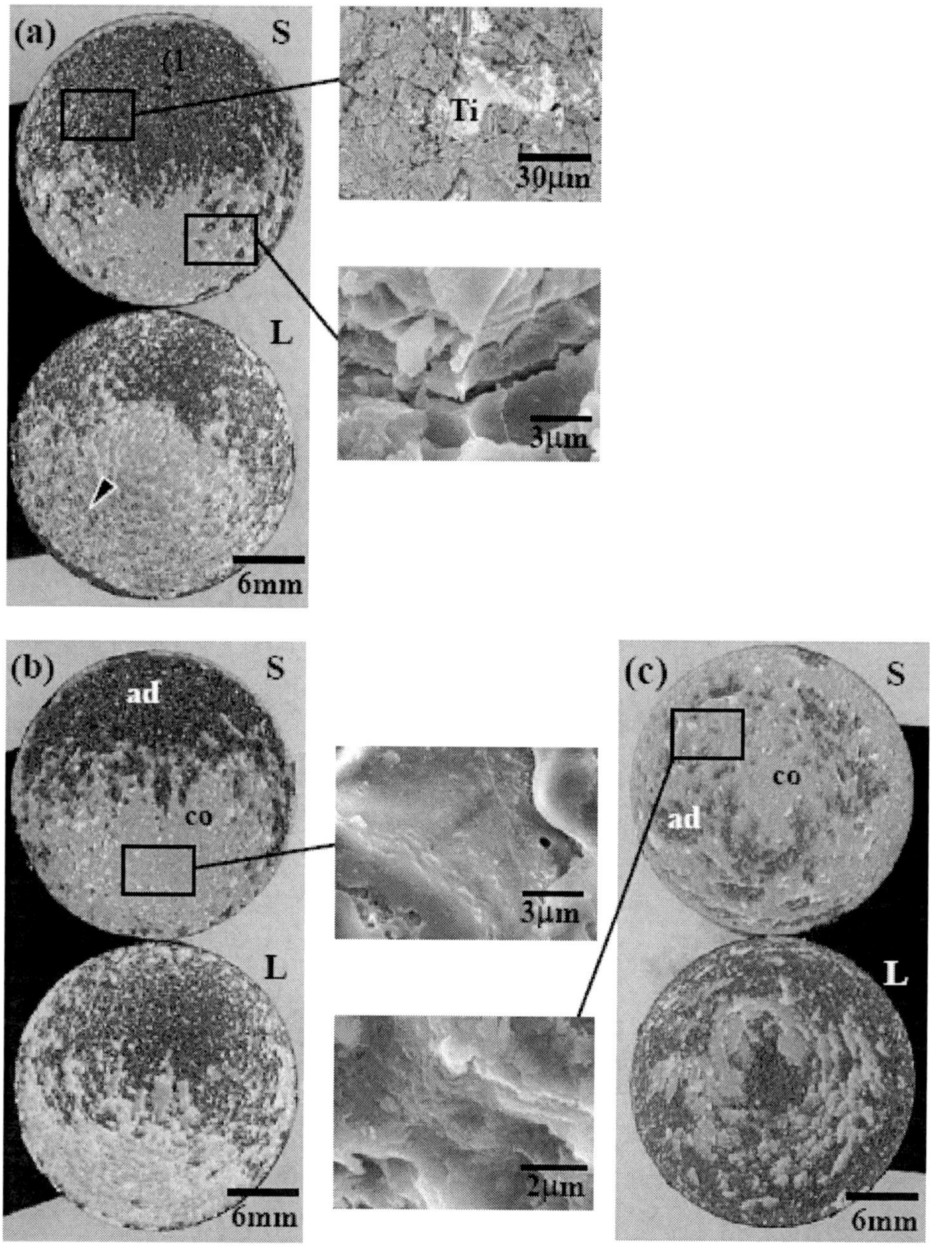

Figure 13. Failure morphologies of (a) AH800, (b) VH600, and (c) HT150 specimens after tensile adhesion tests. The bonding strength of the HACs measured is a manifestation of cohesive (co) and adhesive (ad) strength (L: loading fixture, S: substrate fixture).

Compared with the failures of VH600 specimen, since strengthening coatings resulted from the self-healing effect of hydrothermal crystallization, the failure morphologies of HT150 specimen as shown in Figure 13(c) represents crystallized, homogeneity and displays a larger area fraction of cohesive failure than the others. In contrast, the decreased area fraction of adhesive failure (ad) represents that the adhesion of hydrothermally-treated HACs to the Ti-6Al-4V substrate is further improved after the hydrothermal treatment. Referring to the evidences demonstrated from XPS analysis [40,55], the hydrothermal treatment can help to promote the interfacial crystallization through the replenished and the chemisorbed OH^- groups, which results in a significant chemical bonding between the HA coating and the Ti-substrate interface.

5. Statistical Analysis on Bonding Strength Fluctuation and Failure Mechanism of HA Coatings with the Weibull Model

In 1951, Waloddi Weibull suggested a statistical distribution function to characterize the strength of materials [56], which is widely known as the Weibull model. Owing to the significant data fluctuation of the bonding strength, the Weibull model is applied to evaluate the crystallization effect on the failure mechanism of HACs underwent different heat treatments. Equation (7) shows the general form of the Weibull distribution function:

$$F(\sigma_i) = \int_{\sigma=0}^{\sigma=\sigma_i} f(\sigma)d\sigma = 1 - \exp\left[-\left(\frac{\sigma_i - \sigma_o}{\eta}\right)^m\right] \qquad (7)$$

where σ represents the bonding strength, and at least twenty specimens (n=20) are tested for the purpose of statistical significance of the analysis. $F(\sigma_i)$ is the cumulative failure probability corresponding to a bonding strength σ_i (i is the ranking of the specimens, i=1-20). According to the definition of Eq. (7), the failure behavior of materials is determined by three parameters m, η and σ_o. The Weibull modulus, which controls the shape of function curves, is a measure of the variability of the data. The characteristics strength η corresponds to the strength at which the cumulative failure is 63.2%. The minimum strength σ_o means that the failure probability of HACs lower than this strength is zero.

Fitting the bonding strength data into the Weibull distribution function, the failure probability density function $f(\sigma)$ curves of the as-sprayed HACs, HT150, HT200, VH800 and AH800 specimens are plotted in Figure 14(a). The cumulative failure probability $F(\sigma_i)$ is estimated using the Benard's median rank of Eq. (8) [57], which is a very close approximated solution of a statistical function [57,58]. In addition, the reliability function $R(\sigma_i)$ with a relation of $R(\sigma_i) = 1-F(\sigma_i)$ is defined as the survival probability [59] of the specimens herein for the purpose of following discussion. Through the natural logarithmic (ln) graphs for the cumulative failure probability at each corresponding bonding strength σ_i (i=1-20) of the specimens, it can graphically evaluate the Weibull modulus (m) from the slope of a least-squares fitting method of Eq. (9), and results of the average bonding strength and the Weibull statistical analysis are listed in Table 3.

$$F(\sigma_i) = \frac{i - 0.3}{n + 0.4} \qquad (8)$$

$$\ln\ln\left(\frac{1}{1-F(\sigma_i)}\right)=m\ln(\sigma_i-\sigma_o)-m\ln\eta \tag{9}$$

Table 3. Bonding strength and Weibull model parameters of various HACs

	Bonding strength (MPa), $\sigma^{\dagger}$	Weibull modulus, $m^{\ddagger}$	Characteristics strength (MPa), $\eta^{\ddagger}$	Minimum strength (MPa), $\sigma_o^{\ddagger}$
as-sprayed HACs	32.4 ± 1.5	3.8	33.1	26.8
HT150	38.9 ± 1.0	6.4	39.4	33.3
HT200	28.8 ± 1.6	1.1	29.6	26.8
VH800	30.0 ± 1.7	3.6	30.8	24.1
AH800	23.2 ± 2.6	2.1	24.3	17.7

† Each value was the average of twenty tests (n=20).
‡ Data were calculated from $\ln\ln(1/1-F(\sigma_i))$ vs. $\ln(\sigma_i-\sigma_o)$ plots using Eq. (9).

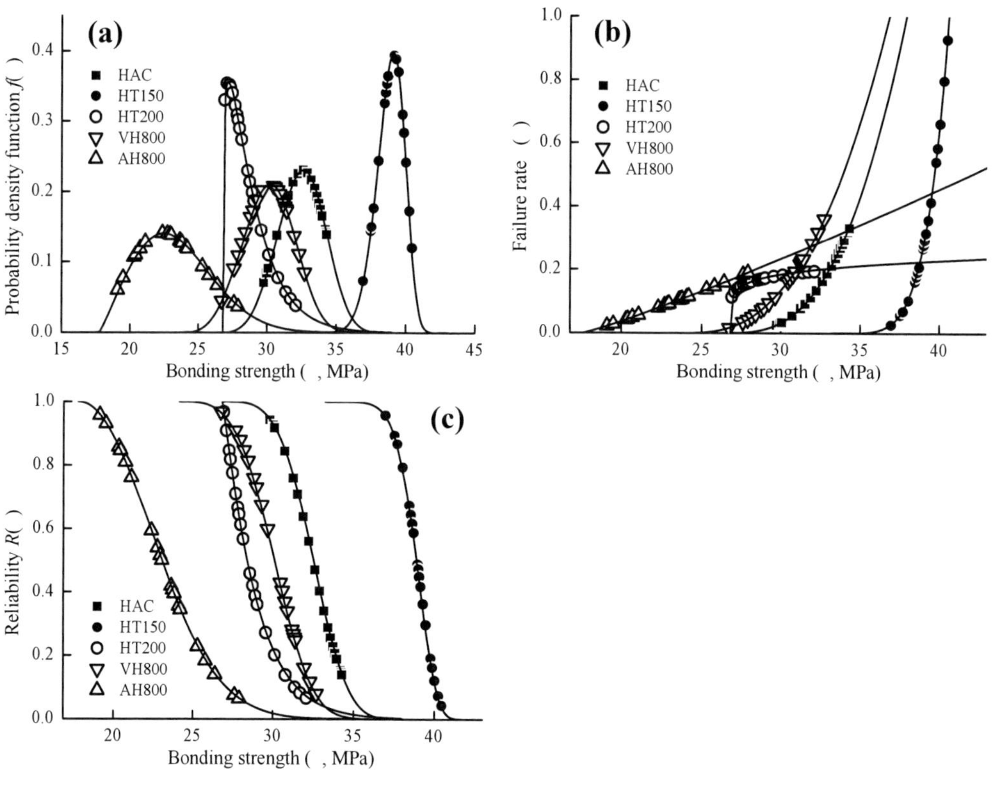

Figure 14. Plots for (a) the failure probability density function $f(\sigma)$ curves, (b) the failure rate function $\lambda(\sigma)$ curves, and (c) the reliability function $R(\sigma)$ curves of the as-sprayed and various heat-treated HACs. These curves start from the minimum strength (σ_o), which is the safety strength for the biological application of the HACs.

Since the Weibull distribution function is used to model the reliability and the failure behavior of materials [35,59,60], a failure rate function $\lambda(\sigma_i)$ shown in Eq. (10) at each corresponding bonding strength is defined for evaluating the failure mechanism of HACs.

$$\lambda(\sigma_i) = \frac{f(\sigma_i)}{R(\sigma_i)} = \frac{m}{\eta^m}(\sigma_i - \sigma_o)^{m-1} \tag{10}$$

Figure 14(b) and 14(c) show the failure rate function ($\lambda(\sigma)$) and reliability function ($R(\sigma)$) curves, respectively. These curves start from the minimum strength (σ_o), which implies the failure probability of HACs less than this strength is zero and the reliability of HACs is 1.0. The minimum strength can be recognized as the safety strength for the HACs herein. Meanwhile, considered with Table 3, knowledge of the Weibull modulus (m) can provide further explanation for the above-mentioned effects on the bonding strength of the HACs. It can be used to determine which material has higher uniformity and reliability. The Weibull modulus is also a measure of the variability of the data, which being larger as the degree of bonding strength scattering decreases (Figure 14(b)). A material with a larger Weibull modulus shows less reliability decrease (Figure 14(c)).

The examination of the Weibull modulus listed in Table 3 indicates that plasma-sprayed HA coating ($m = 3.8$) is reliable material with a wear-out failure model ($m > 1$) of increasing failure rate (IFR) [57,60]. The HT150 specimen also shows a wear-out failure model with a higher Weibull modulus ($m = 6.4$), which represents the hydrothermal crystallization has evident benefits to improve the microstructural homogeneity and to effectively enhance the bonding strength with lower data fluctuation. A material with a higher Weibull modulus is selected as it may be an indicator for clinical use, or lesser technique sensitivity. Yang et al. [61] demonstrated that better biological responses and lower dissolution rate of hydrothermally-treated HACs (HT-HACs) with respect to the as-sprayed HACs after 12 weeks of implantation could be attributed to the low contents of amorphous and impurity calcium phosphate phases. Furthermore, the improved bonding strength of HT-HACs which resulted from its dense microstructure and enhanced HAC/substrate adhesion is possible to provide better mechanical fixation of the implants without dissociation problems. Based on the above-mentioned observations, it can be recognized that the hydrothermal treatment is an available method to improve the mechanical strength, crystallinity, and phase composition of HACs for long-term stable biological and mechanical fixation of implants.

However, the HT200 specimen ($m = 1.1$) should be concluded as a material close to a random failure model ($m = 1$) of constant failure rate (CFR) referring to the failure rate curve in Figure 14(b). The random failure model generally represents a high initial failure probability (Figure 14(a)) and a rapid reliability degradation with increasing applied loading (Figure 14(c)). Lima et al. [62] suggested that Weibull modulus is related to the microstructural uniformity within the coating. In other words, the decreasing of Weibull modulus for heat-treated HACs is essentially due to the degraded microstructural homogeneity. Since the measured bonding strength is dominated by the microstructural characteristics of coating layer such as defects, lamellar texture, and etc., the evident differences in Weibull model analysis for the bonding strength degradation of HT200 specimen can be recognized owing to the appearance of inter-lamellar cracks. The VH800 and AH800 specimens show a wear-out failure model with a decreased Weibull modulus as listed in Table 3. It is resulted from the formation of detrimental crystallization-induced cracks. In addition, the serious interfacial TiO_2 layer at the AH800 specimen induces the debonding of HA coating layer from Ti-substrate (Figure 13(a)). The microstructural degradation phenomena cause a larger bonding strength fluctuation and a lower coating reliability as well as the decreased Weibull modulus.

5. Conclusion

The effect of crystallization with various post-heat treatments on the evolution of microstructural features, bonding strength and failure mechanism has been characterized. The results can be summarized as follows:

1) On the basis of quantitative analysis in crystallinity and phase content, autoclaving hydrothermal treatment is more favorable to eliminate the impurity phases and amorphous calcium phosphate than high temperature heat treatments in vacuum and in the air.
2) The Arrhenius kinetics analysis demonstrates that heating temperature is a controlling factor for HA crystallization. A higher crystallization rate and lower activation energy for the atmospheric heat treatment than the vacuum heat treatment can be recognized as the effect of ambient heating atmosphere with water vapor molecules. The saturated steam pressure of autoclaving hydrothermal treatment is a significant factor which not only accelerates HA crystallization, but also lowers the crystallization temperatures.
3) Hydrothermal crystallization obviously promotes the microstructural homogeneity of plasma-sprayed HACs through the self-healing effect with the grain growth of crystalline HA. Besides the crystallization effect, the microstructural self-healing is another factor which significantly further improves the Young's modulus of plasma-sprayed HACs.
4) The overall dependence of bonding strength on heating temperature can be concluded to be related to the effect of HA crystallization. The inter-lamellar and contraction-induced cracks resulted from the crystallization of HACs may change the fracture behavior and cause the deterioration of bonding strength.
5) According to the results of Weibull model analysis, an increased Weibull modulus (m) represents that the bonding strength and the reliability of HACs are further improved by HA crystallization. However, a decreased Weibull modulus indicates that the crystallization-induced defects and the interfacial oxidation layer are harmful for the plasma-sprayed HACs.

Acknowledgments

This study was financially supported by grants NSC 93-2216-E-006-038 and NSC 95-2221-E-006-118 from the National Science Council of Taiwan, ROC for which we are grateful.

References

[1] Munting, E., Verhelpen, M., Li, F., and Vincent, A. (1990). Contribution of hydroxy-apatite coatings to implant fixation. In T. Yamamuro, L. L. Hench, and J. Wilson (Eds.), *CRC Handbook of Bioactive Ceramics, Vol. II* (pp. 143-148). Boca Raton, Florida: CRC Press Inc.

[2] Jansen, J. A., van de Waerden, J. P. C. M., Wolke, J. G. C., and de Groot, K. (1991). Histologic evaluation of the osseous adaptation to titanium and hydroxy-apatite-coated titanium implants. *J. Biomed. Mater. Res.*, *25*, 973-989.

[3] Holmes, R. E., Bucholz, R. W., and Mooney, V. (1986). Porous hydroxyapatite as a bone-graft substitute in metaphyseal defects. *J. Bone Joint Surg.*, *68A*, 904-911.

[4] Bucholz, R. W., Carlton, A., and Holmes, R. E. (1989). Interporous hydroxyapatite as a bone graft substitute in tibial plateau fractures. *Clin. Orthop.*, *240*, 53-62.

[5] Ducheyne, P., Hench, L. L., Kagan, A., Martens, M., Bursens, A., and Mulier, J. C. (1980). Effect of hydroxyapatite impregnation on skeletal bonding of porous coated implants. *J. Biomed. Mat. Res.*, *14*, 225-237.

[6] Oonishi, H., Yamamoto, M., Ishimaru, H., Tsuji, E., Kushitani, S., Aono, M., and Ukon Y. The effect of hydroxyapatite coating on bone growth into porous titanium alloy implants. (1989). *J. Bone Jt. Surg.*, *B71*, 213-216.

[7] Hardy, D. C. R., Frayssinet, P., and Delince, P. E. Osteointegration of hydroxyapatite-coated stems of femoral prostheses. (1999). *Eur. J. Orthop. Surg. Traumatol.*, *9*, 75-81.

[8] Block, M. S., Kent, J. N., and Kay, J. F. (1987). Evaluation of hydroxyapatite-coated titanium dental implants in dogs. *J. Oral Maxillofac. Surg.*, *45*, 601-607.

[9] Geesink, R. G. T., de Groot, K., and Klein, C. P. A. T. (1988). Bonding of bone to apatite-coated implants. *J. Bone Joint Surg.*, *70B*, 17-22.

[10] Kitsugi, T., Yamamuro, T., Takeuchi, H., and Ono, M. (1988). Bonding behavior of 3 types of hydroxyapatite with different sintering temperatures implanted in bone. *Clin. Orthop.*, *234*, 280-290.

[11] Filiaggi, M. J., Coombs, N. A., and Pilliar, R. M. (1991). Characterization of the interface in the plasma sprayed HA coating/Ti-6Al-4V implant system. *J. Biomed. Mater. Res.*, *25*, 1211-1229.

[12] Galante, J., Rostoker, W., Lueck, R., and Ray, R. D. (1971). Sintered fiber metal composites as a basis for attachment of implants to bone. *J. Bone Joint Surg.*, *53A*, 101-114.

[13] Albrektsson, T., Branemark, P. I., Hansson, H. A., Kasemo, B., Larsson, K., Lundstorm, I., McQueen, D. H., and Skalak, R. (1983). The interface zone of inorganic implants in vivo: titanium implants in bone. *Ann. Biomed. Eng.*, *11*, 1-27.

[14] Ravaglioli, A., and Krajewski, A. (1992). *Bioceramics: Materials, Properties, Applications*. London: Chapman and Hall Press.

[15] Yang, C. Y., Wang, B. C., Chang, E., and Wu, B. C. (1995). Bond degradation at the plasma-sprayed HA coating/Ti-6Al-4V alloy interface: an *in vitro* study. *J. Mater. Sci.: Mater. Med.*, *6*, 258-265.

[16] Yang, C. Y., Lin, R. M., Wang, B. C., Lee, T. M., Chang, E., Hang, Y. S., and Chen, P. Q. (1997). *In vitro* and *in vivo* mechanical evaluations of plasma-sprayed hydroxyapatite coatings on titanium implants: The effect of coating characteristics. *J. Biomed. Mater. Res.*, *37*, 335-345.

[17] Wang, B. C., Chang, E., Yang, C. Y., Tu, D., and Tsai, C. H. (1993). Characteristics and osteoconductivity of three different plasma-sprayed hydroxyapatite coatings. *Surf. Coat. Technol.*, *58*, 107-117.

[18] Chen, J., Tong, W., Cao, Y., Feng, J., and Zhang, X. (1997). Effect of atmosphere on phase transformation in plasma-sprayed hydroxyapatite coatings during heat treatment. *J. Biomed. Mater. Res.*, *34*, 15-20.

[19] Li, H., Khor, K. A., and Cheang, P. (2002). Properties of heat-treated calcium phosphate coatings deposited by high-velocity oxy-fuel (HVOF) spray. *Biomaterials*, *23*, 2105-2112.

[20] Kweh, S. W. K., Khor, K. A., and Cheang, P. (2000). Plasma-sprayed hydroxyapatite (HA) coatings with flame-spheroidized feedstock: microstructure and mechanical properties. *Biomaterials*, *21*, 1223-1234.

[21] Khor, K. A., Gu, Y. W., Pan, D., and Cheang, P. (2004). Microstructure and mechanical properties of plasma sprayed HA/YSZ/ Ti-6Al-4V composite coatings. *Biomaterials*, *25*, 4009-4017.

[22] Lu, Y. P., Li, S. T., Zhu, R. F., Li, M. S., and Lei, T. Q. (2003). Formation of ultrafine particles in heat treated plasma-sprayed hydroxyapatite coatings. *Surf. Coat. Technol.*, *165*, 65-70.

[23] Lu, Y. P., Song, Y. Z., Zhu, R. F., Li, M. S., and Lei, T. Q. (2003). Factors influencing phase compositions and structure of plasma sprayed hydroxyapatite coatings during heat treatment. *Appl. Surf. Sci.*, *206*, 345-354.

[24] Yang, Y., Kim, K. H., Agrawal, C. M., and Ong, J. L. (2004). Interaction of hydroxyapatite-titanium at elevated temperature in vacuum environment. *Biomaterials*, *25*, 2927-2932.

[25] Gross, K. A., Gross, V., and Berndt, C. C. (1998). Thermal analysis of amorphous phases in hydroxyapatite coatings. *J. Am. Ceram. Soc.*, *81*, 106-112.

[26] Yang, C. W., Lui, T. S., Lee, T. M., and Chang, E. (2004). Effect of hydrothermal treatment on microstructural feature and bonding strength of plasma-sprayed hydroxyapatite on Ti-6Al-4V. *Mater. Trans.*, *45(9)*, 2922-2929.

[27] Yang, Y. C., and Chang, E. (2003). The bonding of plasma-sprayed hydroxyapatite coatings to titanium: effect of processing, porosity and residual stress. *Thin Solid Films*, *444*, 260-275.

[28] Cullity, B. D. (1978). *Elements of X-ray Diffraction*. Philippines: Addison-Wesley.

[29] Wang, B. C., Chang, E., Lee, T. M., and Yang, C. Y. (1995). Changes in phases and crystallinity of plasma-sprayed hydroxyapatite coatings under heat treatment: a quantitative study. *J. Biomed. Mater. Res.*, *29*, 1483-1492.

[30] Chou, B. Y., and Chang, E. (2002). Phase transformation during plasma spraying of hydroxyapatite-10-wt%-zirconia composite coating. *J. Am. Ceram. Soc.*, *85*, 661-669.

[31] Yang, C. W., Lee, T. M., Lui, T. S., and Chang, E. (2005). A comparison of the microstructural feature and bonding strength of plasma-sprayed hydroxyapatite coatings with hydrothermal and vacuum post-heat treatment. *Mater. Trans.*, *46(3)*, 709-715.

[32] Yang, C. W., Lee, T. M., Lui, T. S., and Chang, E. (2006). Effect of post vacuum heating on the microstructural feature and bonding strength of plasma-sprayed hydroxyapatite coatings. *Mater. Sci. Eng. C*, *26*, 1395-1400.

[33] Roth, R. S., Negas, T., and Cook, L. P. (1983). *Phase Diagrams for Ceramists, Vol. 5*. (pp. 321-322). Washington DC: American Ceramic Society.

[34] Weng, J., Liu, X., Zhang, X., and de Groot, K. (1996). Integrity and thermal decomposition of apatite in coatings influenced by underlying titanium during plasma spraying and post-heat-treatment. *J. Biomed. Mater. Res.*, *30*, 5-11.

[35] Yang, C. W., and Lui, T. S. (2007). Effect of crystallization on the bonding strength and failures of plasma-sprayed hydroxyapatite. *Mater. Trans. 48(2)*, 211-218.

[36] Chen, J., Tong, W., Cao, Y., Feng J., and Zhang, X. (1997). Effect of atmosphere on phase transformation in plasma-sprayed hydroxyapatite coatings during heat treatment. *J. Biomed. Mater. Res.*, *34*, 15-20.

[37] Tong, W., Chen, J., Cao, Y., Lu, L., Feng, J., and Zhang, X. (1997). Effect of water vapor pressure and temperature on the amorphous-to-crystalline HA conversion during heat treatment of HA coatings. *J. Biomed. Mater. Res.*, *36*, 242-245.

[38] Koutsopoulos, S. (2002). Synthesis and characterization of hydroxyapatite crystals: A review study on the analytical methods. *J. Biomed. Mater. Res.*, *62*, 600-612.

[39] Slosarczyk, A., Paluszkiewicz, C., Gawlicki, M., and Paszkiewicz, Z. (1997). The FTIR spectroscopy and QXRD studies of calcium phosphate based materials produced from the powder precursors with different Ca/P ratios. *Ceram. Int.*, *23*, 297-304.

[40] Yang, C. W., Lui, T. S., and Chang, E. (2007). Low temperature crystallization and structural modification of plasma-sprayed hydroxyapatite coating with hydrothermal treatment. *Adv. Mater. Res.*, *15-17*, 147-152.

[41] Yang, Y., Kim, K. H., Agrawal, C. M., and Ong, J. L. (2003). Influence of post-deposition heating time and the presence of water vapor on sputter-coated calcium phosphate crystallinity. *J. Dent. Res.*, *82*, 833-837.

[42] Chang, C., Huang, J., Xia, J., and Ding, C. (1999). Study on crystallization kinetics of plasma sprayed hydroxyapatite coating. *Ceram. Inter.*, *25*, 479-483.

[43] Huang, L. Y., Xu, K. W., and Lu, J. (2000). A study of the process and kinetics of electrochemical deposition and the hydrothermal synthesis of hydroxyapatite coatings. *J. Mater. Sci.: Mater. Med.*, *11*, 667-673.

[44] Liu, C., Huang, Y., Shen, W., and Cui, J. (2001). Kinetics of hydroxyapatite precipitation at pH 10 to 11. *Biomaterials*, *22*, 301-306.

[45] Zhou, J., Zhang, X., Chen, J., Zeng, S., and de Groot, K. (1993). High temperature characteristics of synthetic hydroxyapatite. *J. Mater. Sci.: Mater. Med.*, *4*, 83-85.

[46] Feng, C. F., Khor, K. A., Liu, E. J., and Cheang, P. (2000). Phase transformations in plasma sprayed hydroxyapatite coatings. *Scripta Mater.*, *42,* 103-109.

[47] Standard test method for adhesion and cohesion strength of thermal spray coatings. ASTM C633-01, West Conshohocken, PA, USA.

[48] Li, C. J., Ohmori, A., and McPherson, R. (1997). The relationship between microstructure and Young's modulus of thermally sprayed ceramic coatings. *J. Mater. Sci.*, *32*, 997-1004.

[49] Kim, H. J., and Kweon, Y. G. (1999). Elastic modulus of plasma-sprayed coatings determined by indentation and bend tests. *Thin Solid Films*, *342*, 201-206.

[50] Yang, Y. C., Chang, E., and Lee, S. Y. (2003). Mechanical properties and Young's modulus of plasma-sprayed hydroxyapatite coating on Ti substrate in simulated body fluid. *J. Biomed. Mater. Res.*, *67A*, 886-899.

[51] Chiang, Y. M., Birnie, D. P., and Kingery, W. D. (1997). *Physical Ceramics: Principles for Ceramics Science and Engineering* (pp. 478-484). John Wiley and Sons, Inc.

[52] Beshish, G. K., Florey, C. W., Worzala, F. J., and Lenling, W. J. (1993). Fracture toughness of thermal spray ceramic coatings determined by the indentation technique. *J. Therm. Spray. Technol.*, *2*, 35-38.

[53] Li, H. Khor, K. A., and Cheang, P. (2002). Properties of heat-treated calcium phosphate coatings deposited by high-velocity oxy-fuel (HVOF) spray. *Biomaterials*, *23*, 2105-2112.

[54] Lo, T. N., Chang, E., and Lui, T. S. (2004). Interfacial characterization of porcelain veneered on the pure titanium under vacuum firing. *Mater. Trans.*, *45*, 3065-3070.

[55] Yang, C. W., and Lui, T. S. (2008). The self-healing effect of hydrothermal crystallization on the mechanical and failure properties of hydroxyapatite coatings. *J. Eur. Ceram. Soc.*, *JECS7024*, in press.

[56] Weibull, W. (1951). A statistical distribution function of wide applicability. *J. Appl. Mecha.*, *18*, 293-297.

[57] Abernethy, R. B. (2000). *The New Weibull Handbook: Reliability and Statistical Analysis for Predicting Life, Safety, Survivability, Risk, Cost and Warranty Claims* (4th edition). Florida: North Palm Beach.

[58] Faucher, B., and Tyson, W. R. (1988). On the determination of Weibull parameters. *J. Mater. Sci. Lett.*, *7*, 1199-1203.

[59] Burrow, M. F., Thomas, D., Swain, M. V., and Tyas, M. J. (2004). Analysis of tensile strengths using Weibull statistics. *Biomaterials*, *25*, 5031-5035.

[60] Lima, R. S., and Marple, B. R. (2003). High Weibull modulus HVOF titania coatings. *J. Therm. Spray. Technol.*, *12*, 240-249.

[61] Yang, C. Y., Lee, T. M., Yang, C. W., Chen, L. R., Wu, M. C., and Lui, T. S. (2007). The *in vitro* and *in vivo* biological responses of plasma-sprayed hydroxyapatite coatings with post-hydrothermal treatment. *J. Biomed. Mater. Res.*, *83A*, 263-271.

[62] Lima, R. S., and Marple, B. R. (2003). Optimized HVOF titania coatings. *J. Therm. Spray Technol.*, *12*, 360-369.

Index

A

B

C

D

E

G

H

I

J

K

L

M

N

O

P

Q

R

S

T

U

V

W

X

Y

Z